# DATE DUE

| | | | |
|---|---|---|---|
| MY 3 0 '96 | | | |
| OC 31 '96 | | | |
| NO 26 '96 | | | |
| OC 1 2 '00 | | | |
| MR 22 04 | | | |
| FE 8 07 | | | |
| | | | |
| | | | |
| | | | |
| | | | |
| | | | |
| | | | |
| | | | |
| | | | |
| | | | |
| | | | |

DEMCO 38-296

*Her Share
of the Blessings*

# Her Share
# of the Blessings

Women's Religions among
Pagans, Jews, and Christians in
the Greco-Roman World

ROSS SHEPARD KRAEMER

*New York*     *Oxford*
OXFORD UNIVERSITY PRESS
1992

Oxford University Press

Oxford   New York   Toronto
Delhi   Bombay   Calcutta   Madras   Karachi
Petaling Jaya   Singapore   Hong Kong   Tokyo
Nairobi   Dar es Salaam   Cape Town
Melbourne   Auckland

and associated companies in
Berlin   Ibadan

Published by Oxford University Press, Inc.
200 Madison Avenue, New York, NY 10016

Oxford is a registered trademark of Oxford University Press

Library of Congress Cataloging-in-Publication Data
Kraemer, Ross Shepard, 1948–
Her share of the blessings : women's religions among pagans,
Jews, and Christians in the Greco-Roman world / Ross Shepard Kraemer
p. cm. Includes bibliographical references and index.
ISBN 0-19-506686-3
1. Women—Mediterranean Region—Religious Life—History.
2. Women and religion—History.   3. Women in Judaism—History.
4. Women in Christianity—History—Early church, ca. 30–600.
5. Paganism—Mediterranean Region—History.
BL625.7.K73   1992
291'.082—dc20   91-33777   CIP

1 3 5 7 9 8 6 4 2

Printed in the United States of America
on acid-free paper

*To*
*Herman Seymour Shepard*
*1916–1989*
*and*
*John A. Hollar*
*1942–1989*

*In Memory*

*Bacchae of the City, say*
*"Farewell you holy priestess."*
*This is what a good woman deserves.*
*She led you to the mountain and carried*
*all the sacred objects and implements,*
*marching in procession before the whole city.*
*Should some stranger ask for her name:*
*Alcmeonis, daughter of Rhodius,*
*who knew her share of the blessings.*

> —Epitaph from Miletus
> Third/second century B.C.E.
> *Maenads,* **8**

# *Preface*

In 1974, while writing my doctoral dissertation and teaching a course on women and religion at Franklin & Marshall College, I first began to envision a collection of sources not on what men thought about women, but rather on what women themselves did in religious contexts in the Greco-Roman world. Fourteen years later, *Maenads, Martyrs, Matrons, Monastics: A Sourcebook on Women's Religions in the Greco-Roman World* (Philadelphia: Fortress Press, 1988) appeared in print.

Once *Maenads* was in the works, my friend and editor at Fortress, John A. Hollar, encouraged me to write a companion volume to provide the context and commentary that the sourcebook deliberately lacked. With a fellowship from the National Endowment for the Humanities, I intended to write just such a work.

As I began my leave, I learned that the eminent anthropologist Mary Douglas was teaching a graduate seminar on women and religion in the Department of Religion at Princeton. She graciously allowed me to sit in on the course, and thereby unknowingly altered the very nature of this book. I had read some of Douglas's work before, and applied it profitably from time to time in my teaching and some of my research, but the semester I sat at the other end of a seminar table from Douglas allowed me to rethink some of my major theoretical concerns regarding women's religions. By the time I began to write this book, it was irrevocably different from the handbook I had initially planned. It no longer follows the organizational order of *Maenads*, and it now primarily covers material from only three sections in *Maenads* ("Observances, Rituals, and Festivals," "Researching Real Women" and "Religious Office") with some discussion of texts from "New Religious Affiliation and Conversion." The sources from "Holy, Pious, and Exemplary Women" and "The Feminine Divine" largely fell victim to the constraints of space and time. Nevertheless, I hope this book provides some of the analysis absent from *Maenads*.

Writing this book has in many ways been a lonely endeavor, though not for any lack of sympathetic friends, family, and colleagues. Rather, I have found myself interested in issues that intrigue few other scholars. The feminists I know are primarily interested in feminist theology, or else they disavow the study of religion altogether, considering religion to be a hopelessly patriarchal institution that oppresses women. At least for the chronological and geographic periods in which I work, the study of women's religions as distinct from theology sometimes seems a field I have carved out myself. My nonfeminist colleagues who share those general parameters of time and culture all too often lack much interest in the study of women's religions. Ann Loades, in her brief review of *Maenads*, was kind enough to refer to me as "one of the most distinguished feminist historians of religions," but it quickly occurred to me that to be one of the most distinguished among a very small company was at the least somewhat depressing.

Since my doctoral dissertation at Princeton, I have argued against the insufficiency of theories, models, and explanations based more or less exclusively on what men do. In its attempt to lay out alternative approaches, this book in many ways represents new territory: efforts to ask new questions of old sources and overlooked sources, and very occasionally even of new sources. In this regard, even though it is the product of almost two decades of work, it is a preliminary venture.

By now, I think I have asked many of the crucial questions and proposed at least a few compelling answers. With this study, I hope to shift the focus of the discussion about women and religion in antiquity away from the contemporary theological concerns that have dominated the scholarly and popular literature to date toward a better understanding of the way religion functions in the lives of women. Ultimately, I wish to facilitate a fuller understanding of religion as a human phenomenon, an understanding that cannot be achieved without careful consideration of the differences between the religious dimensions of women's and men's lives.

But the very nature of the enterprise I have undertaken warrants humility and caution. In response to Ann Loades's praise that I had judiciously refrained from providing more than minimal commentary to the sources in *Maenads*, a colleague sanguinely remarked that I could not possibly provide adequate commentary to the sources I had collected, for who among us could know enough to do that. As I wrote this book, and found myself embroiled in textual and contextual issues ranging from analysis of the archaic Greek *Hymn to Demeter* to early fifth-century C.E. epigraphical evidence for women presbyters in Christian churches, I was acutely aware of the limits of my own knowledge and my debt to and dependence on many other scholars far more versed in various aspects of these materials than I. Doubtless, readers who know these materials intimately will have some corrections of facts to offer; those expert in the theoretical models I have employed will critique my application of the models or my emendations based on ancient examples.

The actual writing of this book bore at least some superficial resemblance to Dionysian possession. At times I know I seemed, at least to my family, to alternate between ecstatic pleasure and borderline insanity. My deepest gratitude goes then to my husband, Michael, and my daughter, Jordan, for their patience, love, pride, and encouragement.

I owe thanks to many other people for their support and assistance. In Cynthia Read of Oxford University Press, I have gained not only a superb editor, but a wonderful friend, whose humor and candor tempers all her insightful criticism. Mary Rose D'Angelo, Robert Kraft, and Susan Niditch each read virtually the entire manuscript; Judith Baskin, Stephen Benko, Daniel Boyarin, Richard Lim, Amy Richlin, and Judith Wegner read various chapters. All were candid, kind, and constructive in their critique. They have saved me from many blunders: for those that remain, I take full responsibility.

The National Endowment for the Humanities awarded me my second year-long fellowship to write this book, for which I am honored and grateful. Over the last nine years, the Department of Religious Studies at the University of Pennsylvania has continued to extend to me the privileges of a visiting scholar, which has quite simply made my research possible. The collegiality of Ann Matter, Robert Kraft, and the graduate students in early Judaism and early Christianity has enriched my life immeasurably. The Department of Religion at Princeton has welcomed me back home numerous times, most recently in 1989–1990 as a visiting faculty member. To my teacher John Gager and to the latest generation of late antiquity graduate students, I am especially grateful for a year of replenishment spent in their midst. Finally, I would like to thank all those who participate in the Philadelphia Seminar on Christian Origins, where my preliminary research presentations have received sympathy, support, and scrutiny.

Although writing this book was often a source of great pleasure and satisfaction, I will always associate it with two personal tragedies. On Friday, August 18, 1989, I delivered the first draft of this work to John A. Hollar. That evening, my father died in his sleep, peacefully but unexpectedly after many years of a debilitating psychiatric and physical illness. Two months later, on Friday, October 20, John Hollar died suddenly on the way home from the Frankfurt Book Fair, leaving a wife and two small children and a vast number of colleagues and friends stunned and bereft. I trust no one will find any irony in my dedication of this feminist study on women's religion in antiquity to my father and my friend, as a modest memorial to two good and decent human beings, both of whom deserved far more from this life than they received.

R.S.K.

*Philadelphia*
*February 1992*

# *Contents*

*Her Share
of the Blessings*

# — 1 —

# *Introduction*

It was a commonplace in Greco-Roman antiquity that religion was women's business, and it was not a compliment. Strabo of Pontus in Asia Minor said it most clearly: "Women [are] the chief founders of religion (*deisidaimonia*) . . . women . . . provoke men to the more attentive worship of the gods, to festivals and to supplications, and it is a rare thing for a man who lives by himself to be found addicted to these things."[1]

As with all generalizations, Strabo's could be qualified. *Deisidaimonia* (literally, "fear of the gods") was neither the only term he used for human reverence for the divine, nor the most complimentary. Not surprisingly, the stance he favored, *eudaimonia*, was far more typical of elite, educated men.[2] Strabo knew that religion was hardly the sole domain of women, and so did every other ancient male author who wrote on the subject. In many quarters, from the Roman priestly colleges to the Jewish study house to Christian church councils, religion could be, and was, a male exercise. But the association of women with religion seemed so obvious that it called forth little explanation in antiquity. Modern scholarship cannot be said to have done much better.

This then is a book about women's religions in the world of Greco-Roman antiquity and about the connections between women's religions and the social constraints under which women lived their lives. Although the phrase *women's religions* tends to conjure up notions of secretive women's fertility rites, I intend a much more straightforward definition: women's religions are simply whatever women themselves do and think in religious contexts.[3]

The task I have set myself here is twofold. I wish in the first place to recover and reconstruct the range of women's religions from about the fourth century B.C.E. through about the fourth century C.E. in the geographic region that ultimately comprised the Roman Empire. Within these parameters fall not only a broad range of what we call pagan religions, but also diverse forms of early Judaism and early Christianity.

Beyond the challenge of this descriptive enterprise, I seek some framework for interpreting women's religions in antiquity that does justice to the fact that women were not homogeneous: distinctions of ancient social categorization together with stages of women's lives figure significantly in the diversity of women's practices and beliefs. I also consider some analytic perspective for the ways in which women's religions were both similar to and different from those of men.

Recovering information about women's religions from the sources for Greco-Roman antiquity is a complex, often frustrating endeavor. It is first of all without significant scholarly precedent. The women's movement of the last century and the early portion of this century did generate some study of women in antiquity, particularly in connection with nineteenth-century theories of primal matriarchy.[4] But until the last two decades, women's history and women's religion were largely ignored by scholars of late antiquity, as they were by scholars of most disciplines. In retrospect the reasons seem painfully obvious: bastions of male scholars rarely even thought about the study of women. Behind their disinterest, which was largely shared by the small number of women scholars, lay the insidious, significant, and unarticulated assumption that human religion and human history were identical with men's religion and men's history.

Now and again, some male scholars glimpsed a different reality. Adolph von Harnack, writing in the early part of this century, observed the large numbers of women evident in early Christian communities, but could not integrate that observation into his conceptual framework for interpreting early Christianity.[5] George Foot Moore, in his massive work on Judaism in the first three centuries of the common era, could casually assert that women constituted the majority of pagans attracted to Judaism and an even larger proportion of formal converts or proselytes, and then go on to devote thirty or so pages to a discussion of proselytes, all of which assumed that proselytes were male. He even went so far as to define a proselyte as a "man who has adopted Jewish law."[6]

The resurgence of a women's movement in the second half of this century reopened the question that had lain largely dormant for centuries, of women's leadership roles in Christian churches. Since Christians sought the answers to this question in ancient precedent and practice, scholars and laypersons alike began to investigate those aspects of women's history that might bear on the contemporary dilemma. Did women in fact hold religious office in early Christian churches? How much were early Christian attitudes toward women's offices rooted not in divinely ordained prescription, but in ancient social reality?

At the same time, feminist historians, including historians of Western antiquity, were beginning to raise major questions about women's history. Their increasing awareness that sources written, translated, and transmitted by elite groups of men were not inherently reliable sources for women's history had significant ramifications for the once seemingly straightforward question of women's roles in early Christianity.

Nevertheless, mainstream scholarship did not warm quickly to the study of women in antiquity. Many male scholars continued to hold the untested assumption that women's experiences were either the same as men's or else aberrant, irrelevant, or both, and in any case, not worthy of scholarly attention. Others routinely asserted that there was no evidence about women worth pursuing.

Feminist research has enabled us to see the fallacy of this assertion, which had two intertwined components: (1) there is not any evidence, and (2) any evidence there is, is not worth pursuing. In reality the second component is probably determinative. What we consider evidence is largely a function of what we consider worth pursuing, of what questions we choose to ask and what issues we consider important.

The study of religion in late antiquity, particularly early Judaism and early Christianity, has been directed and distorted by a disproportionate concern for theological questions. This is perhaps less true for classicists who study pagan religions, but even they have not been exempt from the conditioning of contemporary theological agendas in the kinds of questions posed about religion in antiquity.

Concern for theological issues has led to an emphasis and a dependency on literary works, particularly those works that are themselves concerned with theological discussions. Further, many scholars chose to focus on texts that had become normative and authoritative within various branches of Christian and Jewish tradition, creating a kind of scholarly canon alongside the scriptural canons of Judaism and Christianity. Nonliterary materials such as commercial documents, private correspondence, burial and donative inscriptions, magical amulets, as well as archaeological evidence without anything written on it, received short shrift from many scholars, in large part, I suggest, because their bearing on theological issues was not immediately apparent. Classicists engaged in the study of pagan religions would never dream of ignoring nonliterary sources, but in the study of Judaism and Christianity, archaeologists rarely spoke to theologians and historians and vice versa.

I do not think it wholly accidental that the texts and documents outside the scholarly canon have turned out to be precisely those which figure significantly among the evidence for Jewish and Christian women and their religious beliefs and practices in late antiquity. To a lesser extent, evidence for the study of pagan women is also more prevalent in sources traditionally neglected in the canon of classical scholars.

Traditional scholarship did more than selectively screen the extant sources; it often misread and mistranslated all kinds of evidence from the New Testament to epigraphical data and more. Elisabeth Schüssler Fiorenza has adduced many instances of biases in translation and interpretation in New Testament studies. Particularly telling is her observation that scholars claim on the one hand that masculine plurals encompass both men and women, and then conclude that terms like *the disciples* must mean only men.[7] Fiorenza and others have also suggested that ancient manuscripts themselves reveal androcentric editing of the texts.[8]

Inscriptions offer another example. In his collection of Jewish inscriptions from the Greco-Roman period, Jean-Baptiste Frey found many with fragmentary names. In the absence of any information from the remainder of the inscription, Frey frequently restored the name as male.[9] Such methods are guaranteed to reduce the amount of evidence for women and to mislead in particular those not versed in the original languages.

Clearly there is far more evidence available for fruitful investigation than generations of previous scholarship would have had us believe. But there are still significant difficulties studying women in antiquity, particularly if what we wish to know is not men's attitudes, but what women themselves did, thought, and believed. We must also not lose sight of the important problems posed by the sources for men in antiquity. Scholarship has often tended to assume that the extant sources are sufficiently representative of Judaism, Christianity, and other religions in Greco-Roman antiquity, rarely asking whether what has survived to the twentieth century might only represent the perspectives and interests of a small minority. Not only are women ignored, but distinctions of social and economic status, geographic location, and chronological differentiation are often made neither for men nor for women.

In reality our sources for the study of religion in late antiquity are quite fragmentary, especially when compared with those available for many other periods of Western history. Many texts survive only in fragmented form. Others that once existed are now just names on an ancient list; many texts must have been written which are wholly lost to us. Burial epitaphs commemorate only a tiny fraction of those who lived during this period. We have only to think how much the fortuitous discoveries of the Dead Sea Scrolls in Israel and the codices recovered near Nag Hammadi in Egypt[10] in the late 1940s have changed our reconstructions of Judaism and Christianity to realize how tenuous and presumptuous is the assumption that our current sources are sufficiently representative.

The processes by which the ancient materials have reached us may illuminate the limitations of our sources. To the best of our knowledge, we possess no original manuscript of any ancient literary text. Although we have ancient copies of some texts, including the manuscripts from the Judean desert and the region of Nag Hammadi and a variety of complete manuscripts from the fourth century C.E. onward, the majority of literary texts from antiquity have survived in the form of Byzantine and medieval copies made and stored in the libraries of Christian monasteries.[11]

The processes by which those copies were made have considerable ramifications for the study of women. There are the usual problems of scribal errors: misspellings and mishearings, notes scribbled in the margins that other scribes then incorporate into later manuscripts, scribal efforts to correct what they consider mistakes in earlier manuscripts, and so forth.

There are also the political realities. If texts are dependent for their survival on being read and copied, *who* reads and *who* copies becomes paramount. Male monastics thus gain considerable control over what survives to modern times, coupled, of course, with fortuitous events like natural disasters that destroy whole libraries and accidental discoveries that reveal whole collections not otherwise preserved. Given the demonstrated misogynism of many individual monks and monastic communities, it is hard to avoid the conclusion that what they transmit is likely to be affected in some degree.

Whether conditions were any different in the libraries of women's monasteries in medieval Europe remains uncertain. The surviving plays of the tenth-century nun Hroswitha are based on some of the Apocryphal Acts of the Apostles with their plethora of tales about women, so obviously the library in which she worked had copies of those texts.[12] We know that her library also comprised many classical authors, but it would be important to survey what else those libraries contained, whether they preserved and transmitted different ancient materials from those in male monasteries.

Of course what was not written cannot have been transmitted. A handful of women writers are attested in Greek and Latin literature. At best, only fragments of their work are extant; in some cases, only a brief notice that a woman wrote survives in the work of a male author. Nevertheless, women are known to have composed poetry, epigrams, letters meant for public circulation, philosophical treatises, and so forth.[13]

Around the year 203 C.E., a recent convert to Christianity named Vibia Perpetua wrote a diary detailing her imprisonment and visions while awaiting her martyr's death in Roman North Africa.[14] After her death an unknown editor[15] combined her diary with the visions of another martyr, Saturus, and added the details of their deaths, together with a narrative framework. In the late fourth century, Faltonia Betitia Proba composed a Christian cento, and in the fifth century, a pilgrim named Egeria composed a pious travelogue of her journey to the holy land.[16]

Numerous papyri demonstrate that women routinely utilized letters as a means of communication with distant relatives.[17] Letters of a more literary nature were clearly composed by women such as Olympias, the friend and patron of John Chrysostom, or Paula, a disciple of Jerome's, both in the late fourth century, but we can infer their contents and sophistication only from the replies by the men, which not surprisingly are all that have survived.

Accounting for the dearth of women's writing is far more complex than it might initially seem. The composition of works did not depend solely on the ability to read and write. Susan Cole has pointed out that Aspasia, the mistress of Pericles, was reputed to have been an accomplished orator in Athens of the fifth century B.C.E., yet may not have been literate.[18] It only would have required a professional scribe to transform her orations into written works. In the Greco-Roman period, many authors employed scribes and secretaries, who were often slaves.

While relatively few persons in antiquity of either gender could read and write, some women, though perhaps not large numbers, did possess sufficient education to write with or without the help of professional secretaries, increasingly so after the classical period in Greece. They also had sufficient free time at their disposal. The standard processes for the dissemination and transmission of literature in antiquity may, however, have severely limited the amount of women's writing that ever circulated beyond a very small readership.[19]

For early Judaism and early Christianity our problem is compounded because a significant portion of the texts that have come down to us are either anonymous—without any ascription of authorship—or pseudonymous— attributed to someone unlikely if not impossible to have been the author. In a few cases, the author's name may be known but virtually nothing else. Most of the New Testament, except for the undisputed letters of Paul, falls into these categories, as does virtually everything else in Jewish apocrypha and pseude-pigrapha. We know practically nothing of the authors of most works common-ly classed as New Testament apocrypha, nor of treatises found near Nag Ham-madi, nor those found in the caves of the Judean desert. Only recently have scholars begun to take seriously the possibility that women may have com-posed some of these texts, and at the very least to refrain from assuming that the authors must in all cases have been male.

Although we are by no means certain why some authors in late antiquity wrote texts under the cloaks of anonymity and pseudonymity, it is easy enough to surmise why women might have done so. In societies far less restrictive than those of late antiquity, women have found it far easier to gain acceptance and credibility when writing under a male pseudonym.

One enigmatic Christian text conventionally dated to the fourth century C.E. addresses the legitimacy of women writers directly. In *The Debate between a Montanist and an Orthodox*, the Montanist inquires why the Orthodox refuses to accept the teachings and authority of two female prophets who are widely credited as founders of a prophetic revival movement commonly known as Montanism after the name of its third founder, Montanus.[20] The Orthodox speaker contends that women may indeed prophesy, since Mary the mother of Jesus did so, but that women may not write books in their own names. Pressed for scriptural proof, the Orthodox can do no better than offer an admittedly allegorical reading of letters in the name of Paul. For women to write books in their own names is the same thing as praying and prophesying uncovered (1 Cor. 11:5) and having authority over men (1 Tim. 2:12).

Fictitious correspondence probably also composed in the fourth century between one Mary the proselyte and Ignatius of Antioch (who actually lived two centuries earlier) demonstrates that not all Christians subscribed to this interpretation. In her letter, the fictive Mary advises Ignatius on a sensitive ecclesiastical matter, and concludes with a direct allusion to the prohibitions of 1 Timothy: "For I have set forth these matters, not as if I were teaching you, but merely bringing them to the remembrance of my father in God [Ignatius]. For I know my own place, and do not compare myself with someone like yourself."[21]

Women who wished to write in those circles that shared the views of our anonymous Orthodox speaker could clearly have deduced that pseudepigraphy would have to be the order of the day. Even in the case of anonymous texts, it is worth considering whether the authorship of the text was once known, but has been lost precisely because the author was female.

The attitude exemplified by *The Debate between a Montanist and an Orthodox* may have invidious implications not only for Christian literature, but for works by Jewish women as well. Copies of women's writings, if identified as such, might simply never have progressed beyond women's circles into the male mainstream of transmission. Male monks who were subsequently responsible for the transmission of much of ancient literature might not have bothered to copy women's works, for malevolent or other reasons. Since Jewish literature in Greek has depended for its survival to the present day not on Jews (who relinquished, if not repudiated much of Greek Judaism) but on Christians, only that Greek Jewish literature which Christians found acceptable or adaptable was preserved. Thus even writings by Jewish women would have had to pass muster by Christian standards, which appears not to have been conducive to the transmission of ancient writings acknowledged to have been authored by women.

Refusing to be daunted by this situation, a few scholars of late antiquity have suggested that attitudes toward women and men differentiate writings by women and men and allow us to suggest female authorship for certain texts. In particular, it has been suggested that nonmisogynist literature in which women figure as key players and men only as secondary may come from the quills of female authors.[22]

Appealing criteria though these are, they assume, problematically, that men in antiquity could not or would not write nonmisogynist, nonandrocentric literature, even if women were intended as the recipients, as may be the case for collections such as the Apocryphal Acts of the Apostles. They suggest that the key to female authorship lies in the centrality of female characters and the absence of misogyny. But we must also consider the possibility that women wrote works in which men were the central characters and women were portrayed in a negative light.

Diversity and conflict in early Judaism and early Christianity may also have skewed the sources available for a reconstruction of women's history. In the centuries under consideration, there was serious dispute among both Jews and Christians of late antiquity over acceptable behavior and beliefs. Certain perspectives emerged as predominant by the end of the Roman Empire, both within Judaism and within Christianity, although it would be a vast oversimplification to say that diversity ended with the empire. Texts which represent the points of view that did not survive are much less likely to have been copied and transmitted to the twentieth century.

This is especially significant for the study of women because it is frequently within those texts and communities that women now turn out to have been prominent, which as we shall see was probably not accidental. We know that the prominence of women in gnostic Christian communities, in Montanist circles, and among other groups opposed by numerous heresiologists was very much a point of contention. From the so-called orthodox perspective, attitudes toward women and roles of authorship and leadership held by women may have been sufficient to render a community heterodox. This in turn exacerbates the likelihood that the literature of that community will not survive. Even those texts that do survive, such as the Christian *Acts of Thecla* or the Jewish *Conversion and Marriage of Aseneth*, may undergo subtle and not-so-subtle changes in order to tone down their radical implications and render them palatable within communities less hospitable to women.

Thus ancient sources must be approached with great caution to extract reliable information about what women did. In the absence of sources from women themselves, what women thought may be beyond our ability to recover. Having demonstrated over and over again the invisibility of women in most historical sources and reconstructions, some feminist historians despair of our ability to use male sources for women's history. Even if we could correct for pervasive male bias, we would still sometimes be faced with the difficulty of evaluating male reports about aspects of women's lives from which they were supposedly excluded. A considerable number of ancient women's religious festivals fall into this category.

Some historians would further caution that even texts written by women may not be representative of women's perspectives and voices. The fact that men largely control the processes of transmission is likely to mean that only texts deemed acceptable by them are likely to survive.

While acknowledging these constraints, I have found it extremely helpful to pay particular attention to nonliterary evidence from late antiquity: burial, synagogue, and other public inscriptions, and papyrus documents such as divorce agreements, personal letters, tax registers, magical incantations, and so forth. Such sources are less vulnerable to the biases of gender that affect the transmission of literature, since they are less dependent on copying. Their survival to modern times has been largely fortuitous. Men are still disproportionately represented in burial and dedicatory inscriptions, papyrus documents, and so forth. But these materials yield substantial information about women, particularly about Jewish women and pagan women, and after the third century or so, about Christian women as well.

Beyond the problems associated with the sources themselves is a crucial methodological dilemma inherent in feminist historiography. While the study of women's lives, including their religions, must initially take into account whatever women have said, done, thought, and believed, it is abundantly clear that many women subscribe to androcentric views of the universe which tend to reinforce the patriarchal structures constraining their lives. To test whether women's perspectives and views of the universe are somehow qualitatively different from those of men becomes inordinately difficult. Hypothetically, the reconstruction of women's experiences makes no difference if women's experiences are so constrained by androcentrism and patriarchy as to deny any possibility of recovering authentically female experience and cosmology, that is, constructions of the universe. It seems to me to beg the question to propose that the test of such authentic experience would be that it does not accord with male views of the universe which oppress women.

If the description of women's religious activities in the ancient Western world appears daunting, to move beyond reconstruction and description to analysis, interpretation, and even explanation seems at least as difficult. It would be simple enough to explain the diversity of women's religious expressions in the Greco-Roman world as a mirror of the diversity of Western antiquity itself, particularly since the evidence under consideration ranges from late classical Athens to the late Roman Empire and encompasses not only a multiplicity of pagan religions, but diverse forms of Judaism and Christianity. But as we undertake the formidable task of extracting and testing the evidence for women's religious activities from the morass of androcentric sources, we begin to see indications that the complexity and diversity of women's religions is not just a function of the complexity and diversity of ancient life, but of the complexity and diversity of women's lives, which often differed significantly from those of men.

For example, in those communities where there is significant evidence for severe sex segregation, for the confinement of women to the domestic spheres and the assignment of public life principally to men, religion often emerges as women's primary socially respectable public activity. This appears to have been true for women in classical Greek cities, although, for reasons we explore, it seems less true for women in the Hellenistic and Roman period. In such sex-segregated communities, women's religious activities are often performed in concert with other women, in the absence or with the minimal and often peripheral participation of men.

It also appears that many of women's religious activities either (1) reinforce the location of women within the social matrices of their communities (the Matralia, devotion to goddesses such as Demeter, Artemis, Hera, Aphrodite, and Venus); (2) legitimize alternatives and afford women a greater measure of autonomy (Jewish monasticism or Christian asceticism); or (3) provide temporary, structured, and socially sanctioned release from the daily oppressiveness of women's lives (Bacchic rites and the ritual mourning for Adonis). Still other forms of women's religion may combine several of these elements at once. For example, devotion to the Hellenized form of the Egyptian goddess Isis reinforced certain basic social constructs such as the centrality of marriage and family while affirming women's personhood and offering them a degree of autonomy and perhaps temporary relief from social obligations and pressures.

These observations are themselves problematic. First, no ancient sources make these functions explicit, although we sometimes find tacit acknowledgment of these aspects of ancient religious behavior. Rather, modern scholars

infer such functions by looking at the effects of religious activity on the participants, analyzing the unarticulated meanings of symbols, language, and behavior, and drawing on anthropological, sociological, and psychological studies of
contemporary religious behavior to detect the patterns in ancient religions that
might otherwise be difficult to perceive.

Second, to suggest that women's religious activities sometimes reinforce
and replicate women's social experiences and other times offer compensation
for those experiences, or release from those experiences and the legitimation
of alternatives, is to challenge some contemporary theories which contend
that religion must do one or the other, but not both, and certainly not both
simultaneously.

Third, such endeavors are liable to suspicion for the challenge they pose,
directly or indirectly, to the truth claims of specific religions. This problem is
least acute for the study of pagan Greco-Roman religion, since there are virtually no pagan scholars or practitioners to take offense, and most acute for
scholars of Christianity, given the latent, if not active, theological concerns that
undergird much research in this field.

Until fairly recently, insights from anthropology, sociology, and psycholo-
gy have received short shrift in the effort to illuminate the character and
dynamics of early Judaism and early Christianity, not only from traditional
scholarship, but from feminists as well.[23] Ironically, although feminist scholars
have amply demonstrated the pitfalls of androcentric approaches when it
comes to describing women's religious experiences and activities, many bring
similar enough theological agendas to share the biases of their male colleagues
against social science approaches.[24]

In my own work I have sought not only to detect the evidence for
women's religious activities but to find frameworks in which to make sense of
those activities that would improve on the superficial and cynical explanations
of generations of male scholarship which found it sufficient to observe that
women were just inherently more religious (by which they often meant more
superstitious, more emotional, and less rational) than men. My concerns have
been twofold: to inquire whether women's religious activities differed from
those of men, and if so, to ask why. Of course, the latter is a tricky question: it
can imply a gamut of answers from explicit personal motivations to highly
implicit communal functions that are more accessible by their nature to
observers than to participants.

Previous work, including some of my own, presumed that a major function of women's religious activities, such as women's ecstatic worship of
Dionysos and women's interest in early ascetic Christianity, was compensation.
Greek women, normally devalued by men, restricted to their homes, and limited in the scope of their social interactions, found in the ecstatic worship of
Dionysus temporary experiences of community and the expression of critical
social and psychological tensions that were sufficient compensation for the
inadequacies of their normal lives. Similarly, ascetic Christianity offered Greco-

Roman women something pagan society did not afford: a set of standards by which to judge themselves that did not depend on their success as sexual and reproductive beings.[25]

The assumption that religion compensates for deprivation raises numerous problems. Early deprivation theory emphasized economic deprivation as the primary motivation for and function of certain kinds of religious movements, especially millenarian movements (including early Christianity).[26] Interestingly, it was the strong presence of women in such movements that forced some modification in deprivation theory, as critics were quick to point out that many women who join such movements do not come from the lower economic strata. It would take the insights of feminism to point out that what has been taken as the economic class of women is much more frequently the class of their fathers and husbands, and so not truly an adequate reflection of their economic status. But the assault on deprivation theory had begun, resulting in a modified theory in which what matters is a subjective and therefore relative deprivation, which occurs when people perceive that they are excluded from the standards by which prestige and rewards are meted out in their society.[27] New religious movements then function to replace those standards with measures of prestige and reward accessible to those previously excluded.

Yet the elasticity of this modified version of deprivation theory generated new criticism, for the notion of relative deprivation tends to find some form of deprivation for virtually everyone, in the end failing to offer a sufficiently controlled standard for deciding who joins which religious movements and why. One of the most cogent critics of deprivation theory is anthropologist Mary Douglas, who has developed a different model of the correlations between social experience, religious behavior, symbolic systems, and perceptions of the universe.

In my own efforts to understand women's religions, I have found Douglas's work to be tremendously important. Although this study (as will be apparent) is much more than an application of her theory to women's religions in antiquity, Douglas's work has been extraordinarily helpful to me in formulating and clarifying theoretical issues. It is important, therefore, to offer a brief summary of Douglas's ideas, together with some indication of how they may illuminate aspects and patterns of women's religions that are otherwise not so easily apparent. Both the language and concepts of Douglas's work may initially seem formidable, but I think their utility will become increasingly obvious in the chapters that follow.

Douglas has returned to Emile Durkheim's foundational proposition that religious systems, including both ritual and belief, reinforce and replicate social experience.[28] Durkheim argued that religion was nothing but the projection of society onto the cosmos. Douglas agrees that religion and society are integrally related, but she abstains from any judgment about whether religious beliefs derive from social experience or vice versa.

Most importantly, Douglas has developed a new model of the relationships between religion and society, according to which specific beliefs and practices are seen to correlate closely with specific and measurable social experience.[29]

Douglas proposes that human social experience is governed by two predominant factors that produce four major combinations (see Figure 1). The horizontal dimension, which Douglas terms *group*, represents "social incorporation,"[30] measuring the degree to which individuals feel themselves to be a part of a group and the degree of pressure the group can bring to bear on individuals to obtain conformity of action and belief. Group can be measured by considering

> how much of the individual's life is absorbed in and sustained by group membership. . . .The strongest effects of group are to be found where it incorporates a person with the rest by implicating them together in common residence, shared work, shared resources and recreation, and by exerting control over marriage and kinship.[31]

Communes and small sects exemplify group at its strongest, as does the military, while the experience of hermits represents group at its weakest. The dimension of group is not necessarily synonymous with "institution" or "society."[32]

The vertical dimension, which Douglas calls *grid*, measures the degree of regulation individuals experience. In Douglas's own words, "for this the possibilities . . . run from maximum regulation to maximum freedom, the military regiment with its prescribed behavior and rigid timetabling, contrasted at the other end with the free life, uncommitted, unregulated."[33]

Grid also measures the degree to which people share a system of classification or taxonomy. Where grid is strong, people not only hold common beliefs about the way things are but they also utilize language and symbols that communicate those shared beliefs in condensed forms. Classification is not only shared, but implicit rather than explicit. Jokes are an excellent illustration: What makes jokes funny is often precisely what does not need to be said, but what is rather immediately understood by those who share a common classification system. Only to those who do not adhere to the system (including perhaps children) must jokes be explicitly explained. Technical terminology shared by various professions provides another example of strong grid. So does the language of teenagers, for whom expressions like "radical" or "dude" carry connotations instantly recognized by peers but often frustratingly opaque to parents for whom "radical" referred to politics or mathematics, and "dude" to ranches. Where things must be spelled out explicitly, the level of grid is likely to be relatively low.[34]

Grid also measures the value of individual autonomy and the strength of free interactions between individuals. At high grid, individual autonomy is neither valued nor present in reality. Strong grid is "visible in the segregated places and times and physical signs of discriminated rank, such as clothing and

| B | minimal individual autonomy<br>implicit classification<br>B/C<br>Axis | C |
|---|---|---|

| WEAK GROUP | INSULATION: ATOMIZED<br>SUBORDINATION | | ASCRIBED<br>HIERARCHY | STRONG GROUP |
|---|---|---|---|---|
| | Status ascribed, fixed<br>Weak group support<br>Strong pressure to conform<br>Cosmos arbitrary<br>No rewards except fulfilling<br>  one's station in life<br>**Women and men usually**<br>**operate in separate**<br>**domains**<br>**Typifies experience of most**<br>**women in traditional**<br>**societies** and of (male) serfs<br>and slaves | *Grid Axis* | Status ascribed, fixed<br>Strong group support<br>Strong pressure to conform<br>Cosmos strongly moral<br>[good rewarded]<br>[evil punished]<br>Hierarchy valued<br>**Women and men rarely**<br>**operate in same spheres**<br>**Typifies experience of few**<br>**women** | |
| | Group Axis | | Group Axis | |
| | Status achieved, *not* ascribed<br>Strong individualism,<br>  egalitarianism<br>Rewards depend on personal<br>  effort and attributes<br>Cosmos neutral<br>**Women and men compete in**<br>**same spheres**<br>**Gender differences least**<br>**significant here**<br>*But* **despite egalitarianism,**<br>**beliefs about innate gender**<br>**differences may exist** | *Grid Axis* | Status achieved, *not* ascribed<br>Strong group boundaries and<br>  identity: all else negotiable<br>Schisms and divisions flourish<br>Cosmos often dualistic:<br>  good and evil vie for control of<br>  persons and cosmos<br>**Gender distinctions** *may* **be**<br>**obliterated or minimized**<br>**Sexual asceticism flourishes**<br>**here** | |
| | INDIVIDUALISM:<br>COMPETITIVE FREE MARKET | | FACTIONALISM:<br>SECTARIANISM | |

| A | explicit classification<br>maximum individual autonomy | D |
|---|---|---|

**WEAK GRID**

*Figure 1.* Grid, Group, and Gender.
Based on the work of Mary Douglas (see bibliography p. 253).

food."[35] In the military, everyone wears clothing that immediately communicates his or her place within the established hierarchy: It is easy to tell the generals from the enlisted personnel. In corporate America, clothing, hairstyles, the size and location of one's office, and so forth, all communicate information about one's status in the organization. When grid is high, people tend to be judged by personal attributes over which they have no control, such as race, social class at birth, age, and gender—attributes we may designate as ascribed, as opposed to achieved. High grid is thus marked by concern for hierarchical structure, for discrimination according to race, class, and gender.

With low grid, by contrast, individuals have great freedom in their transactions with one another, both in theory and in actuality. Low grid is characterized by the absence of distinctions according to ascribed traits of race, class, age, and gender and by the primacy of the individual. In a low grid work environment, office space might be allotted according to individual needs, not according to status, and office dress codes might not permit one to tell the boss from the clerical staff by their clothing or other aspects of physical appearance.

Four basic social locations emerge from the intersection of these two dimensions (see Figure 1). Each of these locations is associated with a distinctive cosmology, or perception of the nature of the universe and of the place of humans within that universe. For Douglas, cosmology includes "the ultimate justifying ideas which tend to be invoked as if part of the natural order and yet which . . . are evidently not at all natural but strictly a product of social interaction."[36]

Weak group and grid (A) combine to produce an environment Douglas dubs *individualism*. This environment closely resembles the competitive free market: Egalitarianism and individual achievement are highly valued, rank and hierarchy count for little, and personal accomplishment counts for much. Those who experience this environment are likely to hold views of the cosmos as fundamentally neutral: What happens to individuals is seen as the result of individual accomplishments or failures, rather as the result of a divine plan, or of punishment or reward for deeds.

Insulation and isolation from sources of authority and influence characterize the combination of weak group with strong grid (B). Consonant with strong grid, ascribed status and hierarchy are central here, but in the absence of strong group few if any rewards are available for maintaining that status and hierarchy beyond the satisfaction of fulfilling one's allotted station in life. The individual's experience of temporal power as remote and impersonal is reflected in views of the universe in which the divine is similarly remote and disinterested, in which arbitrary fate plays a major role, and individual deeds and actions are seen to have little effect, either on the immediate world or on the cosmos.

With strong group and strong grid (C), status and hierarchical distinctions combine with a strong sense of group boundaries, allegiance, and identity.

The cosmos is perceived as essentially moral, rewarding good and punishing evil.

Factionalism, sectarianism, and egalitarianism epitomize the final combination of strong group with weak grid (D). There are strong external boundaries (a clear sense of who's in and who's out of the group), but few if any formal internal divisions. The ambiguous nature of relationships between individual members of the group creates an environment in which divisions and schisms flourish. The cosmos is often seen as dualistic, split into good and evil as society is split into us and them. Evil often reigns temporarily, but good will triumph in the end. The prevailing view of the universe may be deterministic: Whether an individual is good or evil is predetermined—unaffected by one's own deeds and unrelated to social rank and hierarchy. On the contrary, in keeping with low grid's rejection of ascribed status, here the last may be first, and the first may be last.[37]

Rarely if ever does one of these four combinations suffice to describe the experiences and outlook of all members of an entire society. Within a given society, the experiences of various individuals may be distributed, if unevenly, across several of the four combinations. But in fact, Douglas maintains that the social experience of most people in every society is likely to be that of insulation, isolation, and atomized subordination (strong grid and weak group).[38]

Douglas's grid/group theory, as the model is often called, is extraordinarily useful for explaining the correlations between individual social experience and beliefs individuals hold about all kinds of things, including why the universe is the way it is, which of course encompasses those beliefs we usually label *religious.* But like many other useful grand-scale models, grid/group theory doesn't sufficiently factor in gender. Only in a few places does Douglas devote any attention to the specific experiences of women. Never in her published works does she consider whether her model is formulated on the basis of male experience and therefore is vulnerable to falsification should it turn out that women's experiences are vastly different from those of men.

In one of the few places where Douglas discusses women's experiences, she asserts that women as a whole tend to experience strong grid with weak group. Since this assertion has important implications for understanding women's religions, it is worth reproducing her remarks at some length before critiquing them.

> The social division of labour involves women less deeply than their menfolk in the central institutions—political, legal, administrative, etc.—of their society. They are indeed subject to control. But the range of controls they experience is simpler, less varied. Mediated through fewer human contacts, their social responsibilities are more confined to the domestic range. The decisions they take do not have repercussions on a very wide range of institutions. The web of their social life, though it may tie them down effectively enough, is of a looser texture. Their social relations certainly carry less weighty pressure than those which are also institutional in range. This is the social condition they share with

slaves and serfs. Their place in the public structure of roles is clearly defined in relation to one or two points of reference, say in relation to husbands and fathers. As for the rest of their social life, it takes place at the relatively unstructured, interpersonal level, with other women in the case of women, with other slaves and serfs in the case of slaves and serfs. . . . Women, serfs and slaves (especially released slaves) are inevitably pinned only weakly into the central structure of their society. . . . Their options are few. They experience strong grid. Therefore they are susceptible to religious movements which celebrate this experience. Unlike those who have internalized the classifications of society and who accept its pressures as aids to realizing the meanings they afford, these classes are peripheral."[39]

Douglas appears to view the commonalities of women's social experiences as transcending any differences between them. Women, like serfs and slaves, constitute a universally observable class, whose characteristics will not vary significantly over time or from one society to another. All three, women, serfs, and slaves, share in common the social experience of isolation and insulation. This suggests that women's religious experiences will be predictably similar in seemingly disparate societies, because for Douglas religious ritual and symbolic structures replicate and reinforce the social experiences of participants. It also suggests that the model Douglas has formulated is highly androcentric: that it is, in fact, predicated on male experience as normative.

A number of questions suggest themselves here, many of which this study attempts to address. Are Douglas's observations about women and gender accurate? Are the cosmologies she considers so fundamentally linked to social experience male cosmologies only? Do women share these cosmologies with men, or are there distinctive women's cosmologies that can in turn be correlated with differences between the social location of women and men even within the same geographic and historical communities? If Douglas's portrait of women's experiences is inaccurate, or if women's cosmologies are fundamentally distinct from those of men, is grid/group theory itself to be discarded, or can the model be refined and reformulated to account more accurately for the experiences of both women and men?

To begin with, it seems to me that Douglas's sweeping categorization of women is insufficient, raising key questions. Are the characteristics Douglas imputes to women applicable to *all* women, regardless of their economic and political status? Does the classification "woman" override all others? Are classifications of race and class (including ancient social distinctions of free, slave, and freed) more or less significant than gender in determining women's social location, experiences, and consonant religious beliefs and practices? How much is women's experience a product of the "social division of labour" that Douglas assumes to be virtually universal and perhaps—although she is not explicit on this point—the reason women are less enmeshed in the central institutions of (male) society? Is the social division of labor between men and women itself related to grid and group? Even supposing that women assume the responsibility for the initial care and feeding of small children in all human societies, do different combinations of grid and group correlate with subtle differences in the division of labor? If a universal division of labor is based on assignment of child-rearing and related responsibilities to women, what happens to women in ascetic communities, where childbearing is frequently discouraged, disparaged, or even prohibited, such as some early Christian ones?

Douglas herself observes that attitudes toward sexuality ought to be profoundly affected by grid and group but cautions that in all cases, "the effects of the division of labor between males and females, and the extent to which valued property is inherited on lines laid by marriage alliances"[40] will skew things considerably. She surmises that when strong group is coupled with strong grid, we will find attempts to control procreation and inheritance by strong prescriptions for permitted sex, and that the higher the degree of grid with strong group, the more successfully the "web of prescriptions" can be cast over all sexual activity.[41]

When low grid combines with low group, Douglas suggests, women will suffer in spite of the inherent possibilities for equality present with low grid. Men who engage in unrestricted competition are likely to mingle kinship, politics, and commerce and use marriage, women, and children as pawns in the competitive game. Despite the theoretical possibility of gender equality with low grid, Douglas expects that when low grid combines with either weak or strong group "women [will] be sources of dishonor, ritual defilement and shame . . . so long as marriage is such a valuable stake that they have to be kept under control."[42] Theories about innate differences between the sexes will be used as explanation for this state of affairs.[43] Douglas seems to be saying that in all social locations, male control of women is likely to be the norm, but that the explanations used to legitimate that dominance will vary according to the combinations of grid and group.

Despite the great utility of Douglas's theory in analyzing the diversity of religion in western antiquity, her brief discussion of gender demonstrates the need to refine her model further. Grid/group theory must be able to account for distinctions of class and race; it must offer a more nuanced model for the social divisions of labor by gender; and it must recognize that strong group with weak grid does not exhaust the range of women's social experiences.[44]

As we begin to examine the religious experiences of women in antiquity, if Douglas's model is essentially workable, we should expect to find that women hold the same cosmologies as men of the same social location. To the extent that men and women within a given community do not share the same location, their cosmologies should show predictable differences. And to the extent that gender effects further nuancing, so should we expect to find those nuances reflected in ritual and belief, in metaphor and symbol.

We will consider whether different locations on the grid/group map are associated with differing degrees of gender segregation and separation. Douglas identifies the hierarchical sector (high group with high grid) as the location where the spheres of male and female activity overlap little: where men and women do not compete in the same spheres.[44] Compare this to the individualist sector (low grid with low group), where women and men do compete in the same spheres, as the vivid example of late twentieth-century American society makes clear. Douglas's own observations suggest that the women and men who experience low group with high grid should exhibit shared cosmology by virtue of their common experience. This will be true even if the women are there by virtue of their gender, and the men are there by virtue of their relatively low social class or social standing.

But it may be rather that women and men adhere to cosmologies that are similar in their reflection of low group and high grid. In that case, we may expect to find not necessarily identical rituals, but rather rituals that reflect similar social location. Of special interest will then be the factors that lead women and men to hold shared cosmologies and to participate in shared rituals and those that lead men and women to participate in gender-specific rituals and to hold gender-specific cosmologies.

Finally, since Douglas considers that the specifics of gender differences are always culturally constructed, there remains the fundamental question of whether different social locations themselves generate different cultural constructions of gender. The magnitude of this question renders its answer beyond the scope of this study, but it is germane even here, for certainly in late antiquity, religion was one of the many means by which gender was constructed and confirmed.

I envision this study as much more than an attempt to test Douglas's theories on women's religions in antiquity. Still, my debt to her work, and my repeated references to it through this book, require me to say a little more about the kinds of unavoidable dilemmas that this sort of endeavor encounters.

As I hope to demonstrate, Douglas's theory, in a gender-sensitized form, may be exceptionally useful for the study of religious diversity in Greco-Roman antiquity, especially for the analysis of gender-related differences in practices and beliefs. Conversely, of course, Greco-Roman religions offer a significant set of test cases for Douglas's theory. Testing her theory and refining it on the basis of ancient examples is complicated by the difficulties we have already acknowledged in our efforts to reconstruct what women actually did in religious contexts in the Greco-Roman world. The sources from late antiquity rarely facilitate the kinds of research that anthropologists take for granted: When I proposed this kind of approach to Douglas herself, she expressed skepticism that the evidence from the Greco-Roman world would allow me to proceed. Frequently we know much more about ancient cosmology than we do about ancient social structure, and Douglas is adamant that while we can predict cosmology from social structure, the reverse is fraught with methodological minefields, since different social structures can generate similar cosmologies, for different reasons.[46] I am well aware of all these difficulties. I think that in the end the results will justify the risks, including the possibility of distracting the general reader now and again with theoretical discussions and technical language that can only be translated into reasonable English so far before the precision of implicitly language is lost! For all this, I beg my readers' indulgence.

The chapters that follow combine description and analysis of a variety of women's religions. Chapters 2, 3, and 4 consider women's devotion to Greek goddesses such as Artemis, Hera, and Demeter, and to two male deities, Adonis and Dionysos. The rites of Roman matrons are the subject of Chapter 5; Chapter 6 examines women's worship of the Egyptian goddess Isis in her Greco-Roman transformation. Chapter 7 investigates women's religious offices and leadership in Greco-Roman paganism.

Chapters 8 and 9 explore some of the salient differences in the portraits of Jewish women and women's Judaism that emerge from rabbinic sources on the one hand and evidence from Diaspora Jewish communities on the other. A discussion of Jewish women's leadership roles in ancient synagogues is interwoven into both chapters.

Three chapters are devoted to early Christian women. Chapter 10 investigates the appeal of early Christianity to women and the roots of early Christian antagonism over women's roles, focusing on the increased autonomy early Christian sexual asceticism afforded women. Chapter 11 pursues these issues in the emergence of Christian movements ultimately considered heretical, particularly those in which women were prominent or significantly visible, such as the New Prophecy (also known as Montanism). Chapter 12 scrutinizes the evidence for women's religious offices and leadership in Christian communities through the end of late antiquity.

In Chapter 13, I consider why Christians in particular found the notion of female priestly office so divisive, as an entry into the general inquiry about women's religious leadership in the range of Greco-Roman religions. The epilogue expands the theoretical considerations introduced here to develop a more nuanced though still preliminary model of women's religions.

# — 2 —

# Women's Devotion to
# Ancient Greek Goddesses

Although women in ancient Greece unquestionably worshipped both male and female deities, goddesses played a much more central role in their religious lives. With the exception of offerings to Asklepios, the god of healing, most recorded offerings of Athenian women were made to female deities.[1] Much of women's devotion to ancient Greek goddesses coalesces around marriage and fertility, reflecting women's primary roles in Greek society as promoters of life, whether agrarian or human.

At the temple of Artemis at Brauron, for example, on the southeastern coast of Attica not far from Athens, the festival of the *Arkteia*, or Bear festival, was held every four years.[2] According to the myth associated with the festival, a female bear came to Brauron: tamed, she played with young maidens. One day, she scratched a girl, whose brother, incensed, then killed the bear. When a plague descended on their city, the Athenians consulted the oracle of Apollo at Delphi, which ordered Athenian girls to play the role of bears in the rites for Artemis at Brauron.[3]

Participation in the Arkteia was apparently a prerequisite for marriage for proper Greek girls, and the rites had several components.[4] Girls aged about seven to fourteen competed in athletic games and races, probably according to age groups: prepubescent younger girls may have competed in the nude; older girls wore short tunics cut above the knee. Held in public view, these competitions may have provided an occasion for young Greek men to observe prospective brides.[5] Dances and choral performances were also part of the rites.

We know little about the mystery initiation that took place at the festival. Taking into account an Athenian vase that depicts Artemis and Apollo accompanied by a man and a woman both wearing bear masks, together with a much discussed brief reference in Aristophanes's *Lysistrata*,[6] Susan Cole proposes that a bear-priestess instructed the young girls in their enactment of the bear dance.

According to Aristophanes, the young bears of Artemis wore saffron-colored robes. Paula Perlman points out that saffron clothing was especially asso-

ciated with brides and married women, and suggests that their use at the Arkteia indicated the girls' readiness for marriage, having successfully completed
the rites of the festival.[7] New clothing symbolized transition to a new social
status, eligibility for marriage. Cole, too, illuminates the transitional nature of
the Arkteia. By imitating wild bears, young girls engaged in activity "antithetical to that of well-behaved Athenian girls and married women."[8] As Victor
Turner has shown, such temporary ritual reversal characterizes rites of transition from one social status to another, and reinforces the appropriateness of
the roles that are temporarily inverted.[9] Cole observes that the Arkteia placed
young girls under the special protection of Artemis during precisely that
period when their virginity was most vulnerable—the time when they were eligible for marriage but not yet married.[10]

The Arkteia to Artemis may not have been celebrated at Brauron past the
fourth century B.C.E. The sanctuary there is thought to have been severely
damaged by flooding sometime during that century, and the temple to
Artemis Brauronia on the Acropolis in Athens may have subsequently taken
over the functions of the temple at Brauron.[11] A festival of Artemis similar to
the Arkteia was also celebrated by Athenian women at a temple overlooking
Mounichia harbor.[12]

In addition to marking the transition from childhood to puberty and mar-
riageability, devotion to Artemis marked rites of passage for women to marriage itself and to motherhood. According to cult regulations from Cyrene in
the fourth century B.C.E., brides were required to visit a special room prepared
for them in the temple of Artemis during her festival: penalties in the form of
presumably more expensive sacrifices were prescribed for women who
neglected to do so. Prior to giving birth, pregnant women were expected to
make the same visit, offering the head, feet, and skin of a sacrifical victim to
the Bear-priestess. Those who failed to do so had to sacrifice a full-grown animal victim after delivery as a penalty.[13]

At Brauron, women dedicated woven garments to the goddess. Their
descriptions provide a colorful glimpse of women's clothing: one was "frog-
green," another had a broad purple border with a wave design woven in,
another was embroidered and scalloped.[14] Some scholars think that women
presented completed textiles to Artemis in gratitude for the birth of a healthy
baby. Unfinished textiles listed in the temple inventories may have been left by
women who died in childbirth.[15] Other textile offerings may have been made
to Artemis in gratitude for relief from female reproductive ailments. Some
scholars have even proposed that young girls dedicated the rags used during
their first menstruation to Artemis.[16]

Similar activities and concerns underlie Pausanius's description of
women's worship of Hera, written in the second century C.E. Games called the
*Heraea,* named after the goddess, were held at Elis every four years. Unmarried girls competed in footraces, according to three categories of age. They ran
with their hair down, wearing short tunics cut above the knee. The winners

received olive crowns and part of the cow sacrificed to Hera and were allowed to dedicate statues with their names inscribed.[17] According to an explanatory legend, the games were instituted by Hippodameia in gratitude to Hera for her marriage to Pelops.

Since we tend to associate Hera with the protection of married women, and sometimes also with the protection of women in childbirth, it is interesting to speculate whether the Heraea at Elis also functioned as puberty rites. In any case, Pausanius's account may display traces of the division of Greek women into three traditional categories (maiden, mother, and crone—prefertile, fertile, and postfertile). The games were conducted by a group of married women called the Sixteen, who dedicated a robe they had woven for the goddess on this same occasion. According to Pausanius, the oldest, most noble, and wisest women from each of the sixteen ancient cities of Elis were chosen originally to constitute a council to settle grievances between the peoples of Elis and those of Pisa. Although only eight of those cities still survive, Pausanius says, two women from each city are still chosen to constitute the Sixteen. These older women in turn are served by married women. The Sixteen are also said to arrange two choral dances, the Hippodameia and the Physcoa (named after a resident of Elis who is said to have borne Dionysos a son and to have inaugurated the worship of Dionysos). Prior to their ritual services, the Sixteen performed the proper rites of purification. Pausanius's account suggests that well into the Roman Empire, traditional aspects of Greek religion persisted that served to reinforce traditional roles for women.

Of all the Greek goddesses venerated by women, the most compelling may well have been Demeter. Her worship was hardly confined to women: the great mysteries of Demeter at Eleusis were celebrated by both Athenian men and women and were open, at least in theory, to all except barbarians (defined as those who did not speak Greek) and murderers.

The centrality of Demeter in the religious lives of Greek women may be approached first from a consideration of the ancient *Homeric Hymn to Demeter*, which dates at least to the seventh century B.C.E. It relates that Persephone, the daughter of Demeter, was abducted by Hades, lord of the underworld, while gathering flowers. Grief-stricken Demeter, haunted by the cry of her daughter sobbing as though she were being raped, learns that Zeus has assented to the rape-marriage of Persephone to Hades. In anguish and anger, Demeter deserts Olympus for the mortal city of Eleusis where she seeks refuge in the house of the king Keleos, as a nursemaid to his only son. Ultimately, she reveals herself to the people of Eleusis, and instructs them to build her a temple and inaugurate her rites.

But even the veneration of the people of Eleusis brings the goddess no consolation for the loss of her daughter, and Demeter wreaks devastation on the earth, preventing any seed from growing. Cattle and crops alike die, and the human race itself is in danger of extinction. Perceiving that Demeter is the cause of human suffering, Zeus sends each of the immortals in turn to beseech

Demeter to relent, but she refuses all their proffered gifts and honors. Only the sight of her daughter will appease her.

Realizing that he has no choice, Zeus sends Hermes to Hades, with orders to bring Persephone up from the underworld. Hades complies with his brother's orders, but first he forces his unwilling bride to eat a pomegranate seed to ensure that she will not remain forever back with her mother.

Reunited finally with Demeter, Persephone relates the tale of her abduction, receiving comfort and consolation in her mother's company. Zeus consents to let Persephone spend two-thirds of the year with her mother and the other immortals, compelling her to live with her husband only one-third of the year. Demeter restores the fertility of the earth, and her rites are established firmly at Eleusis.

Many varied explanations of the myth of Demeter and Persephone have been proposed. It has long been seen as a mythological explanation for the origins of seasons[18] and of the Eleusinian mysteries, as well as for the prominence in the priesthood of certain families at Eleusis. The worship of Demeter at Eleusis and elsewhere has long been understood as an integral part of Greek concerns to maintain fertility of all sorts, and women's centrality in her cult has been thought to require little investigation or explanation, given the natural associations of women with fertility.

As the women's movement began to make inroads on the study of antiquity, new readings of the myth were proposed that sought to take more seriously the worship of Demeter as women's religion. Marilyn Arthur has suggested that the *Hymn to Demeter* is a paradigm for women's psychological development in ancient Greek society, accommodating women to their inevitable separation from their mothers and their submission to patriarchal order. The myth reflects girls' movement from the idyllic, protected, but almost fantasylike community of women to the "real" world dominated by men into which married women move, and to whose authority even Demeter is compelled to yield.[19] Bruce Lincoln has revived the suggestion of Henri Jeanmaire that the *Hymn to Demeter* reflects early Greek initiation rites for girls from puberty into adulthood and marriage.[20] While neither is without problems (Arthur relies heavily on a Freudian model of psychological development and the evidence for actual initiation rites is largely conjectural), both take seriously the extent to which the *Hymn to Demeter* and many festivals to the goddess express the realities of Greek women's lives in ritual and myth.

Regardless of the specific psychological dynamics of ancient Greek women, particularly free women, there can be no doubt that the myth expresses real tensions in the lives of Greek mothers and daughters. Marriage (depicted here, as in much other myth and ritual, as inaugurated through rape), inevitably separates mothers and daughters, who grieve for one another and long for reunion. Men in general, and fathers specifically, are depicted as the agency for such separation, although marriage is not disparaged in itself, but only lamented for the pain it causes mothers and daughters. Demeter is enraged at

Zeus for permitting the abduction of Persephone, but never does she suggest that Hades is himself an unfit husband.

The Eleusinian mysteries are presented in this text as what mortals must do to assuage Demeter and guarantee the fertility of the earth and presumably their own as well. It also seems likely that the mysteries addressed, at some level, the social and psychological dimensions of achieving fertility, offering some resolution of these tensions not only through the compromises that the myth offers, but through their enactment in ritual, namely the restoration of fertility to the earth and Persephone to Demeter.

Although the actual rites of the Eleusinian mysteries remain unknown (as do those of most other ancient mysteries), we have ample evidence for some women's festivals to Demeter, which are intimately linked to the story of the goddess and her daughter. Such festivals and observances seem to have continued up through the Roman period.

Best known of these is probably the *Thesmophoria*, which was widely celebrated in ancient Greek cities[21] at the time of sowing in the month of Pyanopsion (November). There are numerous problems reconstructing the details of women's rites at these festivals, especially since key aspects were considered mysteries to be shared only with initiates, and since no men were allowed to observe. Where our male sources got their information remains a mystery to modern scholarship. However, it seems[22] that women called *antlētriai*, or those who draw (water), kept themselves ritually pure for three days (including abstinence from sexual contact), and then descended into underground pits called *megara*, where they retrieved the remains of suckling pigs that had been thrown down into the pits at some point earlier. Pine cones and wheat cakes baked in the shape of serpents and penises were apparently also thrown into the megara and similarly retrieved. The decomposed remains of the pigs, pine cones, and cakes were placed on the altar of Demeter and mixed with the seed corn about to be planted, presumably to ensure a fertile harvest.

Several sources associate the Thesmophoria with jesting obscenity. Diodorus of Sicily[23] reports that women at this festival engaged in *aiscrologia*, dirty talk, in imitation of the crude jokes that cheered Demeter when she grieved for Persephone. Similar behavior, including obscene gestures, is attributed to women's celebration of the *Stēnia*, a festival observed on the ninth of Pyanopsion just prior to the Thesmophoria.[24] Interestingly, in the *Homeric Hymn*, the obscenity of Iambe's jokes is not explicit. The late second-century C.E. Christian author Clement of Alexandria transmits a version of the myth that replaces Iambe with a woman named Baubo, whose name signifies a vagina, or a representation of the vagina, and attributes Demeter's mirth to Baubo's exposure of her naked body and to obscene gestures by the baby Iakkhos.[25] According to the first-century geographer Strabo, initiates traveling in procession from Athens to Eleusis crossed a bridge on their way into the temple precinct where observers lined up to shout obscenities as they passed.[26]

The Thesmophoria probably ended with feasts presided over by women from prominent Athenian families. Other rites whose precise timing is unknown included women beating one another with bark baskets, rites of penance, and the release of prisoners. Women were forbidden to wear crowns of flowers (Persephone, after all, was picking flowers when Hades abducted her), to engage in sexual relations, or to eat pomegranate seeds. Allaire Brumfield argues, against older views, that participation in the Thesmophoria was not limited to Athenian women of high standing, but was likely open to most if not all women.[27]

Less well understood but equally fascinating is the festival of the *Haloa* celebrated in the Greek month of Poseideion (December/January),[28] apparently only at Eleusis. At the Haloa, women drank wine and feasted on many foods, excepting certain prohibited ones, such as pomegranates. The priestess of Demeter and Kore (Persephone) apparently whispered obscene secrets in the ears of the women, who shouted obscenities in response and brandished representations of male and female sexual organs, sometimes called phalli and cunni. Those fashioned out of cakes may have been eaten as part of the rites. Scholars have also speculated that these objects were used to simulate sexual intercourse in the interests of fertility.[29] While all this transpired in closed rooms, the archons (magistrates) sat outside and discoursed on the gifts and rites of Demeter to the general populace (presumably male), showing them genitals fashioned out of pastry.

In the second century C.E. Pausanius knew of numerous women's festivals and sacrifices to Demeter. In Corinth, elderly priestesses sacrificed cows in the temple of Demeter by slashing their throats with a sickle. Outside the temple, Pausanius reports there were statues of women who had been priestesses, as well as images of Demeter and Athena. Inside, he saw seats where the women sat while waiting to sacrifice the cows. But Pausanius did not know the central cult object at the sanctuary, since men were forbidden to see it, and he obviously had no women informants who would divulge that information![30]

In Achaia, a seven-day festival to Demeter was celebrated at the sanctuary of Mysian Demeter, where a man from Argos named Mysius was said to have given Demeter welcome (during her earthly wanderings). For the first few days, men and women participated, but on the third day, men withdrew, and women performed some secret ritual. The following day, the men returned, and the women and men took turns laughing and jeering at one another.[31] It is tempting to see these rites as a metaphor for procreation, seen from a male perspective: a man and woman do something together; the man withdraws; the woman does something secretive (e.g., conceive); the man then returns. Whether or not this is the case, the rites in Achaia seem to resemble those of the Thesmophoria and other known women's festivals of Demeter.

In addition to rites promoting fertility and celebrating women's transition from one sociosexual category to another, Greek women's devotion to goddesses also took the form of caring for their sanctuaries and the cult statues within them. Statues were periodically washed, dressed, and anointed. Such acts can be seen as replicating the major daily tasks of women in much of Greek society: washing, spinning, weaving, and preparing food.

To summarize, women's religious festivals and observances, widely acknowledged as their main form of public activity, may thus be seen as addressing more than the concerns of Greek society for fertility and continuity: they also reinforced male expectations for women's proper behavior and roles. Undoubtedly women shared these expectations to a significant degree, yet even so central a myth as the *Homeric Hymn to Demeter* reflects a fascinating critical women's perspective. While ancient Greek men (and modern male scholars) have seen marriage as the transfer of women from the household of the father to that of the husband and downplayed its re-presentation as rape, Greek women, both mothers and daughters, are likely to have seen it as a painful separation from the household of their mothers and a community of women. Women's rites also undoubtedly afforded women some pleasure in each others' company, although they may also have been the arena for status competition between women, about which we wish we knew more. Athletic competitions such as those in honor of Artemis and Hera may well have been ancient versions of beauty contests or marriage markets. John Winkler suggests that women's laughter in the rites of Demeter was not just focused on sexuality and obscenity in general, but was specifically directed at men's sexual adequacy or its absence, and at the faulty male perception that men played a crucial role in human and agricultural fertility.[32]

From a wealth of recent research, it is apparent that the vast majority of women in classical Greece were isolated and insulated from the sources of social control and power (which Mary Douglas calls low group) and under great pressure to conform (high grid).[33] Free Greek women lived most of their lives physically segregated from men, in a part of the house called the *gynaikōnitis*, or women's quarters. They rarely took their meals with their male relatives, who often dined outside the home in the company of other men. They also seldom appeared in public in those places that were the center of male activity, such as the markets, the courts, the schools, and so forth. Women probably attended theatrical performances, but drama was intimately connected with Greek religion, which was the major occasion when Greek women assembled in public, so this is not surprising.

There were, of course, exceptions. Women who were foreigners or slaves had more occasions to appear in public and to mingle with men, whether for sexual or economic ventures (or even both simultaneously). The stereotype of the cultivated foreign woman (*hetaira*) who accompanied men to formal dinners, entertaining them with music, witty conversation, and sexual diversion is well known. For poor free women, seclusion was unaffordable, whereas slaves (both male and female) performed the public and domestic tasks that obviated the need for their mistresses to appear in public.[34]

It should not surprise us to find these distinctions in social location reflected in the religious lives of women in ancient Greece. If gradations of group and grid are significant, they should manifest themselves in religious activities that differ precisely in their reflection of these subtleties. Since, however, we know more about the religious observances of free Greek women, primarily in their devotion to numerous goddesses, let us focus on these.

The rites of goddesses such as Artemis and Hera are illuminating. In their apparent exclusion of foreigners and slaves, they reflect the strong classification of Greek society that divides individuals by categories over which they have little control: slave or free, city or tribe of birth. Restricted to women, such rites reflect and reinforce the separation of free Greek women from men. They categorize females according to their reproductive functions and their relationships to men, past, present, and future, over which women had little control. These cults distinguish between prepubescent girls, girls as yet unmarried but eligible for marriage, married women (still capable of producing children), and perhaps even widowed women, or those beyond their childbearing years. In the triad of Kore/Persephone, Demeter, and Hecate, goddesses themselves reflect the trifold categorization of women as prefertile (the virgin), fertile (the mother), and postfertile (the crone).

I do not doubt that one of the primary functions of women's worship of ancient goddesses was the promotion of fertility and continuity for the community, which required the fecundity of all life, both agrarian and human. But the specific rites accomplish far more than magical guarantees of fertility. They ensure women's transitions from one sociosexual category to the next, even asserting that these roles are fundamental to the preservation of the cosmos: in short, they keep women in their particular place.

But they are not without subversive elements that may have played significant roles for women. As we have seen, the myth of Demeter and Persephone expresses important tensions in Greek society—the centrality of the emotional bond between mothers and daughters and the pain that their inevitable separation inflicts. Demeter and Persephone by no means exhaust the range of mother-daughter interaction in ancient Greece, which was surely more complex than the unmitigated love and affection, untainted by any negativity, which the *Hymn to Demeter* depicts. But I suggest that it is precisely women whose lives are lived in relative isolation, whose contacts with men are limited, whose sons, as in ancient Greece, were taken away from them at an early age (around six or so) and raised in an almost exclusively male society, who are likely to find the universe well reflected in the myths and rituals of Artemis, Hera, Demeter, and doubtless other goddesses as well.

# — *3* —

# *Women's Devotion to Adonis*

Although Greek women clearly venerated male deities, propitiating Asclepius for cures and seeking the vision of Apollo's oracle at Delphi, little of most Greek women's religious devotion seems directed toward the male Olympians. Rather the male gods most popular among women tend to be foreign, or to bear a foreign attribution. Women's devotion to Greek goddesses tended to reinforce roles for women acceptable within the confines of low group and high grid, affirming their experiences as daughters, mothers, wives, and even widows. But this was by no means the case in all instances. Their devotion to two male deities, Adonis and Dionysos, seems much more concerned with expressing and mitigating their frustration and oppression within communities characterized by severe gender antagonism. The worship of these male deities appears to provide women some opportunities to relax the specific constraints of insulation, in particular by rituals and concomitant beliefs and symbols that reduce isolation and increase autonomy. Interestingly, although both Adonis and Dionysos are male, each is closely associated with at least one goddess, and both are presented as foreign deities, although the importation of the worship of Dionysus, if it was that, occurred well before the classical period in ancient Greece, whereas Adonis probably arrived in the fifth century B.C.E.

The worship of Adonis spread through the ancient Mediterranean as one of a number of cults of dying and rising gods, the male consorts of mother goddesses revered in the ancient Near East. The cult was originally centered at Byblos in Phoenicia. In his Greek form, Adonis was mourned as the consort of Aphrodite at rites called the *Adonia*. In Athens, the Adonia was celebrated by private groups of women on the rooftops of their houses on one or possibly more of the hot summer nights in late July. From the fragmented references to the Adonia, John Winkler concludes that Greek women planted seeds of various sorts, lettuce and fennel among them, in portable pots, and placed them in the sun. Once the seeds had sprouted, they were left unwatered to wither and die. Set out on the rooftops, these short-lived plantings were the gardens of Adonis, around which women danced, sang, and jested before disposing of the soil and shriveled plants in a spring or the sea, in a kind of mock funeral of Adonis.[1]

In his brief but fascinating essay, Winkler proposes a novel and intriguing understanding of this festival and the raucous laughter of Greek women with which it, and several other women's festivals, were associated. Common to myths of the lovers of Aphrodite is their shared and (for Winkler) essential fate, in which all become "no longer erect, decisively and permanently so."[2] Winkler points out that for Greek men, such stories were undoubtedly understood as warnings about the dangerous powers of female sexuality. For Greek women, he suggests, the Adonia afforded alternative interpretations. "What the gardens with their quickly rising and quickly wilting sprouts symbolize is the marginal or subordinate role that men play in both agriculture (vis-à-vis the earth) and human generation (vis-à-vis wives and mothers)."[3] The laughter of the celebrating women, long associated with obscenity and instructions for sexual misbehavior, may then be seen as women's amusement at men's misconception of their own importance in fertility and productivity. "The many religious-social gatherings of ancient Greek women, so few of which were noted by men, are the obvious location for sharing knowledge about male adequacy—or inadequacy."[4]

Winkler concedes that his interpretation, like that of Marcel Detienne,[5] may be "overly preoccupied with phallic issues of interest to men." The significant difference is that Detienne found phallic men central to the significance of women's rites, whereas Winkler finds them peripheral and amusing in their pretensions to importance.[6] It is nevertheless an admirable inquiry into the possible ways in which Greek women might have understood their own activities in contradistinction to the interpretations of Greek men.

In the third century B.C.E., the poet Theocritus devoted an idyll to the attendance of two women at a festival of Adonis in Alexandria.[7] Sponsored by Arsinoe II, queen of Egypt, the festivities were held in the main palace, in honor of Arsinoe's dead and now deified mother Berenice. Written under royal patronage, the *Idyll* as a whole is a political paean to Arsinoe, in which the mundane women who attend the festival are the antithesis of the powerful, sophisticated ruler.[8] The work provides a rich detailed description of Alexandrian life, of the festival of Adonis, and of two typical participants.

At the opening of the *Idyll*, two women, immigrants to Alexandria from the Sicilian city of Syracuse, are about to set out to attend the festivities. The constraints of their lives are immediately perceptible. Though fellow immigrants and friends, they have not seen one another in quite some time. Praxinoa lives at the far end of town, separated from her friend, and attributes her isolation to her impossible husband, who provides her a hovel, not a house, and keeps her away from Gorgo out of spite. When she complains to Gorgo bitterly in front of her young son, Gorgo warns her not to talk that way in front of the child, and reassures the boy that his mother is not talking about his father. But Praxinoa and Gorgo resume their complaints about their husbands, who bring back salt from the markets instead of soap and rouge, who are penurious and constantly making work for their wives. As they prepare to leave, Praxinoa's baby begins to cry, but she rebukes him irritably and hands him over to the care of a slave. This last detail suggests that Praxinoa is not quite as impoverished as it might appear, for she has at least two slaves in attendance.

Negotiating the streets of Alexandria, where they walk unaccompanied by slaves or male relatives (encountering numerous rude men along the way), they arrive at the luxurious royal palace. After admiring the beautiful tapestries, they listen to women singing hymns of Adonis in a ritual competition. At the conclusion of the ceremony, sighing for Adonis, Gorgo reminds Praxinoa that her grumpy husband will be wanting his dinner, and they set out for home.

The hymn sung in the *Idyll* depicts much of the ritual celebrated by women. On this occasion, Arsinoe, called the Double of Helen of Troy, lavishes gifts on Adonis, Aphrodite's young lover, as tribute to the goddess, who has immortalized Arsinoe's mother, Berenice. Displayed in the palace are a multitude of offerings: fresh fruits, flowers in silver baskets, sweetened cakes of semolina flour and honey, and soft purple wool blankets for the couch of the goddess and her lover. A garden of Adonis, a bower for the lovers, has been constructed, above which tiny cupids fly.

On the first day, the women celebrate the sexual union of the goddess and her lover. The next day at dawn, though, the women carry the god down to the sea, casting him into the waters. There they mourn him, loosing their hair, unfastening their clothes, baring their breasts, and singing funeral dirges. But the singer's hymn ends with the expectation of the god's return, and Gorgo goes home with the words, "Farewell, beloved Adonis, and may your return find us happy."

Although it would not be difficult to see the appeal of the Adonia for harried married women in ancient Greek cities, Theocritus's presentation heightens the comparisons between the drab lives of the Syracusan women, fraught with gender antagonism, and the opulent, conflict-free paradise of Aphrodite and Adonis.

By comparison with stingy, grouchy, demanding husbands, Adonis appears the ideal lover. He is beautiful and young, no more than eighteen or nineteen years old, in a culture where fifteen-year-old girls often married first husbands twice their age.[9] He makes no demands of his wife, but only makes love to her in a beautiful green garden on a couch covered with soft blankets, after which he dies, to return again bringing love and delight. There is virtually no resemblance except that of male gender between Adonis and the prototypical Greek husband of Theocritus's Alexandria. Similarly, the only baby boys in the ideal world of the Adonis festival are sweet little cupids who flutter about the bridal bower and display none of the obnoxious characteristics of demanding, crying children.

As Frederick Griffiths points out in his insightful essay, the Adonia offered Alexandrian women a rare day of festal freedom, in a cult whose symbols and myth temporarily reversed the powerlessness and boredom of many women's daily lives. In the case of Arsinoe's Adonis festival, the power of the queen reinforced and enhanced the creation of a female cosmos, which provided a welcome respite from the oppressive, obnoxious male society just outside the palace gates. The powerlessness of women is mitigated and contrasted in the awesome power of Arsinoe and her divine counterparts, Aphrodite and Berenice. Whereas Praxinoa and Gorgo must dash home quickly to pacify irritable husbands, Arsinoe, Aphrodite, and Berenice are answerable to no man.[10]

Analyzing Theocritus as evidence for the increased "emancipation" of women in third-century Alexandria, Griffiths concedes that "[t]he flowering of vast new metropolises . . . diminished the male monopoly on the marketplace, the law court, the palaestra, the stoa and the throneroom,"[11] but suggests that the picture of domestic life remains remarkably similar to that of fifth- century Athens.[12] Women were still largely confined to private domestic roles, and the relationships between husbands and wives were fraught with hostility and struggles for control. The constraints on women were a little looser in third-century Alexandria: respectable women could mingle with men in public a little more freely. More importantly, though, the political realities have changed, for the dominant ruler in Alexandria is a powerful woman, powerful enough to bring women's festivals in from the margins of Greek society to the very palace itself.

Further, despite his trenchant depiction of ordinary women's lives, Theocritus was no social critic. Griffiths argues persuasively that Theocritus emphasizes the boorishness of Praxinoa and Gorgo in comparison to the sophisticated Arsinoe precisely to affirm the essential justice of the social stratification. Women like the Syracusan housewives deserve no better than their lot, in Theocritus's view, just as Arsinoe is infinitely worthy of the power, prestige, and divine favor she commands. Thus the Adonia may very well get women out of the house for a little relief, but for Theocritus its real purpose is the demonstration of the inevitable distance between the common women and their almost divine queen, precisely through a temporary relaxation of the boundaries between them.

Leaving aside the specific political motifs of Theocritus, we may see significant distinctions between the Adonia and women's other festivals such as the Thesmophoria, the Haloa, or the rites of Artemis Brauronia. All of them reinforce women's location within the structure of Greek society as guardians of fertility: as daughters, wives, and mothers. By contrast, the Adonia is concerned not with marriage, childbearing, and the transfer of women from one male-defined role to another, but with sweet sexuality, unencumbered by the ancient equivalent of dirty socks and diapers. Griffiths cites with approval the observation of Marcel Detienne that the union of Aphrodite and Adonis symbolizes not marriage, but antimarriage.[13] "If the Thesmophoria celebrates the strength of the marital bond, the Adonis festival explores the fantasy of liberation from that institution."[14]

Yet the worship of Adonis was hardly extreme in its provision of protest, for the critique of husbands and society it offers is modest indeed, the release it provides confined to a brief outing and some carefully orchestrated weeping and mourning. Like other occasions of ritual reversal, the ultimate function of the Adonia is the affirmation of the social order as it exists, not its permanent transformation into something else.[15] Ritual reversal is hardly revolution. But in a culture where women had few options, it was apparently sufficient to enable Gorgo to return home to cook dinner for her irritable husband, consoled by the promise of Adonis's return.

Thus where women's worship of Greek goddesses reflected the low group and high grid that characterized the social experience of most Greek women (that is, isolation or insulation), devotion to Adonis, as portrayed in Theocritus, reveals elements more characteristic of lowered grid and modestly strengthened group. Since the worship of Adonis, the consort of Aphrodite, tended to undercut marriage, it is not surprising that the cult shows little concern to discriminate among worshippers on the basis of sociosexual categories. Clearly, the cult expresses tensions most acutely experienced by married women, but at least in the description of Theocritus, the women who assembled at the palace of Arsinoe did not distinguish between themselves on the basis of marital status or motherhood. Altogether absent are the divisions of women into prefertile, fertile, and postfertile women so characteristic of high grid. Even distinctions of social class, of which Theocritus is well aware, are at least temporarily minimized in the Adonia, where housewives of modest means like Gorgo and Praxinoa gain access to the royal precincts that would otherwise be beyond their social parameters. Queens and ordinary women alike worship at the bowers of Adonis and Aphrodite.

Despite its historical connections with New Year festivals, with their paramount concerns for life and fertility, in its Hellenized forms the cult of Adonis appears to have played little part in the integration of women into socially mandated roles. It does not assist in the transition of women from one sociosexual category to the next, but instead obliterates most of the distinctions between women in a regulated, or high grid environment and brings women together out of their homes for a brief period of time.

The rituals of Adonis are what we would expect for weakened grid. The mourners of Adonis release their hair and their clothes and bare their breasts in public (although presumably only in the sight of other women by the seashore). Given Douglas's hypothesis that control of the body reflects the concern of society to control the individual, these ritual releases of the body express temporary release from the constraints of high grid. The women weep and lament, singing funeral dirges for the dead deity, a licensed expression of emotion that typifies low grid. Requiring women to leave their homes and to enlarge temporarily the scope of their social interactions by joining other women to mourn the god, the Adonia strengthened the degree of group that women experienced.

Participation in the Adonia thus comprised a brief ritual foray into an area of lessened grid and strengthened group, which although not a permanent transition, afforded women an opportunity for the mild expression of resistance or resentment at the constraints of their lives.

Douglas herself might argue here that the diagonal shift to increased group and decreased grid is less a ritual foray into a new location on the matrix than a reflection of the differing social locations of women in classical Athens as opposed to those in Hellenistic Alexandria. I would agree that by the time of Theocritus, much of Greek society had experienced a subtle but significant shift down grid, which reduced some of the insular pressures on women, at least for some. But I also concur with Griffiths that the freedom to walk the streets of Alexandria and mingle, even converse occasionally, with strange men was only a modest reduction in the pressures on ordinary women like Gorgo and Praxinoa.[16] The worship of Adonis and other mourned gods did more than give ritual expression to the usual social location of the female mourners. It provided them with a temporary relief from insulation and isolation that displays the characteristics we would expect from such a shift, even if the severity of that isolation and insulation was somewhat less than in classical Greece. By comparison, a much more radical shift toward strong group and weak grid and a much more severe social critique undergirds women's ecstatic worship of the ancient Dionysos.

# — 4 —

## Women's Devotion to Dionysos

Women's proclivity for the worship of the Greek god Dionysos was legendary even in antiquity. Yet precisely what women did in the service of the god Dionysos is not easy to determine.[1] Women and men alike took part in great public Athenian festivals dedicated to Dionysos, such as the *Anthestēria*, the *Lēnaia*,[2] and others characterized by intense concern for agricultural and human fertility. Most intriguing is the tantalizing evidence for women's participation in ecstatic rites to Dionysos.

These rites are described in literary sources as early as the end of the fifth century B.C.E.[3] In the *Bacchae*, the play that posthumously won him the prize at the Dionysian festival in Athens in 405 B.C.E., Euripides offers a bizarre yet compelling account of the acts of the first women worshippers of Dionysos, called both Bacchae and maenads.[4]

In the opening lines of the *Bacchae*, Dionysos explains that he has come to the Greek city of Thebes, not far from Athens, to introduce his rites to all of Greece. He has chosen the Thebans as the first to receive the rites to exonerate his mother's honor and punish her family for refusing to believe the paternity of her child. Semele, his mother, conceived him by the god Zeus, yet her sisters Ino, Agave, and Autonoe were unconvinced by this explanation of her apparently out-of-wedlock pregnancy, and instead spread the rumor that Semele, having been impregnated by a mere mortal, was prompted by their father, Cadmus, the king of Thebes, to ascribe the loss of her virginity to Zeus. This lie, the sisters asserted, resulted in her death by Zeus's thunderbolt. Dionysos reveals that Semele did indeed die from a divine thunderbolt, but as a result of a plot devised by the jealous Hera, while he was snatched from his mother's womb by his father Zeus and hidden away from Hera's continuing anger. Ancient literature contains numerous variations of the myth of Dionysos's conception, birth, and childhood.[5]

Dionysos's revenge on his mother's family takes the form of a divinely induced insanity that causes Semele's three sisters to lead companies of dancing women onto the nearby mountains. Dressed in fawnskins, with loosened hair and writhing snakes draped round their necks, wearing crowns of ivy, oak, and bryony, and carrying the wand called a thyrsus, the women ran wild over the mountains, tearing apart wild animals with their bare hands, snatching up children from villages, and pillaging and destroying. When furious men from the villages attack the rampant maenads with spears, the women suffer no harm, yet their own wooden wands inflict serious wounds on the men. Prior to their rampage, the Bacchae nurse wild animals with milk intended for the infants they have abandoned at home. When they seek water, a thyrsus wand struck against a rock yields a fountain from the god while another rod produces a spring of wine. Milk wells up from the ground as they scratch it with their fingers, and honey oozes from their wands. Afterward, they wash themselves in the springs provided by Dionysos while snakes lick the blood from their faces.

Yet all this is mild compared with the ultimate revenge Dionysos has planned. Agave, the eldest sister, discovers her son, King Pentheus, hiding in a tree disguised as a woman, spying on the secret rites of the maenads. Blinded by her divinely induced insanity, Agave perceives Pentheus as a wild animal that she and the others murder and dismember. Only when Agave brings the torn remnants of her son before her father Cadmus, triumphantly seeking the rewards of the hunter, does she regain her sanity and realize what she has done. At the denouement of the play, Agave goes into exile, the house of Cadmus is destroyed, and the worship of Dionysos is firmly established of Thebes.

The *Bacchae* suggests numerous features of the Dionysiac rites. A conversation between Pentheus and Dionysos disguised as his own prophet reveals that the rites were restricted to initiates and could not be divulged to outsiders. They were conducted at night, included dancing, and carried with them the (false) suspicion of sexual misconduct. Many foreign lands already participated in this particular form of homage to Dionysos.

From the behavior of the possessed women in the play, other aspects of Dionysiac ecstasy may be deduced. In a state of temporary insanity, women left their homes, abandoned nursing infants, relieving their overflowing breasts by nursing wild animals instead, fled dancing to the mountains in bands led by a chief maenad, wore the ritual garments described earlier, consumed honey, wine, and milk, and engaged in incredible acts like rending apart wild animals. The play itself distinguishes between the rites of Bacchic followers in foreign lands (which include nocturnal dancing, presumably in an ecstatic state, that are observed now in Thebes and all Greece) and the specific acts of the sisters of Semele, whose frenzied destruction of Pentheus is a onetime penance for their failure to believe the divine paternity of Dionysos.

This raises the dilemma that has confronted modern scholars in their evaluation of the *Bacchae* as evidence for actual religious practice. The manifestly mythical framework of Euripides's description leaves us wondering just how accurately he reflects, or even intended to reflect, the ecstatic worship of Dionysos in fifth-century B.C.E. Athens or earlier. In fact, there are at least two problems here: first, whether the *Bacchae* reflects what women actually did in or before Euripides's own time, and second, whether the *Bacchae* itself becomes the source from which subsequent Bacchic ritual is derived. E. R. Dodds denied that the rites of the *Bacchae* were enacted in Euripides's Athens,[6] but thought that Euripides had seen some form of Bacchic rites, perhaps in Macedonia.[7] He also suggested that the impetus for the play was the arrival of similar rites in Athens, in the form of devotion to another foreign god, Sabazios.[8] Henrichs takes the position that ritual reenacts myth—so that the *Bacchae* accounts for later behavior, but is not reliable evidence of prior Bacchic activity.[9]

Unfortunately, the evidence available does not permit us to decide these issues, for we have little corroborating epigraphical, archaeological, or other literary evidence for Bacchic rites earlier or even contemporaneous with Euripides.[10] The *Bacchae* is somewhat ambiguous on the participation of men in the ecstatic, private mysteries of Dionysos. The speech of the god in the opening lines strongly suggests that only women were compelled to worship Dionysos in this fashion; the scene so graphically relayed by the terrified messenger confirms the absence of men. Pentheus's tragic downfall is occasioned by his illicit presence, spying on women's secret rites disguised in female clothing, which reinforces the notion that men had no business on the mountains with the women. Yet in the *Bacchae* two men do participate in the new worship of Dionysos, namely Cadmus, the father of Semele and now the ex-king of Thebes, and the blind seer Teiresias.

In my earlier work, I argued that the men's participation seemed limited to donning ritual clothing, carrying a thyrsus, and intending to dance in honor of the god, perhaps on the mountain.[11] However, a scene at the conclusion of the play does support the interpretation that Cadmus and Teiresias danced on the mountains. As Agave returns to Thebes with the as yet unrecognized head of her son, Cadmus enters carrying fragments of the body of his grandson, which he went to find in the glens of Cithaeron. "For I heard from someone of my daughters' shameless deeds, back in the town, when I had passed within the walls with old Teiresias, coming from the bacchants[12] and turning back to the mountain, I recovered my son who died at the hands of the maenads."[13] From this scene, Kirk and others conclude that sincere male participants were welcome, despite Agave's insistence on the secrecy of the rites: Pentheus's crime would then be his insincerity, not his masculinity. But I still find this interpretation problematic, especially since no men are explicitly said to be present during any of the scenes reported on Cithaeron. There is no indication that either Cadmus or Teiresias are possessed by the god and impelled to perform his rites, like the women, and there is no suggestion that they participate in any ritual sacrifice that may have taken place, such as the rending of the live body of Pentheus.[14]

There is some inscriptional evidence to suggest that private Bacchic initiations included men as early as the period when the *Bacchae* was first presented.[15] A gold tablet found in the grave of a woman buried at Hipponion in southern Italy, dating to the end of the fifth or the beginning of the fourth century B.C.E., contains instructions for the deceased. She is to avoid drinking water from a certain spring and instead to seek water from the lake of Mnemosyne (Remembrance). This water will enable her to walk the same sacred road that was trod by other mystic initiates (*mystai*) and Bacchics (here *Bacchoi*).

From the masculine plural *Bacchoi*, Susan Cole concludes that the initiates of this cult were male and female,[16] and that the cult was unquestionably Dionysian. Reading the tablet in conjunction with the *Bacchae*, a passage from Demosthenes,[17] and other evidence, she considers this inscription as evidence for private Dionysiac mysteries that conferred knowledge of the proper road to choose after death and the blessed state that would accompany such a decision on its initiates.[18] The conversation in the *Bacchae* between Pentheus and the disguised Dionysos may allude to just such initiations.

Cole may well be correct in her interpretation, but it seems necessary to distinguish between these private initiations, whose purpose is to guide the dead in the afterworld, and the ecstatic activities of the maenads in the *Bacchae* and elsewhere. For surely, as Cole observes, mysteries aimed at guaranteeing the blessed state of the dead would not have been confined to women, yet there are aspects of Dionysiac rites that *are* restricted to women and must have other functions and probably other forms as well.[19]

Some sources from the Hellenistic and Roman periods allow us to reconstruct certain aspects of Bacchic ritual and cultic organization and to distinguish between mysteries oriented toward the afterlife and those oriented toward a ritual reenactment of the myth in the *Bacchae* whose connections with the afterlife are by no means certain. They also provide varying degrees of information about the gender composition of Dionysiac associations.

In the fourth century B.C.E., a fragmentary inscription from Methymna on the island of Lesbos refers to mysteries performed only by women, from which men were excluded: the presence of the word *thyrs [us]* may indicate the Dionysiac nature of the rites.[20] But little else about the rites can be deduced from this source.

More intriguing is an aside by the famous Greek orator, Demosthenes, who attempts to impugn his opponent, Aeschines, by accusing Aeschines of assisting his mother with her cultic activities.[21] Demosthenes claims that Aeschines, upon reaching maturity, read aloud from cultic writings, mixed libations at night, purified the initiates, and dressed them in fawnskins. He cleansed them with cornhusks and clay, and raising them up, led them in the chant, "The evil I flee, the better I find." By day, he led the group, called a *thiasos*, through the streets of Athens, wearing garlands of fennel and white poplar. Garlanded in snakes, he gave the ritual cry, "Euoi, Saboi," and danced to the tune of "hues, attes." Old women called him by various cultic titles.

As described by Demosthenes, the cult has some immediate affinities with the rites of Dionysos as outlined in Euripides's *Bacchae*. Initiation is restricted and the rites are performed at night. Ritual clothing includes fawnskins and garlands of leaves. Ritual activities include libations, dancing, and the use of live snakes. The ritual cry, "Euoi," evokes the *Bacchae*. The titles attributed to Aeschines, especially *kittophoros* (ivy-bearer) and *liknophoros* (winnowing basket carrier) have clear associations with Dionysos.[22] The cultic organization is also called a *thiasos*.

But other elements may call a Dionysiac identification into question. The only name given to the worshippers is *Saboi*, which virtually all scholars have taken as a reference to the Thracian god Sabazios, who is sometimes associated or even confused with Dionysos. In fact, as I have argued elsewhere, the limited evidence available supports a stronger identification of Sabos and Saboi with Dionysos rather than Sabazios.[23]

The gender of the members of the *thiasos* is not immediately apparent. That these are rites in which women participate is undeniable, not only because Demosthenes associates them with Aeschines's mother, but because that association (with his mother and with women) is a crucial part of the insult (another part being, no doubt, their foreign character, whether Dionysian or Sabazian or something else). Whatever these rites, they are not the appropriate behavior of Greek male citizens. The question for us is whether any other men besides Aeschines are involved, whatever their rank and status in Athenian society. The term *Saboi* (masculine plural of *Sabos*) may signify the presence of men, in contrast to the feminine plural *Bacchae*, which seems to preclude it. The use of the masculine plural should imply either males or males and females, but not females alone, for which the feminine plural would be used. But aside from Aeschines himself, nothing in the passage resolves the matter. Conceivably, the cultic titles attributed to Aeschines by "old women" point to a level of involvement on his part that goes beyond assisting his mother and her religious colleagues, but it may also simply be part of Demosthenes's rhetoric.

My own inclination is to see this passage as evidence of women's religious activities in fourth-century B.C.E. Athens that bear striking and significant resemblances to the ecstatic worship of Dionysos. If the cult is not actually Dionysian, it is so similar that it warrants scrutiny as a related form of women's cultic behavior. Susan Cole also sees this passage as evidence for Dionysiac mysteries, but connects them with the concerns for the fate of the dead reflected in the gold tablet from Hipponion. She suggests that the chant "The evil I flee, the better I find" alludes to the mystic's choice of the proper road in the afterworld.[24]

It is not inconceivable that a mystery cult oriented at one level to afterlife concerns sustained a second level of meaning for women. Perhaps there were two forms of Dionysiac initiations: one shared by men and women and oriented toward the afterlife and the other initially restricted to women, comprising a ritual reenactment of the birth of Dionysos and the death of Pentheus and the deeds of the first women possessed by Dionysos, with no clear connections to any afterlife mysteries. Yet the two may have been easily confused and perhaps even intertwined. Cole may be right that Dionysos's conversation with Pentheus alludes to rites that confer blessings in the afterlife—rites which are now introduced in Thebes, alongside the maenadic rites. Albert Henrichs may be correct to speak of maenadism as distinct from other forms of Dionysiac initiations. The problem with Henrichs is his inability to consider the possibility that men at any point adopt maenadic practices without compromising the nature of maenadism. How the two sets of rites could then be practically distinguished is part of the problem, for both may be detected in the *Bacchae* and even in Demosthenes. But the later literary references may actually provide some clarification.

By the third century B.C.E. we find irrefutable evidence of maenadic associations in several Greek cities in western Asia Minor, including Magnesia and Miletus. An inscription from the early third century B.C.E. distinguishes between the public rites performed on behalf of the city and private rites of initiation, which were performed only by women.[25] A priestess presided over both sets of rites. The inscription refers to biennial celebrations to Dionysos that are attested elsewhere in literary sources[26] and requires that whenever a woman wishes to perform an initiation to Dionysos, whether in the city, the countryside, or on one of the islands, she must pay the priestess at the biennial festival. From this, Henrichs concludes that any woman could set up her own *thiasos*, provided she paid the appropriate fee to the priestess, and that such *thiasoi* must have been numerous in and around Miletus.[27] If men are involved here in the worship of Dionysos, it is only at the public level.

The epitaph of a priestess of Dionysos named Alcmeonis, daughter of Rhodius, supplements our knowledge of the cult at Miletus. She led the Bacchae of the city to the mountain, carrying unspecified sacred objects and implements.[28] This inscription confirms the existence of Bacchic rites practiced only by a ritual organization of women, up on the mountains outside the city.

A third inscription describes how the rites of Dionysos were brought to the city of Magnesia on the Meander River in western Asia Minor.[29] After a storm the Magnesians found a plane tree split open to reveal an image of the god Dionysos. Uncertain about the meaning of this portent, they inquired of the oracle at Delphi who instructed them to send to the city of Thebes for three maenads to establish the rites in Magnesia. The Magnesians complied, and three women came from Thebes, establishing three separate thiasoi.[30] Cosco led a thiasos named after the plane tree in which the image was found, Baubo led a thiasos outside the city, and Thettale headed a thiasos of the *Katabatai*.

From the name of Thettale's group,[31] Henrichs deduces that its membership must have included both women and men. Despite the apparent feminine form of *Katabatai*, it is in fact a masculine plural, which ought to denote either a mixed group or a group of men. He concludes that this group "celebrated nonmaenadic revels," since for Henrichs, by definition, only a group of women could perform maenadic rituals. He then suggests that we cannot tell whether the other two thiasoi were maenadic, and therefore confined to women or mixed as well.

Several Hellenistic authors testify to the practice of women's rites to Dionysos in various Greek cities. Diodorus of Sicily, who wrote in the mid-first century B.C.E., reports that sacrifices were held in many Greek cities every other year, at which Bacchic bands of women gathered.[32] Diodorus distinguishes between the rites of *parthenoi* (literally, virgins), who carried the thyrsus and shouted ritual cries, and those of the (married) women, who formed groups to sacrifice to the god and enact the part of the maenads of old.[33] Nothing in Diodorus suggests that these rites to Dionysos had an afterlife orientation. Unfortunately, it is not clear whether Diodorus was reporting festivals still celebrated in his own time, or whether he was merely utilizing one of his many sources.

The writings of Plutarch from the second century C.E. contain several significant references to women's ecstatic rites in honor of Dionysos. He confirms that these were celebrated in midwinter, and tells the story of a group of women stranded during a severe winter storm who had to be rescued by a search party.[34] In a work that recounts episodes of bravery on the part of women, Plutarch includes the deeds of the women of Amphissa. Once during wartime a band of maenads was discovered, collapsed from exhaustion in the town square. In their ecstatic wandering, the maenads had apparently strayed behind enemy lines into Amphissa. The women of the town guarded the sleeping devotees from the unwanted attentions of soldiers and procured them a safe escort home.[35] Plutarch dedicated this work and at least one other to a priestess of Osiris and Dionysos named Clea,[36] who may have been a source of his information.

In his life of Alexander the Great, Plutarch also attributes an unbridled passion for both Orphic rites and the orgia of Dionysos to Olympias, Alexander's mother.[37] This may allude to precisely the distinction I have just suggested, between initiations oriented toward the afterworld and mimetic rites reminiscent of the maenads in the *Bacchae*. Plutarch suggests that mountain rites were again involved and is explicit about the use of snakes, ivy, thyrsus wands, garlands, and winnowing baskets, all familiar to us from the *Bacchae*. Particularly intriguing is Plutarch's observation that the handling of snakes by these possessed women terrified the men, suggesting that at some point, men were able to observe at least some aspect of the rites, similar, perhaps, to the dynamic in Demosthenes.

Finally, in his travelogue of ancient Greece written in the second century C.E., Pausanius reports that Attic women, called Thyiades, joined with women from Delphi every other year to celebrate rites to Dionysos on Mount Parnassus.[38]

In the Hellenistic period, Bacchic initiations eventually found their way to Italy and Rome proper. The worship of Dionysos is already well documented in Italy in the fourth century B.C.E.,[39] and the gold tablet from Hipponion suggests the existence of Dionysiac initiations oriented toward the afterlife. But according to the first-century Roman historian Livy, orgiastic rites of Dionysos were first imported into Italy and then Rome at the beginning of the second century B.C.E.[40]

Livy's account of the origins of the Bacchanalia in Rome contains much fascinating information, tainted somewhat by Livy's questionable historiography, by his obvious disgust for the rites, and by his concern to demonstrate the dangers of Romans adopting foreign rites, which was equally an issue in his own time.[41] At the beginning of his narration, Livy imputes the origins of the Italian Bacchanalia to a nameless Greek man, who began by initiating a few people, apparently both women and men, into secret nocturnal religious rites. To lure larger numbers into the cult, he offered wine and outrageous feasts. For Livy, any religious aspect to these rites was simply a cover for extreme debauchery and corruption, including sexual promiscuity, perjury, forging documents and wills, murders and other violence—all concealed from the public ear by the cacophonous sound of the ostensibly religious cymbals and drums.

The rites ultimately came to the attention of the Roman government in 186 B.C.E. when a freedwoman named Hispala Faecenia sought to prevent her aristocratic lover, Publius Aebutius, from being initiated into the rites by his scheming mother and stepfather. The account that Hispala offers to the consul Postumius differs somewhat from Livy's initial description. Originally, she testifies, the rites were restricted to women, who performed initiations on three days each year, always during the day. Matrons took turns serving as priestesses.

But when a woman named Paculla Annia was priestess, she initiated her sons Minius and Herennius Cerrinius, held the rites at night, and expanded them to five days a year. The seemingly innocent and inoffensive rituals of the women now degenerated into sexual orgies in which men engaged more in homosexual than in heterosexual acts, and those who attempted to refrain from such misdeeds were sacrificed as victims. Seemingly in a state of possession, men would utter prophecies and women dressed as maenads ran around plunging burning torches into the Tiber River. Many people succumbed to the lures of these rites.

According to Livy, Hispala's testimony spurred an extensive investigation, which culminated in the banning of the Bacchic rites. Livy claims that more than 7,000 women and men were involved, some of whom committed suicide rather than undergo investigation and punishment for their participation. Those who were found to have undergone initiation but not to have committed any criminal acts were imprisoned; those who had actually engaged in fraud, sexual impropriety, or murder were executed. Livy reports that many women and men were found guilty, and more were executed than imprisoned. Convicted women were handed over to their families for execution of sentence, except those women who were not subject to any male relatives who could impose the punishment. These were remanded to the state.

However, an inscription of the actual decree of the Senate, the *Senatus*

*Consultum de Bacchanalibus*,[42] contradicts Livy's account at several points. The Senate apparently called for the dismantling of current Bacchic sites, forbade the establishment of Bacchic cult places, and prohibited Roman citizens, men of Latin rights, and allies from participating in Bacchic rites. Men were forbidden to serve as Bacchic priests, men and women were prohibited from the office of magistrate, and various other restrictions were enumerated. Violation of these regulations was punishable by death. However, the decree provided that the Senate could approve exceptions for those who appealed to the urban praetor. Livy makes no mention of such exceptions.

Rather, Livy's account claims that the Senate authorized the consuls to seek out Bacchic priests, women and men, wherever they might be, and to inquire into the identity of any Bacchic participants. The *Senatus Consultum* makes no mention of such inquiry; the closest parallel is its provision that anyone who claimed the need to celebrate Bacchic rites could appeal through the praetor to the Senate.

In addition to the conflicts with the text of the Senate's position, the speech attributed to Postumius before the assembled Roman people contains yet another description of the Bacchic rites that contradicts previous testimony within Livy's text. Postumius acknowledges that many Romans are vaguely aware of the Bacchanalia, but do not realize their true form and extent. He knows how alarmed the Romans will be when they learn the great number of participants, so he hastens to relieve their anxiety by assuring them of the disreputable character of the initiates. Most of them are women, who are the source of this disturbance, and those men who have participated are like women, "debauched and debauchers, fanatical, with senses dulled by wakefulness, wine, noise and shouts at night."[43]

Postumius's speech makes explicit some of the threats that such foreign cults posed to the Roman social order. The Bacchanalia encouraged and facilitated illicit sexual relationships. They compromised the efficacy of Roman soldiers, for Postumius asks the crowd, "Will men debased by their own debauchery and that of others fight to the death on behalf of the chastity of your wives and children?" (making clear, incidentally, that the assembled Roman people are mostly, if not exclusively, men!). In fact, if Postumius is to be believed, all the crime and villainy that has taken place in Rome in recent years can be attributed to the Bacchanalia.

But perhaps the greatest danger posed by the Bacchanalia (and by foreign worship in Livy's own time) is the potential neglect and insult to the true Roman gods who want only proper Roman sacrifices, without which the prosperity of Rome will suffer greatly. Lest the Romans worry that the extermination of the Bacchic rites will offend the gods, Postumius assures them that it is the gods themselves who wish this false religion extirpated as soon as possible.

The many agendas that underlie Livy's account of the Bacchanalia make it difficult for us to assess its reliability as a source for women's religious activities in republican Rome. At key points, the information he presents is contradictory. Who, for instance, first introduced Bacchic rites, a nameless Greek or equally nameless women? According to the speech placed in the mouth of Postumius, the vast number of participants were women, and women were the cause of the Bacchic conspiracy, although the testimony of Postumius is rendered somewhat suspect when he declares that his (male) audience will find this reassuring. Only two paragraphs after Postumius's speech, Livy reports that the heads of the conspiracy were four men, who were its priests and founders. One of these was Minius Cerrinius, the son of Paculla Annia, whom Livy earlier blames for the original perversion of the rites. Conceivably Livy finds it as convenient as Postumius to blame women for these offensive anti-Roman rites, and places in the mouth of Hispala the testimony that it was a woman, Paculla Annia, who first altered some harmless women's rites. But if so, he has forgotten himself a few pages later when he reports without critique the central role of the four men.

Although women are accused of being the primary participants in the Bacchic orgies, much of the description points to extensive, if not primary, male participation. Much of the sexual debauchery is committed by men with other men and with young boys. Much of the plotting, forgery, and so forth would also seem to be male activity. Four men are named as the actual leaders. Conceivably, the Roman authorities have seized on the old and known association of Dionysiac rites with women to defame and defuse dangerous foreign rites that have become popular in Rome.

It is exceedingly difficult to sort out unbiased information about the Bacchanalia from this account. Unquestionably, Livy's concern to defame foreign rites colors his presentation. Yet we know both from antiquity and from contemporary experience that fraud and deceit are sometimes perpetrated under the cover of religious piety, and it is quite possible that the episode of 186 B.C.E. does reflect such an occurrence. What information, then, if any, can we extract about women's religious practices?

Even the description of the antecedent rites of women attributed to Hispala is problematic. She claims that women performed Bacchic initiations by daylight three days a year from which men were excluded. Conceivably, this is perfectly accurate, but if so, it represents a departure from the testimony of other sources that women's Dionysiac rites were performed every other year, at night. While we cannot exclude the possibility of regional variations, on balance, it seems safer to regard Livy's entire account with great interest and much suspicion.

Conceivably, the accusations and the offensiveness of the Bacchanalia lie at least in part in its very attractiveness to women and in its licensing of their departure from home, responsibility, and so forth. We should not forget that the episode with the Bacchanalia allegedly took place in the intermezzo between the second and the third Punic wars, which, as we see again later, had major implications for the roles of women in Roman society. It is possible that the accusations about the Bacchanalia reflect some of the same issues that underlie both the festival of the Matralia, celebrated by Roman matrons, and the institution of various cults of women's chastity. That Roman women could participate in such rites reflects their growing autonomy, which is also indirectly acknowledged in Livy's report that women who had no male relatives able to punish them had to be remanded to the state for execution of sentence. The Punic wars would have greatly increased the numbers of such women. The Bacchanalia may have been the occasion of sexual excess, but they may also reveal the fears of the Roman government about what happens when women gain some measure of autonomy. Further, in the wake of the first two Punic wars, the Romans may also have been especially sensitive to perceived threats against Roman worship and terrified of foreign incursions of any kind.

Despite all the complexities of reconstructing women's Dionysian devotion, the ecstatic activities attributed to female worshippers of the Greek god Dionysos offer an excellent opportunity to test the relationship between the social location of Bacchic women and Dionsyian religion and to examine the changes in Bacchic worship, including the eventual inclusion of men. If Mary Douglas's theory is viable, we should find correlations between changes in Bacchic religion and changes in the social location of Bacchic worshippers when plotted on a grid/group matrix.

Leaving aside the unresolvable, complicated historical question of whether women in Euripides's Athens were themselves participants in the kind of ritual he describes for the first worshippers of Dionysos, let us consider whether the rituals of Euripides's *Bacchae* correlate in a predictable manner with the social experience of the women in the play, and what resemblances we can identify between Euripides's female characters and fifth-century Athenian women.

At first glance, the society depicted in the *Bacchae* exhibits clear elements of strong group and strong grid. Status is determined by ascribed characteristics: being the king, or the king's son, or the king's daughter. Male and female spheres are quite distinct. Relationships are governed by rigid social regulations and conventions. Foreign gods are dangerous and anathema. On closer examination, we suspect that the female characters in the *Bacchae* share the high grid of their male relations, but experience relatively less group. This would, in Douglas's view, predispose them toward precisely the kinds of ecstatic worship of a foreign god that the *Bacchae* attributes to them. But for Douglas, the explanation lies in the consonance between ritual expression and social reality, each replicating and reinforcing the other. The specifics of Bacchic possession, though, suggest that something more complex is going on.

In his study of women's attraction to contemporary Caribbean and African sar and bori cult possession, which Douglas also utilized, anthropologist I. M. Lewis argued that the main function of such possession was compensatory, affording women an oblique aggression strategy that allowed them to redress the imbalance of power between women and men without directly challenging the status quo.[44] Despite the persuasive arguments offered by Lewis, Douglas was troubled by his "sliding toward deprivation theory," for lack of a better theoretical alternative. Lewis recognized that this dynamic applied to ancient Dionysian ecstatic religion,[45] although he did not pursue it in any detail. Is there a way to reconcile the two: to utilize both Douglas's model and Lewis's persuasive arguments for the effective compensation that sar and bori cults, and by extension, Bacchic cults, provide their participants?

Clearly, Bacchic possession, and the community into which the worshipper enters, may be viewed as a temporary shift strengthening group and weakening grid, analogous to the schema I proposed for the mourning of Adonis. The community of women who flee the city of Thebes display many of the characteristics of strong group and weak grid. They abandon much of the ascribed status that governs their lives normally, lessening the constraints of grid. The mother of Pentheus, the "princess" Agave, retains some leadership, but the play emphasizes that Dionysus has possessed all the women of Thebes.[46] They dress in similar ritual garb, wild animal skins and crowns of leaves, and they carry a thyrsus wand. Instead of their usual food, they drink water that gushes from a rock, wine that springs from the ground, milk that emerges when the maenads scratch the earth, and honey that drips from their wands.[47]

Several of their activities constitute explicit denials of their socially approved roles and inversions of male/female behavior. Nursing mothers abandon their babies, giving their milk instead to wild animals. The Bacchae release their bound hair, which, again utilizing Douglas's insight that the human body is the symbol par excellence of the social body,[48] we may see as a transparent symbolic expression of their (temporary) loosening of the social constraints that formal hairstyles express. The physical actions of the Bacchae convey a similar message: they run, they dance not the stately controlled dances but wild maenadic ones. These crazed devotees of Dionysus usurp the traditional male role of hunter, admittedly with disastrous results, dismembering animals and perhaps even human infants. They also usurp the role of warrior, attacking village men with their wands: against them, the men's swords are impotent.

They legitimize their departure from home, their abandonment of social responsibilities and relationships (the constraints of high grid) by attributing their possession to an amoral divinity. With the exception of the sisters of Semele, the women of Thebes have done nothing to incur the wrath of the god. This capricious aspect of possession is crucial, Lewis observes, to the social functions of such cults. Only because the women are blameless can they obtain any relief from the oppression of their daily lives through participation in the cult.[49] In short, they temporarily obtain the benefits of strong (female) community and weakened restrictions (grid) by a ritual transition to what Douglas labels egalitarianism. Such an interpretation preserves both the insights of Lewis and the model of Douglas.

This analysis may also illuminate the forms of Bacchic worship described in later sources and enable us to account for the changes observable both in the form of Bacchic rituals and in their constituents. In the Hellenistic period, certain critical features of social life change, which we see reflected in ritual. Clearly there is a lessening of the constraint Douglas calls group. In Athenian society (and Theban society as reflected in the *Bacchae*), we can clearly see the characteristics of strong group, such as common residence, shared work, shared resources, exceedingly tight control over marriage and kinship networks, even shared recreation. In the aftermath of Alexander's conquests and policies of resettlement and military assignment, the fabric of the Greek city-states was severely damaged, if not destroyed, effectively weakening the degree of group that many persons experienced and perhaps also the degree of grid. Widened economic exchange, weakening of kinship ties, increased ease of travel, and other changes contributed to a lessening of group, at least for persons previously experiencing the constraints of high group and grid.

What were the implications of this aftermath for Greek women, especially free women, most of whom lived their lives under the constraints of low group and high grid? Numerous studies have suggested that the rights of women increased in the Hellenistic period, reaching a high point in certain circles of Roman society. Does Hellenization produce a corresponding strengthening of group for women, incorporating them a little more into the male community that has itself experienced a weakening of group?

The net effects of such shifts (weakened group for men, strengthened group for women) ought to have predictable repercussions at the level of ritual and cosmology. In the case of ecstatic Dionysiac ritual, the evidence from the Hellenistic period may be quite pertinent. If many women in the Hellenistic period experienced less isolation and insulation than their classical foremothers, they might be expected to seek less redress in temporary flights to strong group and weak grid, which, it seems, provide relief from the isolation and insulation of their antithesis (weak group and strong grid) through the strong experience of women's community. The truly wild and uncontrolled aspects of Dionysiac worship would then be toned down, while the nature of Bacchic community itself might display more similarity with the community of worshippers, especially relative to gender integration. This seems to be what we find.

In the *thiasoi* for which we have historical evidence, participation seems to be a matter of choice, rather than compulsion by a capricious if not lunatic deity hell-bent on revenge. In the inscription from Miletus that recounts the origins of the *thiasoi* there, the importation of the Bacchic rites and of maenadic leaders is brought about by an oracle, sought to explain the portent of the image of Dionysos found in a plane tree after a storm, a much tamer beginning than the one recounted by Euripides.[50] In the Hellenistic period, too, we begin to have evidence for male participation. This would make sense if the intense gender segregation of the classical period had relaxed, creating more congruence in the social experiences of women and men, and hence in the kinds of religious activities in which they were likely to participate.

Douglas suggests that religious activity can constitute a temporary shift in social location. She recognizes that individuals may move from one such locale to another in the course of their daily lives. Within these different groups, does the extent of exercised control vary so that a mini grid/group axis can itself be applied? Douglas would seem not to think so, since she considers individuals who move freely from group to group in the course of their daily lives to be relatively low group, regardless of their degree of grid. Even this, however, accords with our analysis so far, for it is women whose initial experience is that of insulation and isolation (weak group and strong grid) who make the ritual transition to the diagonal opposite, egalitarianism (strong group and weak grid).

By comparison, in the *Bacchae* the men experiencing strong group and grid, like Pentheus the king (and presumably other aristocratic men), cannot move to the egalitarianism of strong group and weak grid, even in a temporary ritual context. Only two men in the play, Teiresius the blind seer and Cadmus the former king, agree to worship Dionysus. Only two men peripheral to the central power structure of Theban male society, which would place them in a social location similar to that of the women, are able to make the move to Dionysian ecstasy and community.[51] Douglas proposes that diagonal shifts across the diagram, particularly from weak group with strong grid to strong group with weak grid, and from strong group and grid to weak group and grid, are much more feasible than straight horizontal or vertical shifts, especially the latter.[52] This would suggest that some men might have participated in Bacchic ecstasies, but that such men are likely to experience the kinds of constraints shared by the women, namely the insulation and isolation of low group with high grid.

# — 5 —

# *Rites of Roman Matrons*

We have seen that the religious practices of ancient Greek women reflected the degree of social and physical insulation and isolation that characterized their lives. When we turn to a consideration of the religious practices associated with elite Roman women, we should expect to find similar correlations between social experience and religious practices and beliefs.

Given the ancient Roman penchant for classification, it is not surprising that the religious activities of Roman women were frequently categorized according to the various components of Roman social stratification: marital status, rank (plebeian or patrician), and legal status (freeborn, freed or slave, citizen or noncitizen). The study of women's religions in Greco-Roman antiquity always requires sensitivity to the ways in which social categorization separated and segregated women from each other, and this is especially true for ancient Rome. This chapter focuses primarily on the reconstruction and analysis of certain religious rites celebrated by married freeborn women, *matronae*; it by no means exhausts the study of Roman women's religions. We briefly enquire into the range of religious activities not only of Roman matrons, but of other women within the broad spectrum of ancient Roman society, exclusive of Jews and Christians.

Not surprisingly, much of Roman women's private religious ritual and how they understood it lies beyond our historical grasp. From the extant literary sources, we know that a significant portion of Roman religious life was conducted at the household level. On behalf of his entire household, the Roman paterfamilias sacrificed regularly to resident gods who guarded the well-being of the household in exchange for such consistent propitiation. As official Roman religion effected a symbiotic relationship between Rome and her gods, whereby the well-being of both was mutually intertwined, so, too, the well-being of the Roman *familia* was inextricably bound up with the proper veneration of the household gods.[1]

The classic illustration of this parallel may be found in the duties of the Vestal Virgins.[2] Just as ordinary Romans tended the flames of their hearths to ensure the prosperity of their households, so these elite Roman priestesses guarded the flames of the hearth in the house of Vesta to ensure the prosperity of Rome as a whole. Vestals who allowed the flames to be extinguished were subject to scourging.[3]

Yet from the analogy of the Vestals we should hesitate to draw too many conclusions about women's roles in domestic Roman religion. The responsibility for guarding the hearth flame would reasonably have gone to women, who presumably spent more time in close proximity to the fire. But the household worship was itself the purview of the paterfamilias who performed the requisite sacrifices on behalf of those who were legally in his power, which included his unmanumitted children and his slaves. It also included his wife, if they had been married *cum manu*. This form of Roman marriage released a woman from the legal power of her father and placed her in a legal relationship with her husband analogous to that of his children.[4] Freeborn women married without *manus* technically remained under the control of their fathers, provided the father lived. They were therefore not part of their husband's household and theoretically excluded from such worship, although the reality may have been quite different.[5]

Almost certainly, women's religious rites and celebrations were intertwined with certain crucial life experiences, such as childbirth, puberty, marriage, serious illness, and death. How social classification should be factored here remains hard to assess. The licit sexual liaisons of freeborn (and freed) Roman women had all kinds of ritual delineation and celebration. Numerous festivals provided these women ample expression of concern for successful marriages, for fertility (a major component of a successful marriage), and for the health of babies and children. Shrines, too, were a locus of women's devotion. A story in Valerius Maximus suggests that elite women sought omens about forthcoming marriages at special shrines.[6] The same author also reports that women could seek the resolution of marital conflicts at a shrine to Dea Viriplaca.[7] Shrewdly translating this name as "Goddess Husband-Pleaser," Amy Richlin renders explicit the underlying perception that a harmonious marriage was one in which the husband was pleased.[8] According to Aulus Gellius, women in Rome hoping to avoid a breech birth prayed at two altars set up to the goddess Carmenta (or Carmentis) explicitly for this purpose.[9]

Such activities are unlikely to have been limited to elite Roman women, for at least some of these concerns, particularly regarding harmony with one's sexual partner and safe outcomes for childbirth, would have been equally important to nonelite women as well. Inscriptions occasionally testify to slaves, freedpersons, and foreigners who sought the compassion of the gods with respect to precisely such matters. But we know virtually nothing about how the transition into heterosexuality was marked, if at all, for slave women or prostitutes.[10]

Roman *matronae* were clearly differentiated from other Roman women by their free birth, their licit marriage, and their capacity to produce legal heirs. Distinctive clothing and privileges publicly delineated the *matrona*, who wore a long dress called a *stola* and a distinctive headband. Roman matrons as a class, the *ordo matronarum*, were sometimes taxed and occasionally organized themselves for political purposes, as in 195 B.C.E. when they successfully lobbied for the repeal of the Oppian law that had limited conspicuous consumption by women.[11]

Numerous official Roman festivals involved the participation of the *matronae*. March 1 was the *Matronalia*, sacred to Juno Lucina. Various ancient etymologies were given for her epithet: she was connected not only with light generally, but with the light of the moon, and therefore with childbirth.[12] On April 1, elite women worshipped Venus at the *Veneralia*. (Lower class women also venerated the goddess on the same occasion.)[13] A festival of the Bona Dea was celebrated on May 1 at her temple on the Aventine.[14] During the *Vestalia* in the month of June, *matronae* brought offerings to the temple of Vesta, and on June 11, the festival of the *Matralia* was celebrated at the temple of the Mater Matuta in the cattle market.[15] In early December, on a date apparently calculated anew each year, *matronae* gathered in the house of the highest Roman magistrate to celebrate secret rites to the Bona Dea.[16] Elite women also participated in the *Carmentalia*, two mid-January festivals to Carmenta, the goddess to whom women prayed to avert breech births and who was closely linked with childbirth generally. Because the Romans favored April for weddings, H. H. Scullard suggests that the Carmentalia would have coincided with a rise in the annual Roman birthrate approximately nine months later in mid-January.[17]

Both the reconstruction of women's participation in these festivals and consideration of their significance for those women present us with serious problems of method. In the first place, our description of these festivals depends on the literary testimony of male authors, who, as we see, often had their own complex agendas in portraying women's rites. In certain cases, such as the December rites of the Bona Dea, men were forbidden to be present during the preparations or the celebration, which raises questions about the accuracy of their information. Further, scholars have found significant discrepancies if not outright contradictions between the literary testimony and the inscriptional evidence for certain cults, such as that of the Bona Dea, which has important implications for our use of other literary evidence.[18]

Beyond the problem of the reliability of our male authors at the descriptive level lies an interpretive dilemma. Roman writers tended to impute sexual concerns and motives to many of women's actions. Juvenal, for example, sees virtually all Roman women's religious rites as a mere pretext for sexual escapades.[19] His portrait may be more vicious than most, but it differs only in degree and not in kind. We must wonder how Roman women themselves understood their religious activities; to what extent did the self-perception of the participants diverge from that of the observers?

Brouwer's study of the worship of the Roman goddess Bona Dea provides an instructive example.[20] In the surviving literary sources,[21] Bona Dea appears as an ancient Latin goddess who guarded the Roman state. Aristocratic women, including both *matronae* and the Vestal Virgins, were delegated to propitiate the goddess on behalf of the state with secret rites from which men and anything male were firmly excluded.[22]

The extant archaeological and inscriptional evidence collected and surveyed by Brouwer contradicts this portrait at virtually every point. Epigraphic sources suggest that the history of Bona Dea is considerably less pristine, and that the goddess was amalgamated with numerous non-Roman, non-Latin deities.[23] The majority of the dedications to Bona Dea are by individuals, seeking her protection for themselves and those close to them. Some inscriptions testify that Bona Dea was the patron goddess of religious associations, called *collegia*. Contrary to the literary portrait, it is clear that the worship of Bona Dea was limited neither to the aristocratic classes, nor to women. Still, most dedications were made by women, the vast majority of cultic officials were women, and dedications by freedpersons outnumber those of other classes considerably.[24]

Brouwer's efforts to make sense of this apparently stark contradiction are important. He proposes that the worship of Bona Dea always had an individual, personal dimension, and was never solely concerned with the welfare of the Roman state mediated through select women. The excessive emphasis on that aspect of her cult in the literary sources can be traced to the writings and concerns of Cicero, who emerges as the basis of most subsequent literary interpretation of the goddess.

It is from Cicero that we obtain our first descriptions of the December festival of Bona Dea. But Cicero was hardly a chronicler of Roman religion for its own sake, and his testimony is not offered in isolation. Rather, events that transpired at two separate celebrations of the festival proved crucial for Cicero personally.

The December festival of the Bona Dea was apparently celebrated in the house of the highest Roman magistrate for that year. In 63 B.C.E., that was Cicero. At the time, he was embroiled in a fierce struggle for power with an opponent known as Catiline. During the festival, a flame apparently leapt up suddenly from the dying embers on the altar, a miracle which Cicero immediately interpreted as a sign from the goddess that his actions to crush his enemies were necessary for the welfare of Rome. Cicero gave his own account of this event in a work that survives only in fragments, but in his *Life of Cicero*, Plutarch reports that it was the attending Vestal Virgins who immediately recognized the significance of this sign and urged Cicero's wife, Terentia, "to call on her husband as quickly as she could, asking him to do what he thought fit to save the country, since the goddess had given him a light as a sign of salvation and fame."[25]

The following year, a juicy scandal marred the festival, which was held in Caesar's house. One of Cicero's major opponents, P. Clodius Pulcher, unlawfully entered the house disguised as a harpist, allegedly hoping for a tryst with his lover, Caesar's wife Pompeia. He was soon discovered and recognized by Aurelia, Caesar's mother, but apparently escaped with the aid of a servant. Subsequently, though, he was tried for sacrilege, and Cicero testified against him. When Clodius ultimately died ten years later near a shrine to Bona Dea,[26] Cicero considered himself personally and politically vindicated.

From these accounts, and from numerous references to the goddess in Cicero's writings, Brouwer concludes that Cicero viewed Bona Dea as his own personal patron, and interpreted her actions and omens as direct evidence of her support for his role in Roman politics and his vengeance on his enemies. "In each case it is Bona Dea who appears as Cicero's protectress in his struggle for the preservation of the established order."[27] Yet such an interpretation would only be persuasive if Cicero's allies and enemies alike shared the perception that Bona Dea was indeed an ancient goddess whose primary concern was the protection of the Roman state. Brouwer suggests that Cicero found support for that interpretation in elements of her worship available in the December rituals and possibly in the May festival as well. For his own purposes, though, Cicero enhanced, if he did not create, the image of Bona Dea as not simply one link with Roman tradition, but its very embodiment. Subsequent literary testimony, largely dependent on Cicero, perpetuates this distorted representation of Bona Dea; the figure of Clodius "becomes the prototype of everything that clashes with these traditional values."[28] Only the epigraphic and archaeological evidence demonstrates the real continuance of the much more personal devotion of individuals and *collegia*.

Brouwer's work thus demonstrates that concerns about the reliability of ancient literary testimony to women's religions (and to men's religions as well) are justified. It confirms that in the absence of corroborating archaeological and inscriptional evidence, we should view all our descriptions and interpretations as provisional, a conclusion I have reached independently, particularly in the course of my own work on women's Judaism (see Chapters 8 and 9).

All this being said, we may still give some consideration to women's worship of Bona Dea. Brouwer appears convinced that reasonably reliable details about the December and May festivals can be extracted from the extant sources.[29] Early devotion to Bona Dea probably included the sacrifice of a pregnant sow and libations of milk and honey. Eventually women began to drink undiluted wine, which was apparently normally forbidden. As is the case with many deities venerated outside the context of state cults, Bona Dea was associated with prophecy (foretelling the future) and with ecstasy. Myrtle and serpents figure prominently in the mythology and the iconography of the goddess.

Brouwer offers a plausible reconstruction of the December festival. After all males, including slaves and servants, children and adults alike, had left the house of the chief magistrate, his wife or mother decorated the house with plants, flowers, vines, and so forth. Sacred vessels to hold the sacrificial portions for the goddess were set out on a table, probably before a platform (*pul-*

*vinar*) on which the cult statue, borrowed for the occasion from the temple, was set up. The central rite consisted of the sacrifice of a pregnant pig and a libation poured to the goddess by the mistress of the house. Afterward, the *matronae* and the Vestals drank wine, caroused, and were entertained by female musicians, who may not have been aristocratic.[30] Although Brouwer does not comment on this here, the location of the rites in a personal residence, even that of the chief magistrate, suggests that not all *matronae* would have been able to attend.

Less evidence is available for reconstruction of the May 1 festival, which was, however, part of the official Roman religious calendar. Brouwer suggests that it may also have excluded men, and involved the sacrifice of a pregnant sow and the drinking of prohibited wine euphemistically called milk.

Precisely who performed the celebration is uncertain. Brouwer notes, though, that several *sacerdotes* of Bona Dea (almost all women) are known from inscriptions at Rome, and he proposes that the festival was organized by these priestesses. Since we know by their names that all of these *sacerdotes* were freedwomen, the festival cannot have been conducted on behalf of the Roman people, and may not have been restricted to elite women.[31]

Various other goddess cults frequently marked the transitions of women's lives. At puberty, elite young girls dedicated their childhood togas to Fortuna Virginalis, and began instead to wear the *stola*, which publicly differentiated respectable women from prostitutes, who wore togas. At marriage, these same women were transferred to the protection of *Fortuna Primigenia*.[32] In this regard, the rites of Roman women do not seem to differ substantially from those we have considered for Greek women. But as some ancient sources admit, a significant number of religious cults for Roman matrons were instituted with the explicit intention of instilling and reinforcing male aristocratic expectations for women's appropriate behavior.[33] The worship of Pudicitia Patricia, Pudicitia Plebeia, Venus Verticordia, and Venus Obsequens, among others, are presented as obsessively concerned with women's sexual chastity and marital fidelity.[34] Conceivably here, as elsewhere, ancient male writers have failed to present or have misrepresented women's own perceptions of their devotions. Yet there are considerable reasons to see these Roman cults of chastity as complex refractions of Roman gender relations.

Virtually all of these cults are known to us mostly from Roman literary sources, such as Livy's *Annals of Rome*, Ovid's *Fasti*, or Juvenal's infamous sixth *Satire* on women. The minimial inscriptional corroboration for a number of these cults[35] may suggest that some were more significant as religious and social propaganda than as manifestations of authentic women's piety.

Studies by Pomeroy and other classicists have demonstrated that Roman women, at least among the aristocratic classes, had obtained a significant measure of economic and social autonomy by the early Republican period.[36] While a variety of factors contributed to this, the wars between Rome and Carthage were of particular importance. For the periods 264 to 241 B.C.E., 218 to 201

B.C.E., and 149 to 146 B.C.E. (the first, second, and third Punic wars, respectively), large numbers of aristocratic men were absent from Rome, and many never returned.

The consequences for the women at home were considerable, as they have so often been for women during wartime. In the absence of men, many women became actively involved in the management of family estates and fortunes. As their male relatives died in battle, aristocratic women often inherited vast wealth and found themselves freed from the guardianship of those relatives. Although state guardians were appointed for women left without male kin, Pomeroy and others point out that such guardianships were often pro forma and totally ineffective in controlling the activities of women.[37]

The absence of patrician men from Rome was also conducive to the establishment of irregular sexual liaisons between aristocratic women and the remaining men, for at least two reasons. First was the matter of simple demographics: war decimated the numbers of suitable, legitimate spouses. In the absence of offended male relatives, women experienced less pressure to avoid improper relationships and little reprisal for such liaisons. While previously women's sexual indiscretions were considered a private matter to be handled within individual families—the rights of a father (*patria potestas*) gave him the authority to punish, and even execute, his daughters for certain acts[38]—the wholesale absence of male relatives compelled the state to take over these functions.

The methods chosen to limit and regulate women's autonomy combined the passage of legislation with the establishment of religious shrines and rites that expressed and reinforced the expectations of elite Romen men for elite Roman women. During the Punic wars, Roman women who found themselves in control of great wealth are alleged to have flaunted it by wearing extravagant clothing and jewels, riding in costly ornate chariots, and surrounding themselves with a multitude of obviously valuable slaves.[39] Pomeroy points out that men had also engaged in similar displays, but "their lavish dinners and entertainments ultimately had the socially approved goal of furthering their political careers."[40] Women had no such justifications for their conspicuous consumption. In 215 B.C.E., the Senate passed the Oppian law, limiting the amount of gold women could own to half an ounce, forbidding women from wearing dresses trimmed with costly purple, and restricting women from riding in carriages, except under certain circumstances like religious festivals. In 214 B.C.E., the state more or less confiscated the funds of single women, widows, and wards: all of them, as Pomeroy comments, women who lacked the protection of male relatives to defend them against this usurpation of their wealth.

The provision that women might still ride in carriages to religious festivals is easy to see as a deliberate concession to encourage women's participation in certain cults, participation which may not have been so easy to compel. Consider, for example, devotion to Venus Verticordia, Venus the changer of hearts. Valerius Maximus, a Roman historian writing in the first third of the first century C.E., relates that on the basis of a prophecy in the Sibylline books, a statue to Venus Verticordia was consecrated in 214 B.C.E. Chosen for the honor of consecrating the statue of the goddess was one Sulpicia. Daughter of Servius Sulpicius Paterculus and wife of Quintus Fulvius Flaccus, she was elected by other women from a list of 100 as the most sexually virtuous woman in Rome. Valerius is quite explicit that the function of this goddess was to turn the minds (*mens*) of virgins and women away from illicit sexual desire (*libidine*) to sexual chastity (*pudicitia*).[41] The epithet of Venus is double-edged, as Danielle Porte recognizes.[42] In urging the goddess to turn the minds of women to chastity, the Romans acknowledge that it is Venus who directs their minds to sexuality in the first place. Simultaneously, it is the goddess herself who is urged to turn her mind to licit sexuality.

Sulpicia's dedication occurs also in Pliny the Elder, one of whose sources was Valerius.[43] Pliny omits the epithet Verticordia for Venus, as well as the explanation that worship of the goddess was designed to affect the sexual behavior of Roman women. To the story of Sulpicia he appends that of Claudia Quinta, a Roman matron about whose chastity several stories are told in connection with the arrival of the goddess Cybele at Rome ten years later in 204 B.C.E.[44]

While these notices demonstrate a connection between religion and women's social behavior, they tell us virtually nothing about actual devotion to Venus Verticordia. According to Plutarch, however, a temple to Venus Verticordia was erected in 114 B.C.E. under curious circumstances.[45] By consulting the Sibylline oracle, the Romans discovered that the death of a woman named Helvia in a riding accident was a sign of the sexual defilement of three Vestal Virgins, who were found to have had sexual relations with three equestrians. The Vestals were walled up and left to die, and a temple to Venus Verticordia was dedicated to propitiate the gods.[46]

Our only description of women's rites in service to Verticordia comes from the ritual calendar, the *Fasti*, of the poet Ovid, who gives April 1 as the date for women's devotion to Venus.[47] He attributes the institution of the rites to a time when Roman chastity had declined precipitously, including the explanation that the cult was inaugurated in response to a consultation of the Sibylline oracle at Cumae. But it is not clear whether Ovid refers to the dedication of the statue in 214 B.C.E., or the erection of the temple in 114, or perhaps to both. In return for washing and dressing the statue of Venus and drinking a mixture of milk, honey, and poppy seed, Ovid observes, women will be rewarded with beauty, virtue, and good fame.[48]

Ovid's account of the festival celebrated on April 1 is somewhat puzzling, for it combines or perhaps conflates the worship of Venus Verticordia with that of Fortuna Virilis, rites of a somewhat different order! According to a note inscribed on an ancient Roman calendar, from Praeneste, rites for Virile Fortune were celebrated at the men's bathhouse, as Ovid also implies.[49] They were performed only by nonpatrician women, although Pomeroy is uncertain whether all plebeian women were permitted, or only *humiliores*, that is, courtesans and prostitutes. She suggests that after the institution of the new rites to Venus Verticordia, respectable women (presumably both plebeian and patrician) would not have continued to participate in the cult of Virilis. For Pomeroy the juxtaposition of the two cults heightened the dichotomy between respectable women and whores, with "the former worshipping an apotheosis of conjugal ideals, the latter worshipping sexual relationships having nothing to do with wedlock."[50] Eva Stehle concurs that the worship of Venus, both as Obsequens (the Compliant) and Verticordia, had as its function "the preservation of the moral and class distinction between *matronae* (or upper-class *matronae*) and women without rank (including courtesans)."[51]

Several other stories about the origins of women's worship of certain goddesses sound similar themes. While Fortuna Primigenia was clearly worshipped by men as well as women (and antedates the second Punic War), Pomeroy's analysis of the establishment of a new Temple to Primigenia on the Quirinal in 194 B.C.E. is germane. According to Livy, the temple had been vowed by Sempronius in 204 B.C.E.[52] But it was actually dedicated only a year after women were finally successful in obtaining the repeal of the oppressive Oppian law, through massive public demonstrations. Pomeroy suggests that the building of the Temple to Fortuna "served to confirm and advertise the traditional expectations the Romans continued to hold for their women, despite the repeal of the law."[53]

Livy appears to be our only source for another story about the origin of the worship of a goddess called Pudicitia Plebeia, which allegedly took place almost a century earlier, in 296 B.C.E.[54] It seems that a patrician woman named Verginia had been barred from worshipping a goddess called Pudicitia Patricia, Patrician Chastity, when she married a plebeian named Lucius Volumnius. Outraged, Verginia dedicated a new shrine in her own large home, the Vicus Longus, and established rites for Pudicitia Plebeia, Plebeian Chastity. The rites of Plebeia were essentially the same as those of Patricia and restricted to plebeian *univirae* (women married only once). Livy claims, however, that within a short period of time, the cult of Plebeia was defiled by polluted women, both *matronae* and others, and faded quickly into oblivion. Finally, Livy records that only a year later, a number of matrons were convicted of adultery, and the fines they paid were used by Q. Fabius Gurges to build a temple to Venus Obsequens, Venus the Compliant.[55]

Traditional scholarship has tended to take these accounts more or less at face value and to see women's religious activity during the Punic wars as one-half of a complementary strategy to cope with Rome's prolonged inability to defeat Hannibal. Interpreting Hannibal's success as evidence of divine disfavor, the Romans took to fervent supplication of the gods to avert their manifest displeasure. As Alan Wardman views it, if the responsibility for debates and military action fell to the men, the responsibility to propitiate the gods fell to the women. When Rome ultimately triumphed, men and women together participated in the celebration of their joint labors.[56]

But without denying that later Roman authors may have adhered to this view of complementary activity, traces of a different pattern emerge when we consider the specific temples built, deities venerated, and cults created or encouraged.

From at least the early third century B.C.E. on, women in ancient Rome sought and obtained a degree of autonomy and control over their economic and sexual lives sufficient to threaten the traditional Roman values of control over women's resources, both economic and reproductive. Numerous cults were apparently explicitly manipulated or created to shore up values that were under assault. That these attempts had only limited success is demonstrated not only in women's ability to compel the repeal of the Oppian law, but in the repeated reinstitution and rededication of cults affirming the values of marital fidelity (and female submission to male authority).

Significantly our sources for these cults come not from contemporaneous authors or from temple inscriptions, but from writers several hundred years later, whose use of these now ancient stories served explicit political agendas. Pomeroy astutely reminds us that Livy presents these stories as propaganda for the marital legislation of Augustus, which was itself (in part) a response to the autonomy of elite Roman women in the first century B.C.E./C.E. Augustus, too, was not above using religious cults to reinforce his beliefs about the centrality of childbearing, chastity, and familial bonds for the good of the empire.[57] Thus Livy's emphasis on the establishment of cults of morality and chastity in the republic, designed to demonstrate the antiquity of such beliefs and practices, demonstrates instead the extent to which such views were constantly under attack, not only in the republic, but in the early empire as well. Further, it is crucial to realize that the real issues were not the chastity and social class distinctions of Roman women so much as continued efforts on the part of elite Roman women, and possibly other women as well, to wrest a measure of autonomy from Roman men.

That the Romans themselves, both men and women, were aware of this is made magnificently clear in the speech Livy attributes to Cato's defense of the Oppian law:

> Unless you act [against repeal], this is the least of the things enjoined upon women by custom or law and to which they submit with a feeling of injustice. It is complete liberty (*libertas*) or rather, if we wish to speak the truth, complete licence (*licentia*) that they desire. If they win in this, what will they not attempt? Review all the laws with which your forefathers restrained their licence and made them subject to their husbands; even with all these bonds you can scarcely control them. What of this? If you suffer them to seize these bonds one by one and wrench themselves free and finally to be placed on a parity with their husbands, do you think that you will be able to endure them? The moment they begin to be your equals, they will be your superiors.[58]

Thus it is not inappropriate to wonder how well attended such festivals would ever have been, and how much these cults can be taken as legitimate reflections of women's religions. Perhaps our male sources have radically distorted the true nature of these devotions, which meant one thing to the women who performed them and another to the male writers, on whom we are largely dependent for our descriptions. Livy's report that the altar of Plebeian Chastity ultimately fell into oblivion, which he blames on its degradation by polluted women, might be taken as evidence that the cult held considerably less appeal for women than Livy would have been willing or able to admit. Alternatively, though, it could be taken as Livy's judgment that Roman women of all categories were simply incapable of chastity! Then, too, it is not inconceivable that little if any historical truth underlies this account.[59]

But Livy can hardly be singled out among ancient writers whose ulterior motives caution us against taking their presentation of women's religious rites, festivals, and observances at face value. Extracted out of context, Ovid's reports on the festivals to Venus Verticordia, Fortuna Virilis, and the Mater Matuta appear as simple descriptions of women's worship of goddesses concerned with chastity, marital fidelity, and familial piety (where piety is understood as the fulfillment of one's obligations to others). But let us not forget that this same Ovid was the author of numerous controversial poems on illicit sexual relationships, especially extramarital affairs. His poems suggest that he had one or more relationships with respectable married women[60] whether or not the figure of Corinna is a composite or a cover for an actual woman.[61] His elegiac poems, the *Amores*, provide insights into the realities of Roman sexual relationships, albeit from the perspective of a man. Although he knows that extramarital relationships are common among the elite, Ovid expects fidelity from his lover, while acknowledging that he cannot offer the same.[62] He begs his lover that if she cannot be faithful to him, she should at least be discreet.[63] The first two books of the *Ars Amatoria* offer men instructions on the art of seducing women; a third book, added later, offers reciprocal advice to women.

Can we avoid seeing something ironic in his account of women's worship of Venus Verticordia—Venus who turns the hearts of women toward a marital fidelity that contrasted so strongly with Ovid's own life and experiences of Roman society? What do we make of these vastly contradictory accounts of the attitudes and practices of allegedly respectable Roman women? What, too, do we do with this ancient expression of the sexual double standard? Ovid exemplifies male complicity in the sexual dalliances of elite Roman women, and yet there are no known cults of male chastity and fidelity! It may not surprise us to find that aristocratic Roman men saw the marital infidelity of Roman women as qualitatively different from their own sexual dalliances. The point here is not only that Roman men considered it acceptable to sleep with a variety of women other than their legal wives, but rather that they were apparently content to place the blame for their liaisons with women legally married to other aristocratic men solely on the women (or perhaps on the women and the goddess Venus), at least when religion was concerned.

Might there not be something subversive and intentional in Ovid's odd conflation of the worship of Fortuna Virilis and Venus Verticordia—a suggestion, perhaps, that the distinctions between chaste married matrons and sexually indiscriminate *humiliores* were not, in fact, nearly as clear as they seemed. Viewed apart from the rest of Ovid, such an interpretation may have little justification, but in light of Ovid's other work, his accounts in the *Fasti* are much more problematic, and at the very least must be treated with caution as evidence for women's actual religion. Even his exhortation "Go, good mothers, the Matralia is your festival" could be seen as caustic and ironic!

Similar observations might also be made about the bitter Juvenal, who sees in women's devotion to Cybele, for example, a transparent cover for lewd sexual activity. Juvenal's entire sixth *Satire*, a vicious attack on Roman women, was written to deter a friend contemplating marriage. When evaluating Juvenal's descriptions of women's religious activities, we must keep in mind the concern he manifests in all his satires for the pervasive decay and immorality into which he believes all Rome has fallen.

In fact, the religious devotions which appear to have held substantial appeal for women were precisely those that Livy and Juvenal in particular disparage: the ecstatic Dionysian bacchanalia, the wild rites of the Great Mother Cybele, and the worship of the greatly transformed Egyptian Isis.[64] If the rites that Livy and Juvenal uphold, in their efforts to shore up sagging Roman morality and morale, were designed to regulate women and confine them to traditional roles, we will not be at all surprised to find that the rites they disparaged were those which did not affirm traditional roles for women, but on the contrary legitimized alternatives and some measure of autonomy for women. The tension between autonomy and sexuality, which will surface again and again, is accentuated here. From a male point of view, the primary problem with women's "unchastity" is the underlying loss of male control over women, including, but hardly limited to, the loss of legitimate heirs that was so crucial in aristocratic Rome. The absence of any cults designed to guarantee marital fidelity on the part of husbands, or even male abstinence from sexual relationships with women married to other men, is one indication of the real dynamic underlying male offense at women's behavior. It is not accidental that all these offensive religions share at least one characteristic: they are all foreign to Rome and thus, one might suggest, fundamentally uninterested in the legitimation and perpetuation of the Roman social order.

There is at least one example of goddess worship by Roman women, though, which may not only have played a major role in the campaign for the proper socialization and regulation of elite Roman women, but may also have addressed some of the needs of Roman women in ways that others did not: the worship of the Mater (Mother) Matuta.

As with the cults of Fortuna and Venus, literary testimony to the festival of the Matralia in honor of Matuta comes primarily from the first century C.E. and later. The Roman poet Ovid and the Greek philosopher Plutarch provide similar accounts that nonetheless differ in some interesting respects. According to Ovid, the rites of the Matralia were restricted to Roman *matronae*, and female slaves were explicitly excluded.[65] The late second-century C.E. Christian writer Tertullian claims that only *univirae*, women married only once, could participate in the rites of Matuta.[66] Plutarch concurs that female slaves are forbidden, but adds the curious story that the women do bring in a single female slave, whom they slap on the head and beat.[67]

According to Ovid, the worship of Matuta at the Matralia owes its origin to events surrounding the birth of the god Dionysos. A number of Greek myths tell of Zeus's passion for the mortal Semele (daughter of Cadmus, the first king of the Greek city of Thebes).[68] When Semele gives birth to the baby god Dionysos, the outraged Hera maneuvers the death of Semele and seeks to destroy the infant as well. While some Greek sources credit the nymphs/nurses of Nysa with the child's survival,[69] Ovid credits the courageous deeds of Semele's sister, Ino, who fled with the god to the shores of Italy. There is considerable irony here, for according to some other accounts, Ino was less than a model mother. She is blamed for indirectly causing the death of one of her sons and directly causing the death of another, Melicertes, when she forced him to jump with her into the sea. In Ovid, Ino still jumps into the deep with her son, but now the boy is really Semele's child, and both Ino and the boy are rescued.[70] Upon arriving safely in the new country, Ino becomes the goddess Leucothea (the white goddess); Dionysos becomes Palaemon (the wrestler); Leucothea in turn is deemed identical with Matuta.[71] Ovid also attributes the exclusion of female slaves to the treacherous behavior of a female slave who was sexually intimate with Ino's husband Athamas and confided Ino's secret rites to him.

Both Ovid and Plutarch report that at the festival of the Matralia, the worshippers of Matuta sought her protection not for their own children, but for those of others.[72] Ovid is somewhat elliptical in this regard, admonishing women not to pray to Mater Matuta on behalf of their own children (*pro stirpe sua*) but rather to commend the progeny (*prolem*) of another (*alterius*).[73] Since Ovid locates the origins of the Matralia in Ino's rescue of her sister's son, it seems obvious that women are admonished to pray for the children of their sisters. A problematic late source reports that the women prayed *pueri sororii,* which may be translated "for the children of their sisters."[74] Plutarch is more explicit that the children of siblings are intended, but some scholars have questioned whether he refers to the children of sisters only, or to the children of brothers, or brothers and sisters.[75] According to Ovid, a main feature of the Matralia was the offering of toasted cakes to the goddess.

There seems to be no doubt that Matuta was originally an Italian goddess who may have been the primary deity at Satricum, a city not far from Rome that eventually fell into Roman control in 346 B.C.E. A temple to Matuta stood in Rome in the Forum Boarium, the cattle market, for hundreds of years before our authors reported on the Matralia. According to Livy, the temple to Matuta and a temple of Fortuna, which stood nearby, burned in 213 B.C.E.[76]

The following year, a board was elected to undertake the restoration of these two temples, and a third, to Hope.[77] In 196 B.C.E., Lucius Stertinius, returning from Spain, built two arches in the cattle market in front of the temples. In 174 B.C.E., Tiberius Sempronius Gracchus set up an inscription to Jupiter in the Temple of Matuta, commemorating his victory over Sardinia. Livy also relates that the temple to Matuta at Satricum was spared from the flames of the Latins in 377 B.C.E. and from a similar fate in 347–6 B.C.E., but was not so lucky in 206, when it was hit by lightning.[78] There was also a temple to Matuta at Praeneste.

Although scholars assume that Matuta was a goddess of birth worshipped by women, it is interesting that none of Livy's references to her temples, either at Satricum or at Rome, make this connection, and Livy provides no evidence for the rites of the Matralia. Instead, it appears that at least for Livy, Matuta's worship was connected to the military fate of Rome and her enemies.

But was the Matralia celebrated in the form Ovid and Plutarch describe prior to the first century? In the absence of affirmative literary evidence, we are dependent on inferences drawn from the archaeological and iconographic sources. The early iconography of the temple of Matuta at Satricum and/or Praeneste, and the testimony to its restoration by Camillus in the fourth century B.C.E., suggest to Valerie French that the worship of Matuta, with its representations of a female figure holding children in the security of her lap, provided a constellation of symbols and rites that would have held strong appeal for the war-torn Romans. The cult of Matuta would have been a effective way to solidify political unions between Romans and Etruscans in the wake of Roman conquests of Etruscan cities and the establishment of Roman colonies in the region, with special implications for women.

> In an exogamous society in which a daughter left her natal family to become part of her husband's family, the ritual of the Matralia reaffirmed the blood ties of the daughter's natal family. This reaffirmation would have value within Rome itself, of course. But it may have had even more value for marriages between the daughters of the Roman aristocracy and the aristocracies of communities that had joined Rome's growing sphere.[79]

French speculates that if Roman daughters returned home from their new households to worship with their sisters, perhaps bringing their children with them, such activities would have had the very practical effect of weaving together the interests and relations of Roman families with Etruscans and other non-Roman peoples now part of the Roman sphere.

She also interprets the material decline of Matuta temples after the early second century B.C.E. as evidence that the Matralia had waned as an effective means of political and social unification. It is at this time, she suggests, that the figure of Mater Matuta was identified with Ino, the aunt of Dionysos, and with the childbirth goddesses Leucothea and Eileithyia, signifying a shift in the function of the cult.

French's analysis assumes that in the fourth century B.C.E., Roman women became part of their husband's families at marriage, and that a festival enabling women to worship together with other women of their natal family not only reinforced personal ties between women and perhaps mitigated some of the stress of separation that marriage entailed, but also served, during the early period, to strengthen alliances between Roman and Etruscan families whose children had intermarried.

This interpretation is not without its difficulties, particularly when applied both to the early second century B.C.E. and the time of Ovid and Plutarch. In the republic, the most common form of marriage may have been marriage *cum manu,* which, as I noted earlier, made a woman part of her husband's family (legally, in fact, it gave her the same status as one of his legitimate children).[80] But there is considerable evidence that by the late republic and early empire, marriage *cum manu* had become increasingly rare, and that legally, at least, a Roman woman married by the most common form of Roman marriage, the so-called free marriage, remained under the control of her own father as a member of his household if he was still living, or was legally independent. To the extent that French's analysis hinges on this particular aspect of Roman social relations, her theory may be less applicable to the later periods.

However, the effects of marriage on the dynamics of women's relation-ships to their husbands and fathers may hold the key to our understanding of the Matralia, which I explore shortly. In Roman law, children born in a legiti-mate marriage were considered the property of their father. Women married with *manus* may have had no more rights to their children than women mar-ried without *manus,* but the former were at least part of the same family while the latter found themselves in the position of legally being members of a differ-ent family than their own children. In the case of divorce, this may have had some ramifications, although in general, with *manus* or without, children stayed with their fathers in the event of divorce.

Scholars like Hallett have illuminated the significance of the likelihood that elite Roman women remained permanently linked to their fathers and their natal families, and yet tied emotionally and practically to their husbands and their own biological children. The frequency of divorce in ancient Rome meant that Roman mothers were often separated from their children. The sig-nificance of the Matralia may well lie in its ability to express the dilemma that such women experienced, both in terms of their split loyalties and in terms of the implications this had for their relationships with their children. In linking elite Roman women with the children of their siblings, the Matralia guaranteed the preservation of ties with the natal family, as well as assuring that siblings (probably sisters) would continue to care for each other's children in the event that mothers and children were separated.

Thus, while ostensibly reinforcing Roman social distinctions and appropri-ate roles for Roman matrons, mothers, and nurturers, the Matralia may have had the additional function of expressing some of the real social and psychological conflicts of elite Roman women, which would have made it more appealing to women than cults whose subtext was the restriction of women.

The religious rites of Roman matrons offer an instructive counterpoint to the worship of Greek women presented in prior chapters, the more so when considered from the perspective of Mary Douglas's work. Several factors suggest that Roman matrons experienced not only strong grid, but also a stronger degree of group, which would place them closer to their male relatives and compatriots than the Greek women of Thebes in the *Bacchae* or real Athenian women, and perhaps closer to men than were many other women in their society, including slaves and freedpersons.

As the work of numerous scholars has shown, elite Roman women were far more involved in the economic, social, and political life of republican and imperial Rome than their Greek counterparts.[81] Compared to Athenian women, they were far less segregated within their homes, they owned property and managed large estates, often without even pro forma male guardians, they could inherit and bequeath land, and they had substantial influence and indirect power in Roman politics. Roman matrons had much more stake in Roman society: they stood to gain power and prestige in ways that other women did not.

Thus, relative to elite Greek women, elite Roman women were far less isolated and insulated. Unlike such persons, for whom few rewards are available except the satisfaction of fulfilling one's station in life, Roman matrons could affect the prestige and power of their families and could reap the ensuing social and economic rewards.

Why this might have been the case is a question largely beyond the scope of my study, although Judith Hallett's work on the centrality of Roman daughters, which she calls *filiafocality*, goes a long way toward unraveling the structure of elite Roman society.[82] Hallett suggests that once Roman fathers agreed to raise their legitimate daughters, they became heavily invested economically, socially, politically, and emotionally in the well-being of those daughters. She argues that the role of daughter is the most well articulated and defined role of elite Roman women, and defines them throughout life. The centrality of the father-daughter tie explains how, if not why, elite Roman women were so effectively integrated into elite Roman society and has major implications for some of the religious activities of these women.

Yet the evidence does not wholly support the conclusion that the primary experience of elite Roman women was one of strong group and strong grid, which does appear characteristic of much of Roman (male) aristocratic society.[83] The relationship of Roman women to their families was both complex and ambiguous. The Roman *familia* was essentially a conjugal unit consisting of an adult male, his wife, and their dependents, who would have included unmarried children, slaves, freedpersons, and foster children, *alumni/ae*. In many practical regards, the *matrona* was a central and significant member of the household. In certain key respects, however, she was almost always an outsider, particularly if she had been married without *manus*, in which case she remained under the control of her own father (provided he was still living) and was legally not a part of her husband's family.[84] As we observed earlier, women married without *manus* were technically excluded from the household worship of their husbands.[85] They were also, at least in theory, not entitled to burial in their husbands' family plots.[86] Regardless of the form of marriage, the children she bore to her husband belonged legally to him and to his family, not to her. In the event of divorce, which was not infrequent among the Roman aristocracy, children remained with the father. Rawson suggests that this convention might have affected the ties Roman mothers formed with their children, encouraging these women to remain somewhat distanced from their offspring as a protection against the possibility of future separation.[87]

Thus the elite Roman woman remained permanently linked to her father and to her natal family. Such dual loyalties often engendered real conflicts for Roman women, as in cases when fathers motivated by political expediency compelled their daughters to divorce one husband and marry another.[88] The central location of elite Roman women within elite Roman society was primarily in their role as daughters to powerful fathers, secondarily as sisters to (powerful) brothers, and only then as wives to their husbands and mothers to their children, especially sons. As wives, aristocratic Roman women, while in many ways more integrated into Roman society than other women, were yet peripheral to the central social unit of the *familia* in certain key respects, so that they experienced relatively weaker group, though perhaps not weaker grid.

If Douglas's model is viable, we should expect to find that the religious rituals and accompanying myths of elite Roman women exhibit characteristics that accord with these observations. Specifically, we may expect to find that in their roles as daughters, elite Roman women participated in religious activities, symbolic systems, and cosmologies that confirmed their experience of relatively strong group and grid, whereas in their roles as wives, we may expect rituals, beliefs, and symbols expressing somewhat diminished group. It is also conceivable that the religious involvements of elite Roman women reflect a tension between their central roles as daughters and their more ambiguous roles as wives. In the rites of Roman matrons previously described, this pattern is more or less what we find.

Whatever the origins and early social functions of the worship of Matuta, by the time of Ovid and Plutarch, worship of that goddess, especially in the festival of the Matralia, expresses the concerns of aristocratic women functionally tied to their marital households, but legally and (if Hallett is correct) affectively tied to their natal families. Douglas's model allows us to see that the Matralia expresses the dilemma and the somewhat ambiguous social location of upper-class Roman women caught between two families, a full member of neither.

It is interesting to consider the mechanism through which the Matralia accomplishes this. Women who were normally separated from one another, namely sisters living in separate conjugal households, came together, perhaps with their children, to offer toasted cakes to the goddess and to pray for the welfare of each other's children. The temporary constitution of a women's community represents, in Douglas's terms, a shift toward stronger group. But unlike Dionysiac ecstasy, which abolished certain social distinctions (although perhaps not that of free and slave) and thus represented a decrease in grid, the Matralia explicitly affirms certain aspects of high grid. Only women of high social standing in the Roman hierarchy were able to participate: married citizen women, presumably who were already mothers and perhaps only *univirae* at that. Slaves were explicitly excluded. The Matralia was not a festival that obliterated social distinctions: rather it confirmed them, even while expressing some of the ambiguity married Roman women must have experienced.

Through the vehement exclusion of slave women and the aetiological myth offered as explanation, we may also view the Matralia as a convenient vehicle for the expression of yet another dilemma faced by married Roman women, namely their relationships with their female slaves. The myth of Ino and her slave (who in Plutarch acquires the name Antiphera and the ethnic origin of Aetolia) may well reflect the nature of antagonisms between married women and their female slaves, who not infrequently had sexual liaisons with the husband. By Roman law, such liaisons were perfectly licit, and Roman wives were expected to tolerate them. The children of such relationships were illegitimate, and posed, in theory, no threat to the status of the *matrona*'s own legitimate children. In reality, however, things may not have been so simple. There is some evidence that foundling children (*alumni* and *alumnae*) and slave children (*vernae*) raised in the master's house (often alongside his legitimate children) may have been, in fact, the illegitimate children of the man and his slave mistress.[89] The rites of the Mother Matuta, performed only by those Ovid calls "good mothers," may have enabled aristocratic Roman women to vent some of their hostility toward the slave women who threatened not their social status, but their household relationships. "Good mothers" are not slaves, and the children of slaves are not protected by the Mater Matuta. In Plutarch, the account of the one slave who is slapped and beaten may further express the hostility that aristocratic wives felt toward slave women who were sexually intimate with their husbands.[90]

Precisely because the mythological explanation of the Matralia and the origins and identification of Matuta are explicitly derived from Bacchic legend, it is instructive to elucidate the salient differences between the myth in Euripides and that in Ovid. In the *Bacchae*, the sanctioned ritual activity is the ecstatic worship of Dionysus himself; in Ovid it is the invocation of Ino/Matuta. In the *Bacchae*, Ino plays a minor role and receives no credit for the succour of the motherless infant. The opening lines of the play accuse Semele's sisters indiscriminately of failing to believe Semele's claims about her baby's paternity, and it is their failure to believe her, their sororal betrayal, which causes Dionysus to punish them with an insanity that ultimately results in the death of Pentheus, king of Thebes, at the hand of his crazed mother, Semele's sister

Agave. Yet in Ovid, Ino becomes a central and positive figure, one who having previously brought about the death of her own child, now succors her sister's, evoking the dangerous wrath of the jealous Hera/Juno on herself.

Numerous motifs are at play here. In the *Bacchae*, mother-son conflicts in Greek society are of primary concern, as are the restraints imposed on Greek women by Greek men.[91] Sexual liaisons, such as Semele's affair with Zeus, damage relationships within the family: Semele's sisters are apparently more concerned for the family honor than they are with their loyalty to her. In the *Bacchae*, motherhood is ritually rejected and perverted. The Matralia, by contrast, evidences concern for proper motherhood: it *is* the festival of good mothers, even if the best mothers turn out to be aunts.

Sibling relationships, to which Hallett has drawn attention,[92] figure significantly in both the *Bacchae* and the Matralia. In the *Bacchae*, there is tension between the sisters and conceivably competition for the respect of their father. In the Matralia, by contrast, sibling ties are strong. Ino risks death to save the child of her sister, in return for which she is richly rewarded with immortality.

The social tensions most evident in the *Bacchae*, as I noted, are between men and women, whether between Greek mothers and their sons or Greek husbands and wives. The appropriate roles for women are emphasized precisely to the extent that they are violated by the Bacchae when they abandon their homes and babies and assume male roles of hunter and warrior. In Ovid, the social tensions are much milder and focus more on conflict between women: between married Roman women and their female slaves; between the outraged Juno and the maternal Ino. Both these conflicts, we should note, center on sexual jealousy and its social ramifications. If French's interpretation of the Matralia's historical origins is correct, in their republican form these rites may also address and ease conflicts between Roman and Etruscan families.

The differences between the Bacchic story of Ino and that of the Matralia may also reflect fundamental shifts in the social location of elite women in Athenian versus Roman society, especially in relation to their male kin. The mythic tensions between mothers and sons, between men and women, are rooted in the real, historical distancing of adult Greek men and women, and in its effects on mother-son relationships.[93] If it is too much to say that in the *Bacchae*, the only good mother is a dead mother (Semele), it may not be too much to point out that in the *Bacchae*, mother-son relationships tend to be lethal for at least one member of the dyad.

In Ovid's version of the Bacchic myth, Ino, the sister who receives the least attention in Euripides, exemplifies different relationships: namely solicitiousness toward one's natal (male) kin, though still characterized by some distancing from her own children, which accords well with what we know of Roman family relations and tensions.

It is also intriguing that the foreign nature of Dionysos, so crucial to the dynamics of Bacchic possession in Euripides, is greatly mitigated in the Matralia. There, Ino's foreignness is deemphasized and ultimately resolved by

her invocation of Roman gods, her recognition that they will provide her sanctuary, and her ultimate transformation into a deity called by one name among
the Romans and another among the Phrygians, Thebans, and other Greeks.[94]

The diverse portrayal of mothers and mother-child relationships in these
traditions suggests further lines of inquiry. Elite Roman child-rearing
practices[95] may not have been terribly conducive to the establishment of
strong affective ties between Roman matrons and their children. Elite babies
were routinely nursed and raised by women other than their mothers. They
were strapped down and bandaged to boards to ensure their proper physical
development and subjected to all sorts of massage to mold their bodies to
elite physical ideals. They received little attention and comfort when they
cried. All this contrasted enormously with the child-rearing practices of the
popular classes, who, Aline Rousselle writes, "bathed newborn babies day and
night, . . . refused to wrap them in bandages, . . . rocked them in their arms
when they cried and even put them to the breast to soothe them, which was
heresy to the doctors."[96]

We may well wonder how Roman women raised under such circumstances felt about mothers and how they would have construed the worship of
mother goddesses. It seems fascinating that several goddesses who had clear
associations with mothering in other contexts were transformed in the worship of elite Roman women into deities whose primary associations were no
longer with nurturing motherhood, but with chastity, marriage, and fertility,
such as Ino, Matuta, and even Cybele. If such connections can be borne out
by further research, we might also expect to find analogous correlations
between the maternal experiences of nonelite women and their own goddess
devotion.

Hallett's observations about the Vestal Virgins, discussed in detail in
Chapter 7, conform neatly to Douglas's schema. Noting the scholarly consensus that the historical roots of the Vestal Virgins lie in the consecration of the
daughters of the king, she sees this cult as the embodiment of elite Roman
filiafocality.[97] In Douglas's terms, the cult of the Vestals reinforces the central
location of Roman women as daughters experiencing strong group and strong
grid, even as the Matralia reflects their somewhat shifted location toward lower
group.

The devotions of Roman matrons to Venus, which as we have seen may
well have been consciously invented and encouraged as a means of enforcing
proper sexual behavior on the part of elite wives, also conform to a Douglasian
pattern. At first glance, the concerns preeminent in the cult of Venus Verticordia, especially as described in Ovid, are typical of persons experiencing high
group and grid, though perhaps not at its extremes. Rank and status are crucial: the participants are described by their status as mothers, brides, and
unmarried virgins. The cult reinforces marriage by ensuring the physical desir-
ability of women on the one hand and their protection from inappropriate lust
(symbolized by the satyrs who unsuccessfully lusted after the naked bathing

Venus) on the other. It guarantees chastity, and by implication, lawful heirs and succession.

While I recognize that the women who worshipped Venus may have seen things quite differently, I propose that a more complex dynamic is at work. The human instigators of the cult are persons, probably elite men, experiencing strong group and grid, who hope that the cult will exert pressure on elite Roman women, whose sexual behavior constitutes a rejection of the rigid stratification of Roman society and whose increased autonomy threatens Roman social order. We should expect the religious concerns of nonelite Roman women to be somewhat different, to express their relative distancing from elite Roman society and their probable experience of weak group and strong grid. Our earlier discussion of women's attraction to the Bacchanalia, detailed and impugned by Livy, seems to bear this out, as does Brouwer's analysis of the cult of Bona Dea and probably numerous other cases as well, as yet unexplored.

# — 6 —

## Women's Devotion to the Egyptian Goddess Isis in the Greco-Roman World

In the Hellenistic and Roman periods, the worship of many foreign divinities spread from their native lands as commerce and military maneuvers moved multitudes back and forth across the Mediterranean. The Phrygian goddess Cybele, the Persian god Mithras, the Syrian goddess, and numerous others established cult centers across the Greco-Roman world, and with the singular exception of the cult of Mithras, all were known (and often mocked by establishment Greek and Latin male authors) for their strong attraction for women. Principal among these was the religion of Isis, which many scholars identify as the primary contender against Christianity for sovereignty in the Roman Empire.

Initially unconnected to other deities, in her ancient Egyptian milieu Isis was apparently venerated early as a mother goddess with her son, Horus. Ultimately, she and Horus form a triad with Osiris, whose cult of death and resurrection was intimately connected with the yearly dying and rising of the Nile, which was crucial to the agricultural sustenance of Egypt. In the ancient Egyptian *Pyramid Texts*[1] Isis appears as the sister of Osiris, who together with another sister, Nephthys, mourns their dead brother and participates in the embalming rites that lead to his resurrection. She is not yet the wife of Osiris, and their son, Horus, plays no part here.

Our fullest version of the myth of Isis and Osiris comes from the Greek writer Plutarch, whose friend Clea was a priestess in the cult.[2] Osiris, the great king of Egypt, had two sisters, Isis and Nephthys, and a brother named Typhon, who conspired against Osiris to gain power. At a feast celebrating Osiris's return from travels throughout Egypt, during which he taught all the elements of refined civilization, Typhon has brought into the room a beautifully ornamented chest, secretly tailored to fit only Osiris. He invites all the men to try the chest on for size, offering to give it to whomever it fits, but when Osiris lies down in it, Typhon slams shut the lid, seals it with nails and lead, and deposits it in the Nile.

71

When the news of Osiris's death reaches his sister, Isis, his lover since their days in the womb together, she immediately sets out in search of her husband. In her wanderings, she learns of the fate of the coffin: she also inadvertently learns that Osiris had had sex with their sister, Nephthys (who had married Typhon), and conceived a child whom Nephthys had exposed for fear of Typhon. Isis rescues the child, the dog-god Anubis, who becomes her guardian and attendant.

Eventually, Isis finds the coffin. Washed up on the shore at Byblos, it was rapidly concealed within the trunk of a large heather plant, which the king of Byblos cut down for a pillar in his house. Presumably at this point, Isis has sex with the body of her husband and conceives a son called Harpocrates, who is subsequently born prematurely, "weak in his lower limbs." Isis temporarily conceals the coffin, but Typhon stumbles on it while hunting and cuts up the body of Osiris into fourteen pieces, scattering them all over. When Isis discovers the fate of the body, she sails the marshes of the Nile in a papyrus boat, finding all of the pieces but the penis, which had been eaten by certain fish. Ultimately, Osiris returns from the underworld to train Horus, the son of Isis and Osiris, who successfully avenges his father against Typhon.

Plutarch also describes the Egyptian rites of Isis and Osiris. When the Nile had completely receded (at the time of the winter solstice), the priests covered a gilded cow with a black linen garment representing the mourning Isis. On the fourth (or perhaps the third) day, the priests go down to the sea at night and bring forth a sacred chest with a small gold container inside. Into this they pour some water, and proclaim that Osiris has been found. They mix the water with fertile soil, expensive spices, and incense, and knead the mud into a crescent shape. Plutarch is explicit that Osiris is a metaphor for the Nile and the story of Osiris shut up in the chest represents the disappearance of the water.[3]

In their Hellenized form, the rites of Isis and Osiris still emphasized death, mourning, and resurrection, although it is not surprising to find a deemphasis of the relationship between Osiris and the Nile, since that river played a unique role in the economy (and cosmology) of ancient Egypt that was paralleled nowhere else in the ancient Mediterranean. The festival *Isia* was celebrated in Rome from October 28 to November 1, comprising ritual reenactment of the mourning of Isis, her search for Osiris, and her joy at his discovery.[4]

Better known than the Isia was the ship-launching festival, the *Navigium Isidis*, or *Ploiaphesia*. Apuleius's *Metamorphoses* contains a glorious description of the March festival, whose roots lay in an Egyptian procession from bank to bank at the time of the inundation of the Nile.[5] At the head of the procession down to the sea came women in white, with flowers on their heads, scattering herbs on the ground from their aprons. Some women carried mirrors; others brought combs for the goddess's coiffure. Women and men carried lights, lamps, candles, and torches. Next came the musicians, followed by young male hymn singers, also dressed in white, and trumpet players. After them, wearing the distinctive white linen, came the initiates: men and women of all ranks and ages.[6] The women anointed their hair and covered their heads with a fine linen veil; the male initiates shaved the crowns of their heads.

After the devotees came priests dressed in white robes bearing sacred objects: a gold lantern, cultic pots, a gold palm tree, an image of a left hand, signifying equity, a gold breast-shaped vessel from which milk poured out; a gold winnowing fan, and a wine vessel. Next came the gods themselves: a person dressed as Anubis and a cow representing "the great goddess that is the fruitful mother of all."[7] A cult official carried a chest containing the secrets of the religion; another bore a strange gold vessel decorated with Egyptian figures containing an asp. Apuleius described this last as "a new invention . . . an emblem ineffable, whereby was signified that such a religion was at once very high and should not be discovered or revealed to any person."[8]

When the procession reached the shore, the high priest (*summus sacerdos*) dedicated a boat, over which he prayed, purifying it with a torch, an egg, and sulphur. The ship was made from citron wood, with a white linen sail, a gold-plated poop, and an inscription for the prosperity of the season. All those assembled poured out libations of milk onto the waters and then launched the unstaffed vessel onto the sea. When it had sailed out of sight, the procession returned to the temple of Isis, returning the various holy objects. A priest then prayed from a book for the welfare of the ruler, the Senate, and the Romans, as well as for all sailors and ships, and pronounced the word *Ploiaphesia*, signifying that it was now lawful for ships to depart.[9] Rejoicing, the people carried off herbs, branches, and flowers, stopping to kiss the feet of a silver image of Isis on their way home.

While the nautical festival is best known from Apuleius's description of its celebration at the port of Cenchreae, the names of "captains" of the ship Isis, including eight women, are preserved in an inscription from Eretrea in Euboea.[10]

The reenactment of Isis's mourning and joy was well known in Greco-Roman antiquity as a public performance.[11] Some devotees were initiated into restricted rituals which, Heyob observes, the Greeks equated with their own mysteries, especially those of Demeter at Eleusis. Interestingly, and perhaps significantly, the only account we have of an initiate in the mysteries of Isis and Osiris is that of a man, Apuleius, whose fictional account of the initiation of Lucius is widely thought to be autobiographical.[12] Initiation was understood as election by the goddess, who conveyed her decision in dreams both to the prospective initiate and to the priest who would perform the rites. Devotees of Isis, both male and female, often slept in the precincts of her temples, hoping to dream the requisite invitation. Formal initiation required a period of abstention and baths of purification. Lucius abstains from meat and wine for ten days; women apparently refrained from sexual contact. As the account of Apuleius makes clear, initiation was a costly venture, whose goal seems to have been personal salvation in the form of immortality.

In addition to the festivals and restricted initiations, daily rites were celebrated at the temples of Isis. In the morning, when the temple was opened, the praises of Isis were sung and the statue of the goddess was washed, dressed, and anointed. Praises were sung again in the evening.[13]

The association of Isis with women is perhaps as ancient as the goddess herself. In pre-Greco-Roman Egypt, women enacted the mourning of Isis and Nephthys. Several Hellenistic aretalogies, or praises, of Isis explicitly identify her as a women's goddess. "I am she who is called goddess by women," proclaims an inscription of the second or third century C.E. found at Kyme in western Asia Minor.[14] The aretalogy also identifies Isis with traditional women's concerns:

> I brought together woman and man . . .
> I appointed to women to bring their infants to birth in the tenth month . . .
> I ordained that parents should be loved by children
> I compelled women to be loved by men
> I devised marriage contracts

Another aretalogy on papyrus found at Oxyrhynchus, Egypt, proclaims Isis as the one who made the power of women equal to that of men.[15]

Yet women's devotion to Isis in the Greco-Roman period defies simple categorization. In a useful study, Sharon Heyob observed that women do not appear to have been the majority of the devotees of Isis, nor the majority of those formally initiated into her mysteries, an observation substantiated by subsequent research.[16] While women held many offices in the cult, including that of canephoros (basket bearer) and priestess, there is no documentation for women holding the ultimate office of high priest.[17]

Nevertheless, Heyob argued strongly that the cult of Isis had great appeal for women in the Greco-Roman world. Claiming that official Greek and Roman religion paid little attention to women's concerns and afforded women little opportunity for participation and cult office, Heyob concluded that worshipping and identifying with Isis filled emotional needs of women that went unheeded in other Greek and Roman religions.[18] In particular, she suggests that in her experience of the joys of marriage and motherhood and the griefs of death and separation, Isis experienced the whole range of human emotions that comprised the life of ancient women, and thus provided a divine figure with whom they could truly identify.

As Heyob and others have recognized, in its Greco-Roman incarnation, the cult of Isis emphasized the goddess as exemplary wife and propounded the sanctity and joys of marriage and the nuclear family. While Isis was frequently identified with the Greek Demeter, the differences between them are significant. Demeter's central role was that of grieving mother, but Isis seeks her spouse, not her child. In the *Hymn to Demeter*, the goddess has little regard for marriage, which separates her from her beloved child. It seems at best a necessary evil, a compromise to be reluctantly tolerated.

The Greco-Roman worship of Isis shifts the focus from woman as mother to woman as wife. It is Isis, after all, who commands women to be loved by men, marriage contracts to be written, children to love their parents. It is Isis who seeks day and night for her missing spouse, who forgives him his (apparently unwitting) adultery with her sister and raises the child of that union as her own, who loves her husband and desires his children so deeply that she engages in sexual intercourse with his corpse, succeeding by her own great power in generating from that encounter a child (albeit a slightly deficient one).

At least one romantic novel from the Hellenistic period testifies that Isis was widely invoked as the guarantor and protector of marital fidelity.[19] Several burial inscriptions commemorate women who were initiated into the rites of Isis by their husbands, suggesting that Isiac initiation was sometimes a form of ancient marital togetherness.[20] Particularly in contrast to Demeter, Isis appears to express significant aspects of the changes that took place in the ancient Mediterranean in the wake of Hellenization. Whereas the cult of Demeter reflected the Greek preoccupation with agrarian fertility and community survival and identity, the cult of Isis, transformed from its Egyptian base, manifests concern for individual salvation and continuity and for the marital couple as the central social unit. While Isis is unquestionably mother and wife, she is wife first and foremost, in marked contradiction to the ultimate mother, Demeter, whose very name conveys the maternal element.

The worship of Isis gave ritual, mythic, and symbolic expression to the changed universe in which many people found themselves in the Hellenistic world. Many scholars have argued that the breakdown of older forms of social cohesion, notably the city-state, and the increased emphasis on the individual were accompanied by an increased sense of anxiety.[21] Although that view has also been much disputed,[22] it is hardly insignificant that one of the key attributes of the Greco-Roman Isis was her mastery over the universe and all its elements, not the least of which was fate. Devotion to Isis guaranteed her worshippers protection from the vagaries of a capricious universe, protection embodied in the symbols of wifely devotion and maternal care.

Heyob's interpretation of Isis's appeal to women poses certain problems. Her conclusion that women were not the majority of Isiac devotees does not cohere with her vision of Isis as the ultimate goddess for Greco-Roman women. If most Isiacs were not women, even though many were, then surely most women were not Isiacs. Heyob's (undocumented) belief that women were largely excluded from Greek and Roman religion proper suggests that the bulk of women's religious activity in the Greco-Roman period was directed elsewhere. Yet she is correct that no other Greco-Roman religion was couched in terms quite so directly addressed to the realities of many women's lives.

Some resolution may come from a mild feminist critique of Heyob's study. For while she admirably undertook to analyze women's involvement in the cult of Isis at a time when few scholars thought to undertake careful investigation of women's religious activities in the Greco-Roman period, Heyob tended to accept the evidence of women's participation at face value. She did not consider that the percentages she found for women's participation in processions, festivals, initiations, and cult offices might not be accurate. There can be little doubt, for example, that the percentage of women commemorated in ancient burial inscriptions from Rome (about 40 percent) does not accurately represent Roman demographics.[23] Feminist historians of many different periods have demonstrated that, given the tendency of male historians to obscure if not erase the history of women altogether, any evidence for women, particularly in the public sphere, is likely to underrepresent their true presence. Thus it is not inconceivable that women constituted a far greater number of Isiac devotees at every level than Heyob concluded on the basis of the surviving data.

Heyob's explanation for the devotion of women to Isis reflects an unarticulated assumption that the same religious system can have different meanings for women and men, a critical issue for the present study. She locates the appeal of Isis for women in factors specific to women's experience, while locating the appeal of Isis to men in factors she concedes were also of concern to women, such as life after death and the provision of security and safety in a universe otherwise dominated by capricious fate. Here, too, a feminist critique may be helpful. Heyob accepted unquestioningly that marriage, marital fidelity, and children were the central concerns of women's lives, and that a goddess whose own experiences mirrored those of human women would most effectively meet the emotional and religious needs of Greco-Roman women. While I have no doubt that for many Greco-Roman women these concerns were indeed fundamental, the presentation of Isis may be viewed with some feminist suspicion as at least subtle propaganda for a particular model of the ideal woman.

Isis, after all, is the wife who puts her husband before her children and both of them before herself; whose true being is expressed in her relationships to brother/husband and son(s); who is separated from her sister by competition over their brother (who is lover to both); who has no significant connections to her parents, mother or father; and who tolerates adultery on the part of her husband, even going so far as to raise the product of that adultery in her own household, begrudging the child nothing, yet who is herself the model of constancy and fidelity. She is supreme and powerful herself, yet she subordinates herself to her husband and ultimately to her son. Plutarch's account of the final scene between Horus and Isis, after the defeat of Typhon, is instructive. When Isis shows mercy to Typhon, Horus, outraged, seizes the crown off his mother's head. Isis apparently does nothing—it is only Hermes (a Greek intruder into this supposedly Egyptian scenario) who responds by placing a cow's head helmet back on the goddess.

I do not wish to suggest here that Isis did not hold strong appeal for many women, or that she did not, in fact, offer approval and legitimation for what many scholars have seen as the increased autonomy and approval for women in the Hellenistic period; I think unquestionably that is part and parcel of Greco-Roman Isiac religion. Nowhere else do we find statements as unequivocal as that of the Oxyrhynchus aretalogy in which Isis has made the power of women equal to that of men.[24] But it also seems to me that the model of Isis is not only reflective of women's experience and perspective, it also presents a program for the ideal woman that was not merely descriptive, but prescriptive as well.[25]

The functions of the cult of Isis for women, then, may be seen as twofold. On the one hand, Isis unquestionably legitimized appropriate roles for Greco-Roman women as wives first and mothers second while downplaying women's ties to their natal families. But the role of woman as wife coupled with the emphasis on marital harmony and fidelity (extended sometimes even to the behavior of husbands) represents a shift from the norms of both classical Greek and Roman society, reflective of the emphasis on the individual and the nuclear family that emerged more strongly in the Hellenistic period. This shift seems to have been accompanied by increased economic and personal autonomy for free women, which the cult of Isis also legitimizes. Thus we may argue that Isis reinforced more or less traditional roles for women, with a somewhat different emphasis, yet also sanctioned increased autonomy and authority for women at an explicit level not seen before (or after) in the religions of the Greco-Roman world.

A most interesting test of the utility of Douglas's theory comes from the worship of Isis, which in its Greco-Roman permutation expresses admirably certain changes in the social structure of ancient society. In the wake of the conquests of Alexander, not only did Greek culture spread throughout the ancient world, but foreign cultures increased their influence on Greeks. The programs followed by Alexander's successors further weakened the significance of traditional Greek tribes and city-states, and gave far greater prominence to individuals. In Douglasian terms, Hellenization weakened both group and grid for most members of Greco-Roman society. The consequences were substantial: increased travel, mobility, decreased classification, increased autonomy for persons in general, but especially for those persons experiencing the constraints of weak group and strong grid: free women, slaves and freedpersons of both genders, and other marginalized men.

The cult of Isis, which took shape and spread dramatically in this period, reflects precisely these concerns, both in the actual details of its transformation from its previous Egyptian form and in its new variants. Conceivably, it had the potential to do this already in its Egyptian form by virtue of its relationship with Egyptian society, which should have been closer to weak group and weak grid than Greek and Roman society: this is what Françoise Dunand explores in a limited way.[26]

Following up on Heyob's suggestion that women played a more significant role in imported religions (often from the East—hence their popular label as Oriental), especially that of Isis, than they did in traditional Greek and

Roman state cults,[27] Dunand suggests that this was not because such religions actively sought women. Rather, she proposes that Oriental cults, "marginalized and perhaps suspect among the Greeks and Romans, offered a structure of acceptance for those whom society rejected, or at least integrated only partially and under conditions: women, foreigners, and slaves, who could scarcely expect to have the opportunity, with a few exceptions, of playing a major role in the official religion."[28]

Dunand then asks whether women played a greater role in the cultures that spawned these imported religions. In a brief but detailed study of Egyptian priestesses, especially in the Hellenistic and Roman periods, she concludes that the role of women in Egyptian religion does seem to correspond to their place in Egyptian society before the Greek and Roman conquests—one of equal rights and benefits with men and relative economic independence, compared to most ancient societies, although she concedes that men still predominated in most arenas.[29] Dunand admits that the priests were always a privileged class in Egypt, which suggests that she may overstate the extent to which the economic independence and rights of Egyptian women in general may be deduced from the status of women in priestly families.

In any case, we may observe that in its Greco-Roman form, the characteristics of Isiac religion are consonant with weakened group and grid. It emphasizes individual salvation and the nuclear family. It offers more or less direct communication between the individual and the divine, with some modest mediation (which the individual and the priest must both verify, thus sharing the power between them). Distinctions of rank and status are not altogether absent, but they often seem based on the achievements of devotees and not on externally determined status criteria. Some degree of egalitarianism exists. Although it would have been difficult for poor persons to afford initiation, as Apuleius's description of Lucius makes clear, rich slaves and freedpersons could have bought these rites/rights and privileges.

Isiac worship bears some affinities with worship along the diagonal axis from weak group and high grid to strong group and low grid, yet only some. It is emotional, focusing on weeping and mourning and also on the joy of finding the god, yet it is not wildly ecstatic. Prophecy figures, but there are some modest controls in the form of the belief in double (simultaneous) revelation. Ritual displays some hierarchy, yet also some obliteration of distinctions, in such phenomena as the white clothing worn by devotees. Initiation, processions, and festivals provided some experience of stronger group, but in general, Isiacs did not have the strong sense of community and group control that early Christians enjoyed by virtue of their faith, or classical Athenians experienced by virtue of their common residence, work, state religion, and so forth.

Dunand suggests that all this reflects the social circumstances of the women, freedpersons, foreigners, and slaves who made up a significant portion of Isiacs. What is interesting here is that they have moved from their extreme isolation to a greater degree of autonomy and a stronger sense of community (that is, stronger group and weaker grid), which seems to have characterized much of Hellenistic culture, though not all.

The appeal of Isis to women can be located within this larger dynamic, which had special, but not exclusive concern for the autonomy of women that Hellenization (and the relaxing of grid) brought. But although the weakening of grid brings, predictably, a theoretical gender equality, expressed in the Isis aretologies, it was not accompanied by a total equality for women (although in cultic terms this is difficult to determine) in part because it did not profess beliefs that severed salvation from marriage and children. But in downplaying children (and seemingly having little to do with property transmission) the potential was there, which seems not at odds with what we do know about Isaic priesthood and cult! At least at the explicit level, the religion of Isis was more favorable to women than any other religion.

# — 7 —

## Women's Religious Offices in Greco-Roman Paganism

Evidence for women's religious leadership in pagan contexts has received short shrift, in part because it has been deemed the least relevant, if not irrelevant, for contemporary theological agendas.[1] Yet women unquestionably held office in a wide range of pagan religions across the ancient world chronologically, geographically, and culturally.

At the outset, it is useful to inquire about the connections between religious office and religious leadership in Greco-Roman antiquity. In their introduction to a collection of essays on ancient pagan priesthoods, Mary Beard and John North offer some wise general observations about the nature of religious offices in Greco-Roman antiquity.[2] Such offices were, as a rule, part time, regardless of social level, and tended to be carefully delineated as to function.[3] Those who carried the sacred water of Isis were different from those who tended the statue. Not all religious offices were priestly, even when we take into account the laxity with which distinct Greek and Roman terms have been translated by the English word *priest*. While Beard, North, and their co-contributors focus on ritual sacrifice as the central component of priesthood, they concede that the concept of priesthood itself requires much more investigation.

The service of ancient deities entailed numerous responsibilities, from the physical maintenance of temples and precincts, to the organization of festivals, to the performance of sacred rites. At small rural temples (or perhaps small Jewish synagogues), one or two persons fulfilled all these obligations; large temples in major cities had grand organizational structures. Religious offices were usually held for defined periods of time, often one year. Service was sometimes by lot, sometimes hereditary.

The concept of religious leadership is more complex. In its modern context, the term *leadership* is laden with connotations of both authority and prestige. Leaders are presumed to have the ability to affect and effect the behavior of others through a variety of means. In Greco-Roman antiquity, how much

80

did the categories of religious office and religious leadership overlap? Were all holders of religious office in fact religious leaders, that is, were they invested with authority and prestige by their communities? Were there individuals who functioned as religious leaders without holding religious office, and how would such leadership be distinct from other kinds of leadership? Are these distinctions equally applicable across the range of religions in the Greco-Roman period? To what extent is gender a factor here: are women any more or less likely to hold office without authority, or authority without office?

In the classical Greek tradition, numerous inscriptions and ancient writers testify to priestesses in service to the Greek goddesses Demeter, Hera, Athena, Artemis, Eileithyia, and many others.[4] Robert Garland asserts that in classical Athens, the gender of the priest generally correlated with that of the deity, but Athena Polias was served by a priest, and Dionysos, Helios, and Apollo all had priestesses in their service.[5] Priestesses (and priests) generally served a particular deity in a particular sanctuary, where they were responsible for the care and upkeep of the sanctuary and the statue of the deity, the performance of rites of purification, and safeguarding the sanctuary treasures and gifts.[6] In payment for their services, priestesses received modest fees and a portion of the sacrifices.

We cannot generalize on the question of how much prestige was attached to such religious responsibilities. Garland proposes that in Athens, priestly offices were determined initially by family membership, but that by the mid-fifth century B.C.E., the only critical factor for selection was appropriate gender for the particular deity. He suggests that priesthoods obtained through familial right, called *gentile* priesthoods, and priesthoods held for life may have conveyed more prestige than those priesthoods to which one was elected or held for a fixed period. Conversely, however, since the method of election for non-gentile priesthoods was by lot, it may well be that such priestesses and priests were considered chosen by the gods themselves, an election conveying a different sort of prestige.[7]

Although the majority of priests for official Roman cults were male and organized into colleges,[8] particularly during the republican period, one of the most famous of all official Roman priesthoods was held by women, that of the Vestal Virgins. Consecrated in childhood, in service to Vesta, they guarded the hearth of Rome that symbolized not only the Roman people but Roman male procreative power.[9]

Many scholars now believe that the Vestal Virgins were originally the daughters of Rome's priest-king, who guarded the fire that represented the father's power to create and sustain life.[10] According to Plutarch there were initially two Vestals, whose number eventually was increased to six.[11] Vestals served for a period of thirty years, after which they were free to marry, although Plutarch emphasizes that few chose to do so, and that those who did rarely had happy marriages. The privileges accruing to Vestal Virgins were considerable, including freedom from any male guardianship and the right to make a will and bequeath property during the lifetime of their fathers. The power of Vestals was such that a criminal who accidentally met a Vestal on the way to execution was spared, according to Plutarch, and anyone who passed under the litter on which a Vestal was carried was put to death. While in office, Vestals were subject to chastisement by the Pontifex Maximus (the head of the college of state priests). Vestals who broke their vows of chastity during their term of office were walled up in a small chamber furnished with a couch, a lamp, and minimal food, and left to die.

Another Roman writer, Aulus Gellius, describes the qualifications for becoming a Vestal, as well as the entry rites.[12] Vestals had to be between the ages of six and ten. Both parents had to be alive, and neither could have been a slave or of a low-status occupation. They had to be free from bodily blemishes or speech or hearing defects, and under the authority of their fathers.

The power and status that accrued to the Vestals was considerable, but its implications even for elite Roman women, let alone Roman women in any general sense, must be carefully nuanced. Mary Beard has argued that in the ritual functions they performed, the rights and privileges they possessed, and their specific dress, the Vestals combined aspects of virgins, matrons, and aristocratic males. Their resulting highly ambiguous status "must be seen as playing an important role in their symbolic position."[13]

Beard observes, for example, that in their virginity and chastity and in their probable origins as the daughters of the priest-king, the Vestals exhibit characteristics of unmarried young women. Their participation in certain festivals with fertility implications, such as the *Fordicidia* in April, where the chief Vestal, the Virgo Vestalis Maxima, burned a fetus torn from a pregnant cow, suggests an identification with married women. Beard points out, though, that this by itself is insufficient, since fertility and virginity are by no means antithetical, especially if virginity is seen as stored-up, potential procreative power rather than sterility.[14]

In significant other ways, though, the Vestals did resemble aristocratic Roman married women. They wore the long robes (*stolae*) and headbands (*vittae*) signifying respectable married women, although some scholars have suggested that the garments of the Vestals were actually those of a bride.[15] The rituals associated with becoming a Vestal bear significant resemblance to Roman marriage rites, particularly in the taking of the Vestal from her family. The authority over the Vestals held by the Pontifex Maximus resembles the power that husbands held over wives married in *manus* marriage. In December, the Vestals participated in the rites of the Bona Dea, that characteristic festival of Roman matrons.

But Beard establishes that the Vestals were not merely part virgin, part matron. In their legal autonomy, in their ability to give legal testimony and to make wills while their fathers were still living, as well as in their right to have a *lictor*, the privilege of a public attendant, which was generally reserved for aristocratic men, the Vestals displayed clearly masculine characteristics.

Utilizing the early work of Mary Douglas on the relationship between ambiguity and sacrality, Beard proposes that the sacredness of the Vestals lay precisely in their combination and symbolic presentation of these three characteristics, and in their perpetual liminal or interstitial condition. That is, the Vestals were perpetually and simultaneously virgin, matron, and male, and always on the threshhold, an interpretation which is strengthened if the judgment is correct that they wore the dress not of married women, but of brides on their wedding day, including a special hairstyle.

Ultimately, Beard suggests that the fire of Vesta with which the Vestals were so crucially linked was itself an ambiguous symbol that mediated between heaven and earth, purity and impurity, life and death, male and female, and that most crucial of polarities in structuralist models, nature and culture.[16]

Although cognizant of Beard's work, Judith Hallett has subsequently returned to the dual aspects of virgin and matron represented by the Vestals. She has, however, emphasized the importance of the Vestal Virgins as the quintessential symbol of Roman daughterhood, a role she argues lay at the heart of elite Roman conceptualization of women. Hallett observes that in elite Roman society, daughter and married woman were not wholly antithetical categories, for Roman wives married with *manus* were in fact in the same judicial relation to their husbands as unmarried daughters to their fathers. For Hallett, the Vestals both symbolized and reinforced these particular Roman understandings of women and their location within the elite power structure.

> The Vestal cult demonstrates not only the metaphoric extension of the daughter role to female conduct in another, state religious and extra-familial realm [punctuation hers] but also shows the ability of the role to coexist with that which daughters were expected to assume literally (and hence the Vestals assumed symbolically). Additionally, the fact that the Vestals were defined symbolically as *both* unmarried daughters and more mature wives helps to clarify why their membership in the order benefited their blood families in the way that it seems to have done.[17]

Hallett also suggests that certain aspects of the Vestals' reported history may reflect conflicts that arose in the wake of war and economic crisis, whose implications for women's religion I have considered previously. While conceding that scholars are not agreed on the accuracy of ancient testimony that the number of Vestals and the length of their service increased over time, Hallett proposes that in times of such crises in the Republic, elite families might have sought Vestal positions for their daughters as an alternative to providing the now even larger dowries that would have been required to obtain a suitable husband in a market diminished by such circumstances.[18] Increasing the number of Vestal positions would have eased the pressure somewhat; increasing the length of service would more or less permanently remove Vestals from the marriage market. Plutarch's testimony that although Vestals could marry after their thirty years of service, few did so, and even fewer successfully, might even be seen as tacit recognition of the second effect. Hallett also wisely remarks that Vestals who married after so many years of relative autonomy might have found it difficult to accept a husband's authority.[19]

As further evidence, Hallett points to the timing of the several instances of Vestals punished for unchastity in the republic. In 216 B.C.E., after the defeat of Rome at Cannae, two of the six Vestals were executed.[20] Again a century later, amid similar tensions, three Vestals suffered the same fate.[21] Sarah Pomeroy interpreted these incidents as evidence of the (male) Roman predilection for associating the fate of Rome with the chastity of Roman women, and assigning responsibility for military defeat to the unchastity of Rome's representative women, the Vestal Virgins.[22] Hallett suggests that additional factors were at work. These accusations may have been sanctioned, if not instigated by "well-born and powerfully situated families [who] coveted for their own prepubescent daughters the places that would be left vacant by such executions."[23] She proposes that if such mechanisms are plausible for the later period, where the historical evidence is stronger, they could also have played a part earlier and may explain the earlier enlarging of the number of Vestals and length of service.

Beginning with the Hellenistic period, evidence for women's cultic offices abounds, not only for traditional Greek and Roman worship, but for new, imported, and transformed cults, including numerous so-called mysteries, and Roman emperor worship. An inscription from Delphi in the second century B.C.E. commemorates a priestess of Athena who took part in a procession to Apollo and who received extensive honors as a result.[24] Tata of Aphrodisias in western Asia Minor was a priestess of Hera and of the imperial cult, who also held the office of *stephanophorus,* "crownbearer."[25] Women held numerous offices in the worship of Isis.[26] Aba of Histria in Thrace was high priestess of Cybele in the second century C.E. and numerous other priestesses of this goddess are attested from inscriptions.[27] Priestesses served in the cult of the Roman emperor as it was celebrated in the Roman province of Asia (western Asia Minor).[28]

Numerous inscriptions record detailed information about some of the responsibilities of religious office. In her capacity as priestess of the imperial cult, Tata of Aphrodisias underwrote the cost of religious festivals and public entertainments for the entire term of her office, probably one year.[29] She supplied oil free of charge for the athletes who competed in public games, a costly venture.[30] She offered sacrifices throughout the year for the health of the imperial family (presumably paying for the sacrificial victims) and sponsored banquets open to the general public. Berenice of Syros similarly funded the public rituals in her capacity as magistrate and as priestess of Demeter and Kore, among others. Aba of Histria not only saw to the proper performance of the great festival of Cybele around the spring equinox, but sponsored a lavish public banquet, surpassing all previous generosity.[31]

In Hellenistic Greek cities and towns, women (and men) who held cultic offices were expected to finance the performance of public religious festivals and entertainments, costs that must have been enormous. Popescu puts the sum of 2 denaria, which Aba of Histria gave to all men, into useful perspective by noting that the annual salary for an auxiliary in the Roman army was 75 denaria at this time.[32] Five lambs for a banquet cost 18 denaria: a day's work in the mines paid four-tenths of one denarion.[33] Not all benefactors gave direct gifts of cash: some distributed free oil and wine in addition to the banquet meal. Women received these donations, if at all, in much smaller ratios. The ratio of one distribution was 30:20:3 for town senators, Augustales, and women.[34]

Several inscriptions from Sillyon in Pamphylia testify to the benefactions of a woman named Menodora, in the early third century C.E.[35] She held a multiplicity of religious and civic offices: high priestess of at least two emperors (probably Septimius Severus and Caracalla), priestess of Demeter, and of "all the gods," hierophant for life of the city's gods, *dekaprotos, demiourgos,* and *gymnasiarch*. Gordon remarks, "As in the case of Cleanax [of Cyme], priesthoods and civic offices are listed together as though there were virtually no distinction between them."[36] Menodora distributed money and corn to the entire populace, 300,000 denaria to orphans and children, financed the building of a temple, and provided numerous other benefactions.

In return for their largesse, Tata, Berenice, Aba, Menodora, and other women like them received substantial public honors and privileges. The inscription for Chrysis of Delphi in the second century B.C.E. is one of the most explicit in this regard: she received the right to consult the oracle at Delphi ahead of others, freedom from taxes, a front seat at all public contests, the right to own land and a house, "and all the other honors customary" for someone so honored.[37] The Jewish community in Ionia awarded similar honors, probably in the third century C.E., to Tation, daughter of Straton, who underwrote major building expenses, for which she received a golden crown and the privilege of sitting in the seat of honor.[38] Not surprisingly, the Jewish community did not award her relief from taxes and other municipal benefits that were outside the scope of its powers. Such honors and privileges were announced by conspicuous and often lavish public inscriptions, sometimes accompanied by a statue or portrait of the person so honored. Three of the inscriptions to Menodora were found on the bases for statues.

It is obvious that religious office in the cities of Asia Minor and elsewhere was inextricably linked with the entire political and social structure of municipal life. Major cultic offices in Hellenistic Greek cities were awarded to women (and men) who already possessed the financial resources requisite for cultic responsibilities, in return for which they received public acclamation and presumably prestige and influence. Although these people undoubtedly had considerable prestige and influence before they assumed these offices, it also seems likely that they garnered additional power and social position from their performance of these cultic offices; otherwise it seems difficult to understand why they would have undertaken the enormous expense involved. Indeed, it became increasingly difficult to persuade people to take on these expenses. Popescu comments that the praise of Aba and her ancestors for their willingness to take on such responsibilities is intended as an implicit contrast to those wealthy persons who would not do so.[39] Ultimately offices were sometimes apportioned by fractions of a year, right down to one day.[40]

Natal family position was an important, if not a determinative factor, in obtaining many prestigious priesthoods. Tata of Aphrodisias (second century C.E.) is explicitly identified as a member of an illustrious family of the first rank.[41] Her inscription first recognizes her patrilineage—she is the daughter of Diodorus, himself son of Diodorus, and holds the title of mother of the city. Only secondarily do we learn the identity and status of Tata's husband, which appears as less of a factor in the cultic offices she obtains. The decree for Aba of Histria similarly places great emphasis on her illustrious family, although she, too, was married. Menodora also came from a most prominent family, and both her children held public office: her daughter, whose name is not transmitted, was *gymnasiarch*; her son Megacles was *demiourgos*.[42]

On the whole, both men and women depended on family connections for public offices, honors, and prestige. Marital relationships appear of secondary importance, but conceivably were factors for husbands as well. Tata's husband, Attalus, held the office of *stephanophorus* at one point, as did Tata, but there is no indication that she attained that office by virtue of his position.

As Bernadette Brooten has convincingly established with regard to Jewish women,[43] little substantiates old scholarly prejudices that women who held prestigious cultic and civic titles did so only by virtue of their marriage to men who held such offices, and that the titles fail to evidence real cultic and civic functions on the part of women.[44] R. A. Kearsley's study of the imperial cult in the Roman province of Asia demonstrates that although a substantial number of women called high priestess of Asia in the imperial cult were married to men who also held titles in the imperial worship, women held their titles and served in that office in their own right.[45]

An inscription from Magnesia on the Meander dated to the mid-first century honors a woman named Iuliane, whose appointment as high priestess of Asia predated her husband's occupation of the high priesthood.[46] Kearsley rightly concludes that since he held the priesthood after her, her title can neither be honorary (designating no actual function) nor derived from his. Although many *archiereia* (imperial high priestesses) were married to men who themselves bore the title either of *Asiarch* or high priest of Asia, some appear not to have been married, and thus unable to derive their titles or their offices from husbands. Conversely, many wives of male *Asiarchs* and high priests of Asia bear no related titles. Similar observations can be made about Jewish women called elders and the titles of women married to men called *gerousiarchs* (head of the council [of elders]).[47] Finally, Kearsley points out that the relative frequency with which husband and wife appear to hold similar offices, or to hold related offices simultaneously, is easily explained by the frequency of

intermarriage between "those wealthy and distinguished families that dominated office-holding in the cities and provinces,"[48] rather than by any direct
correlation between the two.

Particularly in Greek cities of the Hellenistic and Roman periods, women's
priesthoods and cultic offices cannot be separated from the whole system of
benefactions.[49] Municipal life depended to a large extent on the charitable contributions of wealthy members of the community to underwrite the cost of
public buildings, public festivals, and entertainments (including athletic competitions and dramatic performances), and even a modest form of charity for
individuals. Wealthy women were hardly excused from these obligations by
virtue of their sex, but seem to have served in far smaller numbers than the
men of their families. Writes Ramsay MacMullen:

> [we find] the female sex, as such, entirely excluded from no role or aspiration at
> all, in the public affairs of their community, nor required to demonstrate merits
> much different from men's in claiming respect and participation, but yet
> included only in far, far smaller numbers.[50]

Feminist historians should find it hard to know how to construe such numbers.
As a general rule, the tendency to obscure women's history should lead us to
see any examples of women's public activity as the tip of the proverbial iceberg.
But MacMullen points out that many of the cult titles women held were
eponymous magistracies, which meant they were used to date years: "When
so-and-so was such-and-such" is the routine formula for decrees of many sorts.
If we had a sufficient sample of these magistracies, we might in fact have some
reliable idea of the frequency that women held these highest civic offices. Mac-
Mullen did not attempt such a comprehensive study. Instead, he offers the
data from one sampling of coins: in 13 cities, 17 women and 214 men may be
identified as eponymous magistrates.[51] The larger the city, the fewer the
women relative to men.

Such findings have been interpreted as evidence that women were more
likely to hold public office when there were insufficient numbers of men
around, or when men were no longer willing to accept the increasingly costly
burdens of public service, at which point women were invited to fill the gap.
Riet van Bremen, however, offers some cogent arguments that women benefactors are a substantial force long before the decay of such offices.[52] Instead,
she suggests that women become a more prominent force in public life as the
result of social and ideological components of the benefaction system. Partly,
the social distinction between wealthy women and respectable women of
merely modest means would not be as apparent if the latter stayed secluded at
home. Wealthy women would have been both permitted and encouraged to
participate in public life in order to differentiate the rich from the average. But
van Bremen stresses that such participation was made possible by an ideology
which minimized the public nature of this behavior, by depicting and concep-
tualizing the city as an extended family, thus rendering public benefactions

familial solicitiousness writ large. Certainly the language of benefaction bears this out: it speaks of fathers and mothers of the city, of sons and daughters. Wealthy women in public, though, were still praised in terms that posed no threat to the traditional ideology of gender differences and appropriate roles for women.[53]

We see that in Hellenistic cults, as well as in classical Greece and Rome, gender was not a determinative negative factor in attaining and executing religious offices and the attendant honors, privileges, prestige, and power accrued to those who had fulfilled such civic responsibilities. Rather, such offices and their related benefits were available both to women and men who came from the right ancient families. To some extent, as they often have, wealth and social class qualified the restraints that gender generally imposed on women in the ancient world.

Although women undeniably held the highest cultic offices in Greco-Roman religious hierarchies, not all women's cultic service fell into such prestigious categories. Accounts from the first-century C.E. geographer Strabo provide important evidence for the roles women played in the service of goddesses such as Artemis at Ephesus and Enyo at Comana, both in Asia Minor, whose temples were presided over by male priests.[54] At Comana, women called *hierodoulai* (literally, sacred slaves) performed a range of menial tasks and lived their lives in servitude to the temple, almost certainly not by choice. Women designated *hetairae* served as prostitutes, with their fees almost certainly paid to the temple. To what extent such prostitution can be seen within a cultic context, and what that would mean here deserves additional research: Strabo reports that the temples were frequented by soldiers and merchants on holiday who squandered their money on these women.[55]

Judging from the epigraphical evidence, freedwomen played the major leadership roles both at the temple of Bona Dea in Rome and in religious associations under her patronage.[56] With the exception of one male *sacerdos*, an imperial freedman named Claudius Philadespotus,[57] all the *sacerdotes* of Bona Dea at Rome were freedwomen (and no priests are attested for the goddess outside Rome).[58] Similarly, only women are attested as board members (*magistrae*) of Bona Dea collegia.[59] Most of these are freedpersons, but two are slaves owned by private individuals.[60] Four freedwomen and one slave held the title of *ministra*.

The title of *magistra*, which also occurs in the masculine form *magister*, is well attested as an office both in professional guilds and religious associations. Numbering from two to ten, the *magistri/ae* served as the heads of these organizations. In religious *collegia*, which functioned among other things as burial societies and social clubs, the *magistri/ae* were responsible for cultic offerings and the subsequent meal, as well as for financial matters, internal order, and "executing resolutions." The duties of *ministrae* are less well known, but from the existence of freeborn persons with this title, Brouwer argues that it cannot simply designate slaves.[61]

We have scant evidence for women's religious leadership and authority in pagan contexts apart from cultic offices. The evidence we have considered so far does not lead us to expect to find differently, for it suggests that women's religious office was acceptable in the Greco-Roman world precisely to the extent that it legitimized and conformed to expectations about women's roles generally. Conceivably, women's exercise of religious office was perceived as an extension of their domestic cultic responsibilities—the extrusion of the private into the public, still as women's realm.

To inquire about women's religious leadership and authority apart from such offices, though, suggests that such a distinction was meaningful in any of the pagan traditions under consideration, a question requiring more attention than I can give it here. Pagan charismatics or prophets, whose claim to public consideration stems from contact with the divine apart from cultic offices, are not prominent in the extant sources, though male figures such as Apollonius of Tyana or Alexander of Abonuteichus come easily enough to mind.[62]

I would expect to find such persons, male or female, primarily in two realms: in those socially critical philosophical circles and the small communities that sprang up around such, and in that vast and almost undefinable arena of Greco-Roman magic. Sarah Pomeroy points out that in early Greek philosophy, the Stoics and the Neopythagoreans reinforced traditional roles and restrictions for women, whereas both the Epicureans and the Cynics advocated the emancipation of women, at least theoretically.[63] Epicurus, who lived in the third century B.C.E., admitted women to study in his Garden along with men, but no women from his circle are known for their leadership of the community. A number of Cynic epistles are addressed to women, including several to Hipparchia, the wife of Crates, a major early Cynic philosopher, who herself lived the life of a Cynic, dressing in rags, begging for subsistence on the streets, and rejecting social propriety for philosophical pursuits.[64] Regrettably, these writings are likely to date several centuries after their reputed authors lived and died, rendering them considerably more problematic as evidence of women's leadership in Cynic circles.

From the writings of the Stoic Musonius Rufus in the first century C.E., we might expect to find women among his students, for he wrote several treatises advocating education for women, including the study of philosophy.[65] But no such evidence survives. Only for Hypatia of Alexandria, in the late fourth century C.E., do we have strong evidence for a woman who not only studied and practiced philosophy, but who become head of a philosophical academy, namely the neo-Platonic school associated with Plotinus in Alexandria. Hypatia was brutally murdered and dismembered by a mob of anti-pagan Christians in 415 C.E. at the age of forty-five.[66]

How we might interpret the absence of such evidence is not clear. Arguments from silence are always problematic, but especially so when drawing conclusions about women's lives, as much feminist historiography of the last twenty years has taught us. The very nature of the sources we have from antiquity may play a significant role here, for few such male figures are known from literary sources, and public inscriptions by their nature are not likely to yield testimony to persons of either gender who challenged the status quo and claimed authority outside of the usual public channels.

To date, there is little research that would enable us to consider what kinds of authority and prestige women practitioners of magic commanded in antiquity. The practice of magic was by no means restricted to women, as the numerous spells of the magical papyri from Egypt demonstrate, yet clearly women were significant consumers and almost certainly dispensers of magical lore, however we construe the category of magic in Greco-Roman antiquity. Juvenal's perplexing portrait of the Jewish beggar woman, whom he calls "an interpreter of the laws of Jerusalem, high priestess with a tree as temple,"[67] deserves mention here among the numerous literary references. To what extent these kinds of activities can be considered leadership, and what ramifications it might have for other discussions of religious leadership in antiquity remains to be seen.[68]

What conclusions, if any, can we draw from this description of women's exercise of religious office in the numerous traditions of Greco-Roman paganism? Surely women's service as priestesses to goddesses and even gods cannot by itself be taken as any sort of status indicator for women, given the vast evidence from classical Greece, for example, for pervasive misogynism and narrow delimitation of the roles available to women of any social class. We should not be too quick to equate service to ancient deities with power and authority. The evidence we have considered suggests that, under some circumstances, priesthoods and other religious offices are related to the power, prestige, and authority of those who held office. But whether priestly and other cultic offices themselves convey power, prestige, and authority to those who do not otherwise have them, or whether, as in the case of benefactors in Roman Asia Minor and elsewhere, such offices confirm and enhance the already existing power of the social and political elite, deserves further exploration. In any case, it is clear that no simple equation is sufficient.

Once again, Douglas's model may provide a coherent framework for our observations. The religious offices held by women who experienced the social constraints of low group and high grid seem unlikely to have provided opportunities for women to violate norms of social propriety, or exercise authority in ways that fundamentally challenged existing power structures. If anything, the opposite is likely to have been the case.

Women's service as priestesses and other cult functionaries in classical Greece, particularly in the service of goddesses who themselves functioned as paradigms for proper women, provides one example. There is little evidence that such priesthoods extended the scope of women's public activities beyond the specific cultic context to the larger political, economic, and social spheres, even when we acknowledge that the very delineation of the religious from other spheres of human activity is a modern conception.

Certain cultic regulations reinforce these perceptions. For instance, while abstinence from sexual relations seems to have been routinely required of priests and priestesses for limited periods of time surrounding the performance of their ritual obligations, a higher degree of purity seems to have been expected of women than of men. Numerous priestesses were required not only to abstain, but to be virgins.[69] In some cases they had to swear not only to their ritual purity, but to their marital fidelity as well.[70] One literary source reports that the *Basileia*, the wife of the priest-king, had to be both a virgin at marriage and subsequently faithful.[71]

Those women who, according to Strabo, served as *hierodoulai* and *hetairai* in the large temples of goddesses such as Artemis and Ma/Enyo in Roman Asia Minor afford another similar example. Such women obviously experienced the constraints of insulation and isolation reflected in their cultic service.

The religious offices and related benefactions of elite women offer confirmation of a different sort. By virtue of their extreme wealth and social position, these women seem to have experienced far less of the isolating constraints experienced by *hierodoulai*, *hetairai*, and many other ordinary women. These elite women are more likely to have experienced strong group and strong grid: their experiences more closely approximated those of their fathers, brothers, and husbands than did those of ordinary women.[72] But again, as van Bremen's insightful analysis suggests, their public offices and benefactions, however magnificent, were carefully constructed not to threaten the social order and existing definitions of appropriate roles and behavior for women of their social class.

Similar observations may be made about the Vestal Virgins. By their highly anomalous nature, as Beard has shown, the Vestals actually reinforced traditional expectations for other elite Roman women. Furthermore, despite the undeniable power and importance of the Vestals, it seems that elite Roman religion effectively restricted the interpretation of divine will to carefully limited male channels. Beard offers the intriguing argument that at least in republican Rome, it was the Senate alone which could properly interpret omens reported by observers, and thus determine the wishes of the gods.[73] Once ascertained, though, those wishes could be fulfilled by women and men in service to the gods. Interpretive control, then, remained with the ruling male elite, even while women might be deemed perfectly appropriate to propitiate the gods once the proper form of that propitiation had been determined.

The one place where women's religious leadership may have had more revolutionary implications is in communities characterized by strong group and weak grid: among the Cynics and Epicureans, for example. Such philosophical communities often advocated strict control of bodily desires and sexuality to attain philosophical harmony. Through their attention to sexual control, if not abstinence, they provided precisely the necessary mechanism to separate marriage and childbearing from self and communal worth for women. In such environments, we should not be surprised to find that women could not only function publicly, at least within the group, but could exercise meaningful authority within the community, authority that did have the potential to threaten the standards prevailing outside the relatively small community. As we see, this is more or less what occurs in early Christianity.

# — 8 —

## Jewish Women's Religious Lives in Rabbinic Sources

Strikingly different portraits both of Jewish women and women's Judaism emerge from ancient rabbinic sources on the one hand, and inscriptional, archaeological, and neglected Greek literary sources from the Greco-Roman period on the other. Rabbinic writings have led many scholars to conclude that Jewish women led restricted, secluded lives and were excluded from much of the rich ritual life of Jewish men, especially from the study of Torah. Evidence from the Greco-Roman Diaspora suggests, however, that at least some Jewish women played active religious, social, economic, and even political roles in the public lives of Jewish communities.

Accounting for these divergent portraits is a complex and frustrating task. Perhaps the discrepancy is more apparent than real, attributable to entrenched beliefs that rabbinic writings accurately reflect the social realities of Jewish communities in the Greco-Roman period. As more and more scholars are beginning to concede, rabbinic sources may at best refract the social realities of a handful of Jewish communities, and at worst may reflect only the utopian visions of a relative handful of Jewish men. The portrait of Jewish women that emerges from these writings may then be largely discounted in favor of the more persuasive evidence of epigraphical, archaeological, and nonrabbinic writings for Jewish communities both in the Diaspora and in the land of Israel.

While recognizing that rabbinic sources may be used only with the greatest caution to reconstruct the social realities of any Greco-Roman Jews, this chapter explores the portrait of Jewish women's religious lives that emerges from those sources. Chapter 9 juxtaposes the evidence from nonrabbinic sources, especially the archaeological and inscriptional evidence, to compose a richer portrayal of Jewish women in the Greco-Roman world.

At the outset, a few caveats are necessary. For the sake of convenience and stylistic ease, I sometimes resort to general references to "the rabbis" and "rabbinic sources." These expedients are not meant to conceal the considerable diversity of persons, communities, and perspectives represented within the large body of literature collectively called rabbinic. Rather, I mean to suggest that within that literature and those communities, certain themes regarding women appear sufficiently often to warrant such broad sketches.

The division between rabbinic sources, and evidence from Diaspora communities actually suggests far too neat a dichotomy between Jewish communities in the land of Israel and those dispersed throughout the Roman Empire. Recent research has begun to reveal in considerable detail the diversity of Jewish communities not merely in the Diaspora, but also within the land of Israel, before and after the destruction of Jerusalem in 70 C.E. and again in 135 C.E.[1] Nevertheless, I argue that significant social distinctions permeate the kinds of communities reflected in rabbinic sources, and Jewish communities reconstructed from many nonrabbinic sources, and that these distinctions have important ramifications for Jewish women. I do not consider here the evidence for the considerable number of Jewish communities in Sassanid Babylonia, outside the geographic and cultural mainstream of the Greco-Roman world.

This chapter draws primarily on the standard rabbinic compilations of the Mishnah, and occasionally on the Babylonian Talmud, but these are by no means the only sources included in the rubric of rabbinics. The methodological problems inherent in the use of rabbinic writings as historical evidence are exceptionally complex. There is considerable skepticism about everything from who compiled them, to when and where, to whether the rulings and stories attributed to various rabbis are in any way accurate.[2]

Partly as a response to these difficulties, I have deliberately approached these texts as evidence for the mindsets and worldviews, or cosmologies, of their compilers. I am willing to consider the kinds of social structures that would correlate with such cosmologies, but I remain fully cognizant of the tenuous status of any attempts to reconstruct the realities of rabbinic Jewish communities.

Within these constraints, a few words of introduction about the sources are required. The earliest of these is known as the Mishnah, which collects discussions and conclusions about numerous issues in Jewish law, attributed to rabbis who were believed to have lived before the third century C.E. Tradition attributes the compilation of the Mishnah to Judah ha-Nasi, Rabbi Judah the Prince, around the year 200 C.E., but recent scholarship suggests that redactional activity may have continued for sometime after that date.[3]

A supplement or appendix to the Mishnah, called the Tosefta, closely parallels the Mishnah in structure. Even less is known about its compilation and date, although it is viewed as relatively contemporaneous with the Mishnah.

Eventually, the Mishnah was combined with commentary called Gemara into two Talmuds, one associated with Babylonia, the other associated with the land of Israel, known sometimes as the Jerusalem Talmud or the Palestinian Talmud. In their written form, the two Talmuds date no earlier than the fifth or sixth centuries C.E.[4] The Babylonian Talmud emerged as the standard for the Judaism that became dominant in the West, and references to "the" Talmud are to the Babylonian version.

To the extent that religious life can appropriately be segregated from the rest of life in a rabbinic context, the religious lives of Jewish women according to rabbinic tradition were considerably more limited than those of men. According to Jewish rabbinic tradition, three precepts are incumbent only on women: the lighting of candles at the onset of the Sabbath, the separation and burning of a small piece of the dough prior to baking bread, and the observance of the laws of menstrual purity, called Niddah. But various rabbinic sayings make it clear that the rabbis did not regard these precepts in the same category as those religious obligations incumbent on men. Judith Baskin aptly points out that rather than seeing these requirements as "divine commandments whose observance enhances the religious life of the observer and assures divine favor,"[5] some rabbis saw them as God's punishment of Eve visited perpetually on all generations of women:

> Concerning menstruation: The first man was the blood and life of the world . . . and Eve was the cause of his death; therefore has she been given the menstruation precept. The same is true concerning *Challah* (leaven): Adam was the pure *Challah* of the world . . . . and Eve was the cause of his death; therefore has she been given the *Challah* precept. And concerning the lighting of the [Sabbath] lamp: Adam was the light of the world . . . And Eve was the cause of his death; therefore has she been given the precept about lighting the [Sabbath] lamp.[6]

Failure to observe these precepts causes women to die in childbirth.[7]

Other than these three precepts assigned solely to women, religious responsibilities in rabbinic sources do not seem parceled according to gender analogous to the divisions we have seen for Greco-Roman paganism. There are, for example, no festivals restricted to women, with or without corresponding men's festivals. On the contrary, as Judith Wegner points out, the rabbis of the Mishnah forbade women's *havurot*, or commensal fellowship groups.[8] There are no synagogues restricted to men, or to women,[9] nor, not surprisingly, any temples set aside for one gender or the other. The temple in Jerusalem did have a court known as "the women's court."[10] But as the (description by the first-century Jewish historian Josephus makes clear, the Temple layout consisted of progressively more restricted courtyards, culminating in the holy of holies, into which only the high priest could enter, only in a state of purity, and only on Yom Kippur, the day of atonement. Gender segregation, together with other forms of purity delineation, was nevertheless carried out within a single sacred precinct, whose organization mirrored that of ideal Israelite society.

In fact, broadly speaking, rabbinic Judaism subscribed to the belief that there was one set of religious obligations that God had imposed on God's people, Israel, namely the observance of the law that God gave to Moses at Sinai. But if the law given at Sinai contained everything needed to place a person in the proper relationship with God, the rabbis nevertheless understood that not all of the commandments given at Sinai were equally binding on all members of the community. The entire law was binding only on free adult males; all others (children of either sex, free adult women, and slaves of any age or either gender) were exempt to varying degrees from certain religious observances and obligations. Such exemption had an unavoidable consequence: exempted persons could never serve God fully, and therefore could never stand in the same relationship to God as a free adult male who observed all the commandments.

According to the rabbis, God's commandments could be divided into two categories: the positive precepts—things that one had to do—and the negative commandments—things that one should not do. The negative commandments were held to be binding on women and men equally. The positive precepts, on the other hand, were further divided into two categories: those that had to be performed at a specified time, such as eating one's meals in a special booth (*Sukkah*) during the fall festival of Sukkoth, and those that were not linked to a specific time, such as having a mezuzah on the doorposts of one's house. The rabbis believed that women were in theory exempt from affirmative precepts limited to time, since their domestic responsibilities might make it impossible for them to fulfill these obligations, and presumably it would be unfair to hold women accountable for religious obligations whose fulfillment was beyond their control.[11] Modern defenses of the rabbinic position have argued that such an interpretation is compassionate and realistic, without acknowledging that such a system denies to women the possibility of full rapprochement with the divine.[12]

Yet the rabbis recognized that such a general principle did not apply across the board; women were in fact obligated for certain affirmative precepts limited to time, and exempt from others not limited to time. Rabbinic discussions that attempt to explain these apparent violations of the general principle are exceedingly convoluted and suggest that other explanations underlie rabbinic assignments of religious responsibilities and obligations. Judith Wegner astutely proposes that "the likeliest explanation for [rabbinic exemption of women from some religious duties] lies in the pervasive androcentrism of the sages. Viewing woman primarily as man's enabler, they wish to avoid situations that may impede that function."[13]

So while adult males were responsible for all the positive commandments of the Law, with the exception of the three precepts assigned to women as penance for Eve's error, women were only responsible for certain affirmative commandments. Of course, after the destruction of the Temple in Jerusalem in 70 C.E., men and women alike were unable to perform any of the required sacrifices.[14] After the Bar Kochba rebellion of 135 C.E., Jews were forbidden to enter the city of Jerusalem, so that in reality, the fulfillment of many positive commandments, such as that of festival journeys, became impossible for men as well as women.[15]

Men were expected to pray daily, wearing prayer shawls (*talit*) and phylacteries (*tefillin*), and to wear a fringed garment called *tzitzes* underneath their outer clothing. Women were exempt from daily prayer obligations and from wearing these ritual garments. Women were obligated to recite the grace after meals, but as Wegner points out, they were explicitly excluded by the rabbis from forming part of a quorum to say grace in common. Observing that no scriptural warrants, indeed any reason at all, is adduced by the sages of the Mishnah for their position, Wegner concludes that this exclusion of women "reflects a deliberate decision by the sages to bar women from joining in public rites."[16]

Men and women alike were responsible for observing the kosher laws (*kashrut*), which were technically negative commandments—abstaining from prohibited foods rather than eating any required foods. As evidenced by the chores prescribed in the Mishnah for wives, women were responsible for guaranteeing that the food eaten, the methods by which it was prepared, the pots in which it was cooked, and the jars in which it was stored were all in accord with rabbinic interpretations of kashrut. They were not necessarily, however, responsible for the food's purchase, which would have required them to venture out into the public markets.[17]

Men and women alike were responsible for the observance of Shabbat (the Sabbath), the festivals of Sukkoth, Pesach, and Shavuoth, and the other holidays of the Jewish calendar, including Rosh Hashanah, Yom Kippur, Purim, Chanukah, Rosh Chodesh (the new moon), and so forth. But in rabbinic interpretation the specific obligations of men and women differed here as well. Women were responsible for the dough offering[18] and for lighting Sabbath lamps. Men were obligated for festival pilgrimage, whereas women were not, based on Exodus 23:17. Men and women were required to eat unleavened bread on Pesach and to fast on Yom Kippur. Women were exempted from the requirement to dwell in a *Sukkah* during the festival of Sukkoth.[19]

Women were also exempted from the study of Torah, the obligation to procreate, and the obligation to redeem a firstborn son,[20] in spite of the fact that these are affirmative precepts not limited to time, for which women were generally liable.

Not all exemptions always translated to exclusions, although many did. At one point, the Babylonian Talmud includes the blowing of the shofar as one of the affirmative precepts limited to time, from which women are theoretically exempt.[21] But in another discussion it implies that women have the option of blowing the shofar.[22] Similarly, the Mishnah implies the fitness of women to read the scroll of Esther on Purim.[23] The Tosefta agrees in principle, but asserts that in practice this is not done,[24] and the Babylonian Talmud concurs that women should not read from the Torah "because of the dignity of the congregation," a phrase which, Wegner notes, the Talmud does not explicate.[25]

Had the rabbis been internally consistent, they would have considered women obligated to study Torah, since this is not a time-bound precept. The subject is briefly discussed in the Mishnah in an exchange attributed to Ben Azzai, R. Eliezer, and R. Joshua concerning the ritual prescribed in Numbers for a wife suspected of adultery.[26] Brought before the high priest, the accused

woman was obligated to drink a potentially poisonous draught composed
largely of water and dirt from the Temple floor.[27] If she was innocent, the
drink would have no effect, but if she had committed adultery, her "belly
[would] distend and her thigh . . . sag." The rabbis discuss the possibility that
previous meritorious deeds can nullify the effects of the drink, so that an adul-
terous woman who had otherwise behaved in a meritorious fashion might not,
in fact, suffer any ill effects. Ben Azzai proposes that a father should teach his
daughter Torah in case she has to undergo the rite. If she is guilty, but does
not suffer the predicted consequences, she will know it is because merit has
averted the decree.

R. Eliezer responds, "If anyone teaches his daughter the law (torah), it is
as though he taught her lasciviousness." Wegner attributes to Eliezer fears that
"the more a woman knows, the more liberated she may become—above all, in
her sexual conduct."[28] R. Joshua supports the notion that women are predis-
posed to sexuality in the first place: presumably anything that would encourage
this natural tendency to sexual excess is undesirable.

At least to modern sensibilities, Ben Azzai's opinion is a bizarre explanation for why women should study Torah. Wegner proposes that the discussion
is not about the general study of Torah, but rather limited to the specific law
of the accused wife. In the Mishnah, she points out, Torah "generally has the
limited connotation of a specific law or set of rules . . . rather than denoting
'the Torah' in the general sense that talmudic Hebrew later assigns to the
term."[29] Ben Azzai, then, is not arguing for the education of women in general, but for a father's duty to see that his daughter is acquainted with this specific law. However, Wegner also points out that another Mishnaic ruling does
support teaching Scripture to daughters.[30]

It is also conceivable that rabbis who did favor the education of daughters
and their study of the Law, broadly construed, would have to offer an explanation that fit within the patriarchal rabbinic framework of most of their colleagues. What is particularly odd here is that it is virtually inconceivable that
this discussion had any application in reality, for by the time of the Mishnah,
the Temple had lain in ruins for well over 100 years, and the rite of the suspected wife, like all rites performed within the Temple itself, was no longer
possible, at least in this form.[31]

Whether women in rabbinic communities actually studied Torah remains
uncertain. The discussion in the tractate Megillah about women reading from
the Scroll of Esther seems to presume that some women were sufficiently educated to perform such an act, but since the point of the discussion is the inherent eligibility of women, that may not be the case. The Babylonian Talmud
does mention one learned woman, Beruriah, the wife of R. Meir, who is said
to have learned 300 *halakot* from 300 scholars in a single day.[32] Curiously,
although Meir himself is cited extensively throughout the Mishnah, Beruriah
surfaces only in the Talmudic commentary. In recent years, scholars have been
divided about the proper interpretation of the Beruriah traditions. Some argue

that Beruriah demonstrates at least a few women in rabbinic communities could study Torah and issue authoritative rulings; others, including Wegner, suggest that Beruriah is a fiction designed to show the absurdity of such a woman.[33] Certainly the medieval tradition that Beruriah was ultimately seduced by one of her husband's students and subsequently committed suicide out of shame serves as a warning to other women and as proof of the dangers of women studying Torah.[34] Further, it may be an oversimplification to seek only one view behind the Beruriah stories, which may rather reflect divergent rabbinical opinions about the figure of Beruriah and about learned women.[35]

As I indicated at the beginning of this chapter, how reliably rabbinic sources portray Jewish women's lives is a question that extends far beyond the specifics of the Beruriah story, or other stories about specific women, to the broader question of to what extent rabbinic sources accurately reflect the lives, beliefs, and practices of the majority of Jews in the Greco-Roman world. This has significant ramifications when we look at the evidence for women's Judaism. For example, the Mishnah contains a discussion on whether a woman married to a pious scholar (*haber*) may lend her cooking pots to a woman whose husband is not particularly scrupulous in his observances (an *am ha'aretz*—a common man), since by doing so she may run the risk of ritually polluting her own pots.[36] Good neighborly relations are posited as a possible justification for this problematic act. The Talmud addresses a similar situation by ruling that the wife of a scholar may grind corn with the wife of an *am ha'aretz* only if the former is in a state of menstrual impurity, and therefore will need to undergo ritual purification anyway. Presumably the same holds true for any pots or utensils used in the process.[37]

Beyond the meager details of women's lives that these texts provide lies more interesting information. These discussions suggest that the common people, both women and men, do not adhere scrupulously to the kosher laws or the laws of menstrual purity, and that the most the rabbis can do about it is to forbid social intercourse between themselves, their wives, and such persons, or alternatively to limit such intercourse to those occasions when cultic purity is a moot point for other reasons. The rabbis may permit their menstrually impure wives to engage in activities with other women that would ordinarily carry a risk of ritual pollution only because they are already polluted! But this suggests that in their daily relationships with one another, women were more concerned about aid to one another, in the form of lending cooking pots or companionship while grinding corn, than they were with the risk of ritual contamination.

Let us now take a closer look at the three precepts that the rabbis thought were binding on Jewish women: the separating of the dough (challah), the lighting of Sabbath candles, and the observance of menstrual purity. As Wegner remarks, the separating of the dough offering was in fact a ritual obligation incumbent on men, from Scripture, as was the obligation to abstain from ritual pollution. Similarly, the requirement of lighting Sabbath lamps before sundown (to avoid the temptation to violate the law by lighting a lamp after dark) was also incumbent on men, although it has no Scriptural authority.[38] Wegner points out that all three precepts are in fact time-bound, and therefore according to the sages' own logic, women ought to be exempt from them, yet all devolve in fact on women. The separation of the dough offering falls to women because they normally bake bread,[39] and women must light Sabbath candles to prevent men (who have gone off to synagogue) from any inadvertent transgression of the law.[40] Women must assume the responsibility for male ritual purity because only they can know the pertinent details of their menstrual cycles.

Wegner thus attributes the rabbinic assertion that failure to observe these precepts causes women to die in childbirth to the sages' implicit admission that such regulations are not inherently binding on women at all, yet are crucial to prevent men from transgressing. The primary role of religious observance for Jewish women, in the rabbinic view, is to enable men to fulfill their covenantal obligations.

In their insightful analyses, neither Wegner nor Baskin considers the possibility that the passage in *Genesis Rabbah* represents a minority viewpoint, rather than a majority one. It is conceivable that rather than convey a general rabbinic consensus, this vicious tradition is a piece of polemic, designed to counter or neutralize another viewpoint—one which held that women did have significant religious obligations. This passage may exemplify the phenomenon that intensification of prescriptions against women is often a response to the increased autonomy and authority of women. Unfortunately, we cannot establish the historical context for *Genesis Rabbah*, let alone this specific passage, with any reliability. *Genesis Rabbah* is thought to date from the period when rabbinic Judaism was becoming increasingly dominant in the western Mediterranean (perhaps fifth century C.E.). Interestingly, in this period we have significant evidence for Jewish women who played important leadership roles in communities such as Venosa, Italy. Perhaps the passage in *Genesis Rabbah* reflects some rabbinic opposition to the power and prestige of women in Jewish communities previously outside the influence and authority of rabbinic traditions.[41]

Given all this, we may wonder how much these three regulations were actually observed by women, and how many women would have accepted the rabbinic explanation of their theological weight. It is interesting that the first two requirements have clear parallels, at least to some degree, in the religious observances of women in the various forms of paganism with which Jewish women were likely to have come into contact. The lighting of lamps and candles is associated with numerous ancient religions. Small oil lamps were frequently left in temples as offerings: many are catalogued in connection with Cybele, for example.[42] At a festival to Isis called the *Lychnapsia*, the goddess sought her missing husband with torchlights.[43] Bacchic devotees carried torches on their nightly encounters.

The separation of the dough may have its historical antecedents in Israelite sacrificial practice, but the baking of special Sabbath bread suggests other connections. Women baking cakes to be offered to ancient goddesses is documented in multiple sources.[44] Especially close to home are the practices attributed to Jewish women in service to the Queen of Heaven in the book of Jeremiah.[45] Triangular cakes filled with a dark prune mixture or poppy seeds were originally offered to the Babylonian goddess Ishtar. In the Purim festival, which commemorates the deeds of the Queen Esther (whose name closely resembles that of Ishtar, as the name of her uncle Mordecai closely resembles that of Ishtar's consort, Marduk), these sexually suggestive cakes have become sanitized as the tri-cornered hat of the villain Haman.[46]

It is exceedingly difficult to trace the evolution of cakes offered to the goddess Ishtar, with clear fertility implications, to their transformation into the Jewish festival of Purim. It is equally problematic to detect what happened to the Israelite custom of sacrificing to the Queen of Heaven. But it is surely noteworthy that in Jewish sources, ultimately the Sabbath is represented as a heavenly Queen, who descends once a week into the homes of pious Jews and is greeted with the lighting of candles and the offering of a fragrant Sabbath challah. Then, too, we must wonder whether the notion of the Sabbath as a bride is not itself a remnant of fertility concerns, transformed into a metaphor more aptly suited for male concerns. One rabbinic text suggests that it was considered laudable (a Mitzvah) to have marital intercourse on Shabbat for the purpose, of course, of procreation.[47]

The constellation of symbols and language associated with the Sabbath is reminiscent of women's worship of goddesses designed to ensure fecundity, both in other religious traditions in antiquity and sublimated within Jewish tradition itself. This raises the suggestion that while the rabbis, and perhaps other men, may have seen observance of the Sabbath from a male vantage point and accounted for women's ritual activities as their weekly reminder of the sins of Eve, women may have understood the Sabbath as a more sympathetic expression of concerns relating to sexuality and fertility. It is absurd to assume that all women's religion may be reduced to such concerns, but to accept the rabbinic explanation without further analysis is equally absurd, and misses the intriguing connections between Shabbat and other known forms of religious devotion for women in antique societies.[48]

The observance of menstrual purity regulations is the most problematic of the three precepts imposed on women, partly because it appears to have been unique to Judaism. Various pagan cults sometimes required women and men to abstain from sexual relations for prescribed periods of time in order to purify themselves for some form of worship,[49] but none of these requirements approaches the structure and regularity of the Jewish Niddah regulations. Interestingly, some Christians shared the perception that sexual impurity (encompassing both menstrual blood and seminal discharge) could interfere with one's ability to receive the Eucharist, as we see in Dionysius of Alexandria's third-century C.E. *Epistle to the Bishop Basileides.*[50]

Rabbinic menstrual purity regulations, which have their basis in biblical prescriptions,[51] but which the rabbis greatly expanded, have been a sore subject for feminist scholarship. Their obvious concern to confine and regulate women and their spheres of activity and their horror and denigration of women as almost continuous sources of ritual pollution for men is sufficiently offensive to most modern feminists (with the exception of Jewish feminists concerned to rescue Jewish tradition on this point) to make disciplined analysis difficult.[52]

We are faced with several questions. Did women actually adhere to the strict ritual requirements set forth by the rabbis? How might they have understood such adherence? Did they share the rabbinic perception of themselves as dangerous sources of pollution to their husbands, or might they may have understood such activities from a different perspective?

No sources from antiquity shed any direct light on these questions. We have very little evidence from real Jewish women in the Greco-Roman period, and the few documents we have say nothing about women's observance of menstrual purity laws. Marriage and divorce deeds are silent on the subject, as is the little personal correspondence that has survived. No epitaphs praise women for their scrupulous observance of Niddah, and, perhaps more interestingly, no tales of virtuous women are explicit on this point. All this may simply be evidence of a sense of delicacy about such matters. But it is interesting to note that the one ancient Jewish writing which might be expected to address the subject, the *Conversion and Marriage of Aseneth*, is absolutely silent.[53] This anonymous Greek fiction, which was probably composed prior to the mid-second century C.E., purports to explain how the biblical Joseph married the daughter of an Egyptian priest. Aseneth's premarital conversion is offered as the explanation. When Aseneth does accept the God of Joseph, the angel who serves as her spiritual guide and instructor tells her nothing about the necessity of such observances.

A burial inscription from Rome, commemorating a Jewish synagogue archon, his wife, and their three children, might be construed as evidence that Jewish women in Rome did not observe at least certain aspects of Niddah, for two of the three children are said to have been exactly nine months apart in age.[54] According to Leviticus 12:1–5, a postpartum woman was ritually unclean for 40 days after the birth of a son and 80 days after the birth of a daughter, making it highly unlikely that she could conceive a second child on the heels of the first without having sexual relations while ritually impure. Shaye Cohen, however, has suggested to me that such postpartum ritual impurity was not considered the equivalent of menstrual impurity and did not affect marital relations.[55] Thus we can draw little if any information from this inscription concerning women's practice of menstrual purity regulations, whether or not there is any evidence that Jews in Rome knew of rabbis or rabbinic interpretations of Jewish law.

There is also a brief reference to menstrual purity regulations and postpartum sacrifices in John Chrysostom's invective against the Jews in fourth-century Antioch. In one of a series of sermons delivered in the fall on the occasion of the Jewish holidays of Yom Kippur, Rosh Hashanah and Sukkoth, Chrysostom attempts to prove to his congregants that the Jews continue to observe these festivals in defiance of God.[56] All Jewish ritual was intimately tied to the Temple in Jerusalem. Since God has now destroyed that Temple, clearly such rituals are no longer God's desire, yet the Jews stubbornly persist in their observance. Part of Chrysostom's problem, of course, was that Christians, too, were attending these festivals and finding the rituals of the Jews (and their arguments) appealing. Among the examples Chrysostom cites are the sacrifices required of women after their menstrual periods and after childbirth. Since we know that Chrysostom's polemic against Jewish festivals was firmly rooted in the social realities of fourth-century Antioch, it is conceivable that Chrysostom knew that (some) Jewish women still adhered to the menstrual purity regulations of Leviticus in some form. Regrettably, we cannot be certain.

If Jewish women did observe menstrual purity regulations to whatever degree, we must also consider the possibility that their perceptions of their actions differed from those of the rabbis. Judith Wegner points to allusions in rabbinic literature that Jewish women (and their husbands) engaged in various forms of sexual behavior designed to avoid conception, such as anal and oral intercourse.[57] The rabbis blamed women who initiated such behavior, apparently because they recognized it as a contraceptive ruse—for Wegner also points out that the rabbis were not inherently offended at such sexuality, which they considered permissible under the law if desired by the husband. If Wegner is correct that women burdened by multiple children and multiple pregnancies were especially likely to employ strategies to meet their marital obligations without risking conception and further children,[58] we might also pursue what other means of contraception were available to such women. In particular, I would suggest that menstrual purity regulations allowed women an opportunity to decline sexual relations altogether, or at least to decline vaginal intercourse, on the grounds that they were ritually impure, even, perhaps, when they were not.

How this meshed with prevailing theories of conception is problematic. According to a fine study by Aline Rousselle, "[t]here was no doubt in ancient times that the most fertile phase in the female cycle was that which immediately followed the menstrual period."[59] Since rabbinic law forbade sexual intercourse precisely at this time (a woman had to be "clean" for seven days), this would have presented some interesting problems. Perhaps the rabbis and their followers had different and even more accurate opinions on these matters. My point here is not to pursue the theories of conception that were likely to have been held by the rabbis or other Jews at this time. Nor am I suggesting that Jewish women of the period possessed an accurate knowledge of their menstrual cycles and of reproductive biology sufficient to allow them to abstain from intercourse only at the crucial times, for there is ample evidence that this is unlikely to have been the case. Instead, I am suggesting that menstrual

purity regulations could have been utilized by women seeking relief from pregnancy and childbearing, within the confines of a system both women and men
accepted.

The last example of rabbinic circumscription of women's lives we need to
consider here concerns women's religious leadership. Whether we construe
that leadership as the performance of cultic obligations on behalf of others
(which the rabbis considered permissible for certain obligations), or as the
study of Jewish scripture and oral tradition in order to arrive at interpretations
binding on the behavior of rabbinic Jews, the Mishnah accords women virtually no capacity and no authority for leadership.

As Wegner elucidates, the rabbis of the Mishnah arrived at their conclusion that women were unable to perform cultic obligations on behalf of others
(an act that adult males could perform under certain circumstances) by some
rather curious logic. They reasoned that since women are not *required* to perform these obligations, for example certain daily prayers, they cannot discharge
the obligations of those who *are* so required, namely adult Israelite males.[60]
Wegner points out, though, that the exemption of women from certain cultic
obligations is nowhere explicit in Scripture, but is an invention of the rabbis
themselves, who derive it in a most bizarre and inconsistent fashion.[61]

Rabbinic exclusion of women from the authoritative interpretation of
scripture is similarly derived from the position that women are exempt from
the obligation to study Torah, exemptions that Wegner aptly notes almost
always translate into exclusions.[62] The Mishnah offers no examples of women
who study Torah and engage in discussions of its correct application to Jewish
life. Only within the discussion about the biblical ordeal prescribed for women
suspected by their husbands of adultery, do we find an argument for teaching
women Torah, which we have already considered.[63] Aside from the problematic stories of Beruriah in the Talmud, and the oblique intimations that women
might, under some circumstances, blow the shofar and read the book of Esther
in public, rabbinic sources would lead us to expect no women leaders in Jewish
life during the Greco-Roman period. As we see in the next chapter, such
expectations are not borne out by evidence from nonrabbinic sources.

Jewish women's lives and experiences were undoubtedly far more diverse
and complex than rabbinic sources allow us to see. In the next chapter, I
explore the insights that Douglas's work affords us into some of the differences. Here, though, I want to point out just a few of the implications of rabbinic prescriptions for women's appropriate religious behavior. Judith Baskin,
Jacob Neusner, and Judith Wegner in particular have demonstrated that the
rabbis' interest in women was limited to those occasions when women threatened to violate the categories of society and the cosmos so carefully constructed by the rabbis.[64] In the rabbinic view, to the extent that we can legitimately speak of a univocal position, the appropriate place and behavior of
women was worth little articulation, for it was widely assumed. So long as
women fulfilled their expected roles as daughters, wives, and mothers, the rab-

bis said little about them. Only when women deviated from these categories, when they became emancipated minors, divorcees, or widows, or threatened male relationships with the divine through the transference of ritual impurity, did the rabbis devote any extensive discussion to them.

Both those few obligations binding only on women and the many obligations binding on men from which women were exempt must be examined in light of this perspective. The requirements of Sabbath rituals and Niddah reinforce rabbinic views of women's proper roles within society and the cosmos: they limit women to roles defined by relationships to men, by separation from men, and by separation from the larger society as well. Women's subordination to men is justified by the volitional acts of Eve, and reinforced weekly in Sabbath rituals, and monthly, if not daily, in the observance of Niddah. To the extent that women were expected to ensure family adherence to the kosher laws as well, women were also made active participants in the separation of Jews from non-Jews, and, as we have seen, in the separation of "pious" Jews from the *am ha'aretz*. Such distinctions would have narrowed the circle of permissible relationships for women and possibly contributed to their isolation and segregation, especially for those women from "pious" families who lived in close proximity to undesirable neighbors.

Those obligations from which women are exempt may be seen to function similarly, circumscribing separate and unequal spheres for women and men. For within the rabbinic system, it is precisely those activities from which women are exempted, and therefore in most cases excluded, that are the most valued—certainly by the rabbis themselves, and perhaps even by women as well! Study of the law is by far the most prominent of these, but clearly others, such as daily prayer, should not be overlooked. Certainly, in such a cultural universe, women who exercised authority and leadership over men and women would have been anomalous in the extreme.

# — 9 —

# Jewish Women's Religious Lives and Offices in the Greco-Roman Diaspora

We may glimpse a fuller though scarcely complete picture of women's religion within Jewish communities by turning to nonrabbinic sources, which often provide evidence significantly at odds with that derived from rabbinic sources.

Consider the story of the unnamed widow who each day traveled a considerable distance to pray in the study house of R. Yochanan, rather than in her neighborhood synagogue.[1] In the Babylonian Talmud, this story is offered as the exception that proves the general rule: women did not participate in public Jewish religious life. But in fact a multitude of sources, from the rabbis themselves (perhaps inadvertently) to the New Testament to numerous inscriptions, testify to women's attendance at synagogue services and their participation in synagogue life across a broad chronological and geographic spectrum of Jewish communities in Greco-Roman antiquity.[2] Bernadette Brooten has convincingly demonstrated that some women even held positions of synagogue office, evidence that we consider in detail later.[3]

The prominence of women as financial supporters of ancient synagogues is widely documented by inscriptions, especially from the cities and towns of Roman Asia Minor. At Apamea in Syria, the bulk of the donative inscriptions that have been found for the synagogue there are made by women, although with reference to a household.[4] At Hamman Lif in North Africa, a beautiful stone mosaic now in the Brooklyn Museum commemorates the donation of Juliana, who financed the entire mosaic floor of the synagogue.[5]

From Kyme in western Asia Minor comes an inscription announcing the gift of a woman named Tation, who paid for the assembly hall and the enclosure of the open courtyard with her own funds. In gratitude, the synagogue honored her with a golden crown and with the privilege of sitting in the seat of honor.[6] Tation's inscription appears to contradict the assumption prevalent among many scholars that women and men were separated in the ancient synagogue as they are in modern Orthodox Jewish practice.[7]

If the seating in ancient synagogues was gender segregated, which is by no means certain, gender-segregated seating in public was hardly unique to Judaism. Rawson points out that from Augustus's time on, different groups sat in different sections of the Roman theater. Married men, married women, and schoolboys all sat separately; at the circus, families sat together to watch the horse races.[8] The Christian *Acts of Thecla* depicts women sitting separately from men in the arena at Psidian Antioch where Thecla is brought to fight the beasts.[9] Tertullian alludes to gender-segregated seating in church,[10] and the *Constitutions of the Holy Apostles* command an elaborate segregated seating order in church based on gender and other status criteria.[11] An inscription from the theater at Miletus in Asia Minor designates a specific seating area for Jews.[12] Yet no extant evidence confirms that Jewish women sat apart from Jewish men in synagogues. Brooten demonstrates that none of the ancient synagogues excavated to date sustain this interpretation or support the use of upstairs galleries or adjacent rooms for women.[13] Perhaps more to the point, the large synagogue excavated at Sardis, also in western Asia Minor, has a large semicircle of marble seats facing back toward the assembly hall, which archaeologists theorize were reserved for important members of the synagogue: elders, donors, and so forth. In precisely such a seat, conceivably, Tation would have sat, together with men and other women honored by the community.

Ironically, except for the prescriptions of the rabbis themselves, we have no evidence for women separating dough, baking Sabbath bread, lighting Sabbath candles, or observing kashrut or Niddah, the laws of menstrual purity.[14] A burial inscription from the Roman catacombs, dating perhaps from the second century C.E., praises the piety of a woman named Regina in terms whose precision, if any, evades a modern reader.[15] Her husband commends her chastity, her love for her people, her observance of the Law, and her devotion to her marriage, but gives no clue about the specific ways in which Regina might have demonstrated her love for her people or observed the Jewish law.

The Greek Jewish book of Judith potentially offers fascinating insights into women's practices. The tale is ostensibly set during the reign of Nebuchadnezzar of Babylon, who destroyed the Temple and the city of Jerusalem in the early sixth century B.C.E. But it is manifestly fictitious and replete with historical anachronisms, such as calling the Babylonians Assyrians. The pious widow, Judith, manages to save her people from the besieging army by feigning seduction of the commanding general, Holofernes, and decapitating him at the appropriate moment.

Unfortunately, virtually nothing is known about the date and origin of this work, which was included in the Greek Jewish scriptures but is absent from the Hebrew canon.[16] It seems to have been composed in Hebrew, perhaps no later than the early second century B.C.E. In this story, Judith fasts every day except the Sabbath, the day before and the day of the new moon (Rosh Hodesh), and the day before and the day of festivals, presumably Sukkoth, Shavuoth (the spring festival of weeks), and Passover. She convinces the besieging Assyrian army that it is her custom to go outside of the camp nightly to pray, which provides her a convenient alibi and means of escape when she ultimately murders Holofernes in his bed. After her courageous deed saves her people, Judith leads a celebratory procession of women dancing to Jerusalem, with the men following behind, bearing garlands and singing hymns to God.

Whether, when, and where Judith represents the actual piety of Jewish women is probably beyond our historical grasp. Solomon Zeitlin thought that the figure of Judith was intended as an antidote to the example set by Esther, whose tactics for saving her people included sex without benefit of marriage with the Persian king and seem to have excluded any acknowledgment of divine intervention.[17] It is probable that the author's description derives at least partly from depictions of women's behavior found in (other) Jewish scriptures including Jael and Deborah in Judges 4–5 and Miriam in Exodus 15:20–21.[18] But it is also likely that this description was credible to the author's audience, and significant that Judith was later held up as a model of piety and virtue for Christians. Finally, Judith's public cultic activity contradicts Mishnaic restrictions on women's public participation in Jewish religious life.

From the vituperative sermons against Judaizing Christians by John Chrysostom, we surmise that certain Jewish festivals held particular appeal for women, at least in Antioch in the fourth century C.E. Angrily, Chrysostom (whose epithet means the golden-mouthed) implored his (male) congregation to stop their wives from celebrating the upcoming Jewish festivals of Rosh Hashanah, Yom Kippur, and Sukkoth, but especially the holiday he calls the Feast of Trumpets, which is probably Rosh Hashanah.[19] Chrysostom is here denouncing the behavior of Christian women, but it seems unlikely that Christian women would have been drawn to a festival in which Jewish women were not also actively involved. Beneath Chrysostom's polemical characterization of those who flock to the festival as sexually disreputable women, effeminate men, and theatrical types, we may detect evidence that pagan women were also attracted to Jewish observances in fourth-century Antioch.

Exactly what women and men did at such festivals is hard to extract from Chrysostom's invective. He asserts that Jews danced barefoot in the marketplace during the fall festival cycle,[20] which he deemed dancing with demons. He also attacks the audacity of Christians who dare to return to church to partake of the body and blood of Christ after partaking with those who shed the blood of Christ, suggesting some sort of common meal in the Jewish observance: the verb he uses to describe this activity is the same in both cases, *koinōnai*. Numerous times in this cycle of sermons he denounces the Jewish fasting associated with the fall festivals. Chrysostom does not report whether women blew the shofar on these occasions.[21]

The piety of Jewish women doubtless included the use of what we would call magic, in rural Egypt (where the majority of "magical" papyri have been found) and elsewhere, although Jewish magic was hardly restricted to women. Chrysostom, for example, viewed it as a major drawing card in what he perceived as the battle to lure Christians away from the church, and one which was especially effective in times of illness.

One papyrus contains the adjuration of a woman seeking to attract a husband, invoking Aoth Abaoth, the god of Abraam, Iao of Jacob, Iao Aoth Abaoth, and the god of Israel to accomplish its ends.[22] A Greek magical amulet from Syria, dating perhaps to the fourth century C.E., seeks the protection of Iao, Michael, Gabriel, and others for the newborn daughter of an unnamed woman.[23] In the early second century C.E., Juvenal's infamous sixth Satire attacking the religious corruption of Roman women depicts an old Jewish beggar woman who interprets dreams for small sums of money: he calls her "an interpreter of the laws of Jerusalem, high priestess with a tree as temple, a trusty go-between of high heaven."[24] Mary Rose D'Angelo sees this passage as evidence that the woman "thinks of herself as a prophet and an interpreter of the law (scribe?), the very role most contemporary scholars deny to women in antique Judaism," and that Juvenal assumed that Jewish women thought "of themselves as students of the law." This interpretation is intriguing and worth further consideration, although I am not sure we can move easily from Juvenal's rhetoric to the self-understanding of a mendicant soothsayer.[25]

Rabbinic sources, too, allude to women's practice of magic.[26] The Talmudic saying, "a widow who runs hither and yon . . . destroy[s] the world," is explicated by the story of a widow named Yohani, who is alleged to have employed witchcraft to make childbirth difficult for women and then to have prayed for them (presumably for a fee?). Such an accusation reminds us of the legend of Lilith, the first wife of Adam, who refused to subordinate herself to him, and who, having run away from Eden, now threatens newborn babies.[27] It reminds us, too, of the frequency with which midwives were accused of witchcraft when the outcome of a labor was unfavorable. Yet the rabbis themselves utilized magical means to cure illness when it suited them. The rabbinic texts offer us a polemical judgment of what was likely to have been widespread popular religion, and worth far more exploration as evidence of women's piety.

The Greek *Testament of Job*, whose authorship and date also elude us, contains a fascinating section on the daughters of Job that has no parallel in the biblical account of Job. On his deathbed, Job has distributed his earthly goods to his sons, whereupon his three daughters complain. He assures them that their inheritance is a finer one, and gives them each mysterious, indescribable shining bands. When they complain that such bands cannot sustain their lives, Job counters that not only will the bands sustain them, they will lead the women to life in the heavens. Wrapping themselves in the bands, the daughters chant angelic hymns, which are said to have been written down by the sisters and by Job's brother. If such books did in fact exist, they are lost to us now. It seems possible, however, that the account of Job's daughters may reflect the prophetic, ecstatic activities of some Jewish women in the Greco-Roman period.[28]

A compelling, problematic picture of women's Judaism may be extracted from the enigmatic *Conversion and Marriage of Aseneth.*[29] This tale, as I noted earlier, purports to explain how Joseph, the son of Jacob and Rebecca, married a woman whose father was an Egyptian priest, as we read several times in Genesis without further explanation. Most scholars are agreed, however, that it emanates from the Greco-Roman period, probably before the early second century C.E.

The story of Aseneth and Joseph is set in the time of the seven years of plenty, when Joseph traveled Egypt collecting corn against the forthcoming famine. In his travels, he comes to Heliopolis, the city of the sun, where Aseneth lives with her father, Pentephres (an Egyptian priest), her mother, and their large household of servants. Aseneth is a virtuous virgin who has had no contact with any males outside her family and has spent all of her eighteen years in the family compound, notably in a high tower sumptuously decorated and appointed. Her only fault seems to be her worship of Egyptian idols.

When Pentephres learns from the advance party that Joseph is coming to his household seeking rest and refreshment, he calls his dutiful daughter and proposes that she marry Joseph, whom he calls "the Powerful One of God, the ruler of all the land of Egypt."[30] But Aseneth refuses, recounting the local gossip that Joseph is an abandoned son who, sold into slavery, had sex with his master's wife—and who is therefore clearly an unsuitable husband for the virtuous and virgin Aseneth.

But when Aseneth actually sees Joseph, she is thunderstruck with his glorious appearance and with the power of God that emanates from him. Perceiving the error of her judgment, Aseneth renounces her worship of Egyptian gods and flees to her high chamber, where she spends the next seven days in weeping, ashes and sackcloth, and general repentance. At the conclusion of her self-mortification, a figure resembling Joseph, but who is clearly an angelic being, appears miraculously in her chamber. Informing her that her repentance has been accepted, he instructs her to rise up from the floor, to wash herself, and to put on clean, fresh clothing. When she does so, he shows her a mystery involving bees, a honeycomb, and so forth. When the angel finally departs, Aseneth returns to her family and marries Joseph, her preordained spouse, and they live, more or less, happily ever after.

*The Conversion and Marriage of Aseneth* suffered substantial scholarly neglect for many years, I suspect partly because its protagonist is female. In recent years, it has become the focus of greater attention, although curiously few scholars have taken its feminist implications seriously.[31]

From the first, *The Conversion and Marriage of Aseneth* presents some daunting technical problems. The existing manuscripts differ from each other substantially at some points, and scholars by no means agree on how the earliest version is likely to have read. Out of those discussions, a peculiar difference has emerged between the two major reconstructions proposed in recent years.

The text as reconstructed by Christoph Burchard may be read as considerably more androcentric and sexualized than that proposed by Marc Philonenko, even allowing for the anachronism of applying the term *sexist* to ancient texts. For example, when Aseneth and Joseph first meet, Aseneth's father instructs her to come forth and kiss Joseph as though he were a member of her family, an act she finds repugnant. Philonenko's version reads simply: "And as she came forth to kiss Joseph, Joseph stretched out his right hand and placed (it) on her chest and said. . . ."[32]

Burchard's text, on the other hand, contains titillating details, rendering explicit the sexual component which remains subtle in Philonenko: "And as Aseneth came forth to kiss Joseph, Joseph stretched out his right hand and placed it on her chest between her two breasts and her breasts were already standing up like ripe apples."[33]

In Philonenko's reconstruction, Aseneth was chosen by God before her birth, which would place her in a small elite circle that previously included only men. Burchard, on the other hand, prefers a reading in which it is God's people that have been chosen before all things came into being.[34]

Father imagery is much stronger in Burchard's reconstruction than in Philonenko's. In Philonenko's text, Aseneth's confessed sins consist of blasphemy against Joseph, the son of God, and idolatry, the repudiation of which has cost her the love of her parents. In Burchard's version, Aseneth confesses not only her prior idolatry, but also her hatred of men and her refusal to marry, in prayers that have no real counterpart in Philonenko's text:

> I have sinned, Lord, I have sinned; before you I have sinned much.
> And I had come to hate all who had asked my hand in marriage, and despised them
>     and scorned them.
> I have sinned, Lord, I have sinned; before you I have sinned much.
> And I spoke bold words in vanity and said,
> "There is no prince on earth who may loosen the girdle of my virginity. . . ."
> I have sinned, Lord, I have sinned; before you I have sinned much,
> until Joseph the Powerful One of God came.
> He pulled me down from my dominating position
> and made me humble after my arrogance,
> and by his beauty he caught me,
> and by his wisdom he grasped me like a fish on a hook,
> and by his spirit, as by bait of life, he ensnared me,
> . . and I became his bride forever and ever.[35]

The differences clearly have contextual ramifications. The Burchard version more consistently denigrates Aseneth and brings her into conformity with a patriarchal view of women. Burchard's text emphasizes Aseneth's sexuality, and attributes to her not only a heightened use of father imagery, but intense guilt and remorse over her initial hatred of men and rejection of marriage. Philonenko's Aseneth is a far more autonomous woman, who represents God's chosen and sees her past sins primarily in terms of idolatry and blasphemy against God's physical representative on earth, Joseph. I find it easier to envision redactors who would sexualize the text and bring Aseneth into conformity with classically patriarchal visions of women than the reverse.

It may be that the quest for the earliest text here is misguided. Both versions may be relatively early and reflect differing perspectives about Aseneth that are of central importance for our efforts to reconstruct women's religions and women's view of the cosmos. I do not think it pushes the evidence too far to suggest that Burchard's reconstruction, whether it more accurately represents earlier versions of the text or later revisions, represents the concerns of a male audience and a male author; the text proposed by Philonenko is a much better candidate for female authorship or editing.

In her evaluation of work known to have been written by women in antiquity, Mary Lefkowitz suggests that women's writing is characterized by a reflection of the minutiae of women's lives, and the story of *Aseneth*, though perhaps not *The Testament of Job*, is replete with such details.[36] The author provides elaborate descriptions of interior spaces, of Aseneth's many rooms, her clothing, and her food. Lefkowitz observes that known writing by women exhibits strong affection for other women and for childhood companions.[37] In *Aseneth*, the author pays significant attention to the seven women who were born with Aseneth on the same night, who have been her constant companions since then.[38] When Aseneth retreats in grief and misery to her chamber, the seven companions attempt to comfort her, but she turns them away kindly, feigning a headache.[39] After her mystical initiation with the angelic figure, Aseneth seeks and obtains blessing for the seven.[40] Particularly in Philonenko's version, the text emphasizes her love for both her parents. Interiority is a key aspect of this text, set largely within the house of Pentephres and mostly within Aseneth's own chambers, except for the view from her balcony. All of this accords well with Lefkowitz's observations about the perspectives and concerns of known women writers, and suggests that *The Conversion and Marriage of Aseneth*, particularly in Philonenko's reconstruction, is a likely candidate for female authorship or redaction.

Two aspects of Aseneth's tale are particularly significant for our purposes. First, Aseneth's conversion differs in many ways from what rabbinic sources would lead us to expect. Most interesting is that Aseneth receives no instruction from the angel about Jewish ritual observances, but only a bizarre mystery which, to a modern reader, if not to an ancient one, seems incredibly arcane. Second is the angel's description of the figure of Metanoia, the daughter of the Most High:

> For Metanoia (Repentance) is a daughter of the Most High, and she appeals to the Most High on your behalf every hour, and on behalf of all those who repent, because he is the father of Metanoia and she is the mother of virgins, and at every hour she appeals to him for those who repent, for she has prepared a heavenly bridal chamber for those who love her, and she will serve them for eternal time. And Metanoia is a very beautiful virgin, pure and holy and gentle, and God the Most High loves her, and all the angels stand in awe of her.[41]

This is the most explicit description of a feminine aspect of the divine to be found in a Jewish text from late antiquity, other than some accounts of Wisdom/Sophia. What is even more fascinating here is the unavoidable connection between the heavenly daughter of the Most High, Metanoia, and her earthly counterpart Aseneth, who, as the angel tells her, will be the refuge of all who repent, like Metanoia herself. Whatever many difficulties surround the story of Aseneth, to my mind it remains one of the most intriguing portraits of a Jewish woman's religious beliefs in the Greco-Roman world.

While much of our knowledge of the religious life of Jewish women in Diaspora communities must be extracted from elusive references in anonymous and pseudonymous literature, and from inscriptional information whose context is often equally elusive, a brief treatise by the first-century Jewish philosopher, Philo of Alexandria, offers us a more substantial glimpse into the religious lives of a handful of Jewish women.

*On the Contemplative Life* describes in detail a Jewish monastic community living on the shores of Lake Mareotis outside Alexandria.[42] The Therapeutic society consisted of both men (Therapeutae) and women (Therapeutrides) who were equally devoted to the contemplative philosophical life.[43] Contemplative monastics could be found outside many cities, but the greatest community was outside Alexandria. Members lived in simple houses. Daily they prayed at dawn, studied the scriptures through allegorical interpretation until evening, and prayed again at night. They composed and wrote down hymns and psalms. After their evening prayers, some ate a modest meal and otherwise tended to the minimal needs of their bodies. Others put off eating for intervals of up to six days.[44] The Therapeutics were vegetarians who often ate little except bread and water, even on special occasions.

On a daily basis, the Therapeutics lived in solitude. On the Sabbath, however, they came together for a communal assembly and a modest meal.[45] Men and women attended together, but they sat in discrete portions of the sanctuary, separated by a wall that extended only partway to the ceiling, allowing women to hear everything, but preventing visual contact. It is this description, by the way, which many scholars have used as evidence for the physical segregation of men and women not just in the monastic Therapeutics (whose concern for gender segregation might well be rooted in their ascetic lifestyle) but among Jewish synagogues generally. Philo emphasizes that the women "customarily form part of the audience with the same zeal and the same sense of calling."[46]

The crucial passage occurs in the course of Philo's description of the society's major festival, which seems to have been Shavuoth.

> The feast is shared by women also, most of them aged virgins, who have kept their chastity not under compulsion, like some of the Greek priestesses, but of their own free will in their ardent yearning for wisdom. Eager to have her [wisdom] for their life mate, they have spurned the pleasures of the body and desire no mortal offspring, but only those immortal children which only the soul that

is dear to God can bring to the birth unaided because the Father has sown in
her spiritual rays enabling her to behold the verities of wisdom. . . . [47]

The festival included a modest banquet meal, preceded by a learned yet
unpretentious discussion by the society president of questions arising from the
Scriptures, according to the accepted allegorical methods. The highlight of the
celebration was an extraordinary evening of singing. Individuals sang favorite
hymns, which might have been of their own composing, while the whole community, men and women, joined in the closing lines and refrains.[48] At the conclusion of the festival, the men first formed one choir, the women another, and
they sang in turn and together. Ultimately, though, the two choirs intermin-
gled to reenact the deliverance from Egypt. Filled with ecstasy, they sang until
dawn, when they returned to their individual houses to resume the contemplative life.[49]

From Philo's tantalizing description we may draw a reasonable portrait of
the female Therapeutics. Like their male counterparts, the women who came
to Mareotis were already favorably disposed to the allegorical interpretation of
Jewish scriptures (which they, like Philo, read in Greek). Presumably, they had
also received a classical Greek education similar to Philo's.[50] Few women, Jewish or otherwise, would have been so highly educated.[51]

Like their male counterparts, the Therapeutrides may also have had significant financial resources over which they had some control prior to entering the
community.[52] In fact, there seems to be only one crucial distinction between
the male and female Therapeutics: the men left wives to join the community,
but the women left no husbands, either because they never had them, or
because they waited until they no longer had husbands to take up such a life.[53]

Further support for this interpretation comes from Philo's brief yet enormously rich description of the Therapeutrides. Most of them are virgins,
advanced in age, whose chastity is by choice. Instead of human mates, they
have sought Wisdom, personified in the female form, as a spouse; they have no
children, preferring immortal, spiritual offspring to the ordinary kind.[54]

This description is, I believe, rooted in social reality. For Philo, however,
the details of that reality were just that much more grist for his allegorical mill.
For Philo, female figures, especially those in Jewish scripture, and the feminine
usually symbolized the lower, sensate part of the soul, and so served as symbolic representation of the sensible world, all of which carried negative connotations. Male figures and the masculine usually symbolized *nous* and *logos*
(mind and reason), the higher state of the soul, and the imperceptible world.[55]
For Philo, it was crucial for the soul, which was feminine, to become virgin in
order to attain the capability of union with the divine. Becoming male was similarly a transitional spiritual accomplishment.[56]

The implication of this is straightforward but startling. For Philo, the
Therapeutrides were female in form only. In other respects, like good philosophers who aspire to mystical union with the divinity, they had purged their

souls of their female elements and become male and/or virgin. All this was evident in the fact that they were childless, unmarried, and probably post-menopausal—the state in which Sarah, here the paradigm of the soul, finally becomes capable of conversing with the divine (and ironically, conceiving Isaac!).[57] Unattached to husbands and children, beyond the restrictions of menstruation and potential fertility, they became suitable candidates for mystical union with the divine: they have left behind the feminine and attained the masculinity/virginity necessary for the soul to unite with the divine in the mystical bridal chamber.

This interpretation allows us to understand why Philo could write so favorably about the Therapeutrides when, in general, his writings display disparaging attitudes toward women characteristic of the androcentric culture in which he lived.[58] Interestingly, Philo's depictions of Jewish women in Alexandrian life and his personal attitudes toward women do not conflict with his portrayal of the Therapeutrides. For example, to the extent that Philo thought women ought to be removed from public spheres of activity and confined to private areas,[59] the Therapeutrides conformed to his values. After all, they spent most of their time in seclusion not only from their male counterparts, but from each other as well. When they did come together for communal worship, the women and the men maintained some physical separation, except for the final chorus of Pentecost/Shavuoth. There is nothing in Philo to suggest that he took offense at the Therapeutrides' education or to their devotion to philosophy. Much of Philo's criticism of women is directed at their behavior as wives in relation to their husbands, from which the Therapeutrides would have been exempt.

The cluster of characteristics that Philo attributes to the Therapeutrides gives us some insight into the women themselves and their participation in the society, as women's religion. Their childlessness emerges as a central element in their choice of the contemplative life. In a society that valued women primarily for their roles as wives and mothers, women who did not marry and did not bear children would have found themselves somewhat suspect.[60] The values of the Therapeutics, who prized knowledge, philosophy, chastity, the solitary life, and the spiritual quest of the individual soul above all other human endeavors, legitimized, if unintentionally, women who did not fulfill the usual social expectations.

How much this was an issue for Greek-speaking Jewish women is difficult to gauge. There is ample evidence that among the Roman aristocracy the birthrate had declined enough by the end of the first century B.C.E. to warrant legislation that penalized male celibacy and encouraged aristocratic women to bear children.[61] The Augustan *lex Julia* (18 B.C.E.) and the *lex Papia Poppaea* (9 C.E.) encouraged legal marriage and the production of legitimate heirs among Roman citizens, primarily in Rome, by prohibiting unmarried childless men and women from receiving certain inheritances and by decreeing that freeborn women with three legitimate children (four for freedwomen) would no longer be under the authority of a male guardian (the so-called *ius* [*trium*] *liberorum*). These regulations affected only Roman citizens, who must have married legally permissible spouses and produced legitimate heirs.[62] Aline Rousselle suggests that the decline in the aristocratic Roman birthrate was a function of many factors, including a widespread belief among upper-class Roman men that ejaculation was detrimental to male well-being and errors about the fertile period in the menstrual cycle.[63] Sarah Pomeroy concurs that the desire not to have children was an upper-class phenomenon.[64]

For a limited number of upper-class Roman women, in control of adequate financial resources, childbearing may not have been a crucial measure of social worth and therefore not a critical factor in their religious choices. But this legislation was directed specifically at the Roman aristocracy, and so it is unlikely that more than a handful of women in the Greco-Roman world would have been exempt from the social pressures of marriage and childbearing. Further, these laws would have put pressure on certain women to produce legitimate heirs.

That Philo's own society, however narrowly we construe it, regarded childless women with some suspicion may be deduced from an expansion on the Septuagint that Philo offers in the course of his discussion of the wife accused of adultery. "For if she has been falsely accused, she may hope to conceive and bear children and *pay no heed to her fears and apprehensions of sterility or childlessness.*"[65] Philo's elaboration, here italicized, suggests that if childlessness was not viewed as evidence of adultery, it was at least seen as evidence of divine displeasure. Philo also addresses the plight of the married man who has no children, expressing sympathy for the man who chooses to remain married to a woman who has failed to produce children,[66] when Jewish law allows him to divorce her.

Clearly, becoming a Therapeutic was an alternative available only to a very few Jewish women in the first century C.E.—those who already had the requisite education, which included not only literacy, but also favored the allegorical school of interpretation. It is unlikely that most childless Jewish women became Therapeutrides. Most would have lacked the sophisticated education necessary (as would have most men). Of course, it was precisely unmarried women and married women without children who might have had more opportunity to obtain the education necessary to join the Therapeutic society. Of those women who already had the requisite training, childless women would have been particularly drawn to a community that utilized so extensively the metaphors of Philo. For them, contemplative asceticism would have legitimized a permanent change in their lives and the attendant standards of worth.

Women may have been drawn to the Therapeutics as an opportunity to live the contemplative life within a community that both symbolically and actually provided refuge from a world critical of childlessness. They also may have found inviting the metaphoric representation of the mystical union of the soul with the divine not only as a spiritual marriage but as a reunion of the child with the mother. Although Philo does not address this directly in *On the Contemplative Life*, elsewhere in his works divine union with the soul is explicitly described as rebirth through union with the mother.

The critical text is extant only in Armenian, here paraphrased by Goodenough.[67]

> Rebecca is, like Sarah, Virtue or Sophia interchangeably. . . . She is so exalted a figure that her bracelets are sufficient to represent the entire cosmos which the immaterial Stream of God similarly wears. When she gives the servant to drink at the well it is the Logos itself which he receives. . . . Rebecca-Sophia approaches Isaac who has gone out in the evening to meditate in the field. . . . Rebecca comes down to him, and gets down from her camel as Sophia comes down to the mystic. She is veiled as are the inner secrets of the Mystery. At last they are united in the wedding chamber, which is itself the house of Sarah, also the type of Sophia, and here Isaac is united with the eternal Virgin, "from whose love," Philo prays, "may I never cease." Now Isaac is consoled for the loss of his mother, for in Rebecca he has found Sophia again, not as an old woman, but as one who is eternally young in corporeal beauty. . . .
> Again the soul is reborn by union with its own mother. . . .

Goodenough obviously found his own reading of Philo sufficiently provocative to require immediate defense, for in his very next sentence he writes:

> This is not madness. It is the cry of the soul to be at one with the greater thing from which it has come, to unite itself with the universal spirit which for all its cosmic and hypercosmic motherhood is eternally virgin. For sexual as are the experience and language, the soul has become united with something whose own essential superiority can be expressed only by assertions of its eternal virginity.[68]

Given Philo's extremely high opinion of the Therapeutics, his testimony that they read Jewish Scripture allegorically, and the possibility that he was at one time a member of the community, it seems feasible that the Therapeutics might have shared this understanding. Further, if for women the bearing and raising of children offers one possibility for regaining one's lost mother through the recreating of the mother-child relationship, childless women might have found the opportunity to regain the lost mother through a spiritual experience extraordinarily compelling.

Finally, some scholars have suggested that both *The Conversion and Marriage of Aseneth* and the *Testament of Job* may have been written by Therapeutics.[69] While we have no way of establishing more than the plausibility of such hypotheses, it is important to note that the Therapeutic authors could equally have been women.[70]

From Diaspora communities comes the most important evidence for the participation of Jewish women in the public life of Jewish communities, including the synagogue. In her now definitive study of women leaders in ancient Jewish synagogues,[71] Bernadette Brooten established that women who held

titles of synagogue office in burial or donative inscriptions are most likely to
have held those offices in their own right and not to have derived their titles
from offices held by their husbands or fathers. Virtually all of the inscriptions
that Brooten analyzed have been known to scholars for many years, even centuries, but until her study, they were routinely explained away as honorary, that
is, denoting no functional leadership activity on the part of Jewish women.

Brooten documents women who held the positions of head or president
of the synagogue, elder (most probably meaning a member of the council of
elders), and leader (or perhaps founder) in at least eight synagogues throughout the ancient Mediterranean. Two heads of the synagogue are documented
for western Asia Minor: Rufina of Smyrna in the second or third century C.E.
and Theopempte of Myndos in Caria, whose dedication of a chancel screen
post of white marble dates no earlier than the fourth century and possibly as
late as the sixth.[72] On Crete in the fourth or fifth centuries C.E., a woman
named Sophia was both head of the synagogue and an elder.[73] An inscription
from Thessaly of uncertain date commemorates a woman named Peristeria,
who held the otherwise unattested title *archegissa*, which Brooten suggests
means either leader or perhaps founder.[74]

Several women elders are known from the town of Venosa in southern
Italy, where some members of the Jewish community were prominent in civic
affairs, probably during the third to sixth centuries C.E.[75] Just recently, a new
inscription testifying to an elder named Murina, whose husband Pedonious
was a scribe or secretary, was discovered in southern Italy.[76] Women elders are
also attested in Thrace, in North Africa, and on the island of Malta, in roughly
the same period, probably in the fourth or fifth century C.E.[77]

We know relatively little about the structure of Jewish community life and
synagogue organization in the Greco-Roman Diaspora, so that it is difficult to
say exactly what these offices were or what they entailed.[78] The term *synagogue*
is itself misleading, for it can designate not only physical premises, but discreet
congregations (as in Rome) and whole Jewish communities as political and
social entities.[79] The structure and function of Jewish communal organizations,
though, bears significant resemblance to that of other religious *collegia*, and
many of the titles known from Jewish synagogues are also documented for
pagan associations.[80]

Brooten offers some plausible reconstruction of the offices known to have
been held by women.[81] She suggests that the *archisynagōgos*, or head of the
synagogue, was the leading office, although there could be more than one per
community. In some cases, the office seems to have been hereditary, and could
be held for life, but other persons appear to have held the position for a finite
term.[82] The duties of the head of the synagogue may have included "the exhortation and spiritual direction of the congregation,"[83] including teaching and
inviting other members of the congregation to preach. From rabbinic references, Brooten concludes that synagogue heads only read from Scripture if
they could not find anyone else to do it. They had various financial responsibil-

ities, including building, restoring, and maintaining synagogue facilities. She considers it likely that ancient synagogue heads also represented Jewish communities in their relations with other Jews and with non-Jewish neighbors and civic authorities.

Synagogue elders are likely to have been members of a council of elders, headed by a *gerousiarch*. Although the specific duties of elders probably varied from community to community and over time, they are likely to have included the collection of monies to be sent to the patriarch (in the late Roman period), administration of worship services, and communal decision making, including judicial functions. Noting that the title *elder* has its origins in political, civic terminology, Brooten considers it possible that elders also had political representative functions.[84]

It seems reasonable, then, to assume connections between the economic and financial status of synagogue leaders and their ascension to those offices, analogous to what we saw for other Greco-Roman religious and civic offices. While Jewish leaders may not have been responsible for the public entertainments and athletic events financed by women such as Tata of Aphrodisias, it seems clear that economic resources and synagogue office were not unrelated in many instances, and that this applied to women as well as men (if not more so).

The system of benefaction, or euergetism, which was an integral part of Greco-Roman religious office, extended to Jewish life in the Greco-Roman period. Major donors to ancient synagogues in particular may be presumed to have had considerable prestige and influence, which their contributions both reflect and enhance. There is no reason to view things any differently when the donor was a woman.

Intriguingly, although some Jewish women clearly made lavish gifts to local synagogues, not all the female benefactors of Jewish communities may have been Jews. We have already considered the fascinating inscription of Tation of Phocea, whose privilege of sitting in a seat of honor violates the assumption that men and women sat separately in ancient synagogues.[85] Scholars have routinely assumed that Tation was Jewish. But I am puzzled by the phrase in her inscription stating that Tation built an assembly hall and enclosed an open courtyard and gave them as a gift "to the Jews." Does the use of this phrase suggest that Tation is not included in the category? Conceivably, the synagogue of "the Jews" is meant to distinguish one congregation from another within the same region, as was the case in ancient Rome.[86] But alternatively, the language may subtly reveal that Tation is an outsider.[87] One might argue that a Jewish community would not acknowledge the gifts of a non-Jew with privileged seating, but this is a judgment we cannot base on any evidence, and conceivably Tation herself is the proof of such behavior.

The woman whose benefactions to the Jewish community have received far more attention and discussion is Julia Severa,[88] who, like Tation, also constructed a synagogue.[89] Hers was for the Jewish community in Akmonia.

Whereas Tation's gift was commemorated with an inscription, we know of Julia Severa's role only through a later inscription that recognizes the refurbishing of the building by three male synagogue officers at their expense.[90] Julia Severa appears in another inscription as a high priestess (*archiereia*) and as a judge of festival games (*agōnothetis*). From coins minted at Akmonia during the time of Nero, she and her husband, Lucius Servenius Capito, are known to have been either archons or priests of the imperial cult several times.[91]

The judgment of Lifshitz that "a Jewish woman could not have been a priestess of a pagan god"[92] represents the conclusion of virtually all scholars that Julia Severa was not simply a wealthy Jewish woman who supported her local synagogue. Interestingly, however, William Ramsay did think she was Jewish, and harmonized her imperial priesthood with little difficulty, arguing that it "was compulsory on those who wished to engage in the imperial service that they should freely accept the forms of that cultus, for it would have been a mark of disloyalty disqualifying an officer to refuse to participate in the established forms."[93] Most scholars have taken a compromise position, seeing Julia Severa as another example of an upper-crust Roman woman attracted to Judaism and favoring Jews with various forms of support, including political intervention.[94] Few have considered the possibility that Julia Severa's patronage of the Jews had no implications for her personal religious affinities—that friendly Gentiles might have donated buildings to the local Jewish community out of social, political, or economic ties with no specifically religious sympathies.[95]

Ironically, we have virtually no way to adjudicate between the assessment of Ramsay, who took the example of Jewish accommodation to late nineteenth-century English and French society as his model, and interpreted antiquity accordingly, and the assumptions of Lifshitz and others. No ancient sources from Diaspora Jewish communities confirm that Jews would not, on principle, serve in pagan offices which were at once civic and religious, such as magistracies, and the assumptions we bring govern the interpretations we produce. However, if Julia Severa did indeed donate the physical premises of a synagogue to the Jewish community in Akmonia in the mid-first century C.E., it seems inconceivable that she would not have expected prestige, privileges, and honors in return, even if the inscriptions commemorating her initial gift have not survived.

The benefaction system must also underlie those inscriptions that designate women as mothers of specific synagogues. The title, which is also found in its masculine correlate, father of the synagogue, is attested two or three times in Roman inscriptions, including that of Veturia.[96] All are difficult to date with precision, but are likely to be no later than the third century C.E. It is also attested in an inscription from Venetia in Brescia (Italy) of uncertain date.[97]

Of all the titles Brooten analyzes, this is the only one that may conceivably honor the person so called without connoting any specific duties or obligations. But as she also points out, if the title is "honorary" it must be so both for fathers of the synagogue and for mothers.[98] In fact, the primary reason scholars traditionally interpreted "father of the synagogue" as honorary was precisely because it was the male correlate of "mother of the synagogue," and their myopic perspective required them to consider "mother of the synagogue" honorary. Regardless of whether synagogue parents had specific functions, the title must be related to influence and prestige within the community, and probably to financial considerations as well. Veturia Paulla, who converted to Judaism at age seventy and became the mother of two synagogues, may conceivably have come from a wealthy Roman family or had substantial personal resources to allow her to play the kind of patronage role that would have earned her such a title.[99]

The Jewish catacombs from the city of Venosa in southern Italy have produced not only inscriptions of three women elders, but also one in which a woman is designated *pateressa* (literally, fatheress) and another in which a woman is called "mother" and her husband is designated "father and patron of the city."[100] Brooten observes that ten Jewish men are designated "father" in the inscriptions from Venosa, in addition to the two women noted here.[101] Seven of the men are named Faustinus; one of the women is named Faustina, and Brooten suggests that the "Faustini" may have been a wealthy and influential Jewish family many of whose members attained public offices and titles.[102]

Unfortunately, we do not know precisely what the title patron of the city entailed, or whether the phrase "father and patron of the city" should be construed as "father of the city" and "patron of the city," or as "father" and "patron of the city," and if the latter, what the title "father" might entail.[103] Similarly, we cannot tell whether the titles "pateressa" and "mother" refer to civic offices in Venosa, or to synagogue offices, or even how separable the two might have been. Tata of Aphrodisias was called "mother of the city," but it is not clear how relevant that might be for Jewish women in Venosa, probably several centuries later.[104]

A recurring motif in several of these inscriptions is the factor of conversion to Judaism, which Brooten did not address at any length, since her primary concern was to refute old scholarly dismissals of the epigraphical evidence for women synagogue leaders. Veturia Paulla, mother of the synagogue of Campus and mother of the synagogue of Volumnius, both in Rome, was clearly a proselyte.[105] As we have seen, both Tation in Phocea and Julia Severa might have been converts to Judaism, although the evidence remains tantalizingly ambiguous.

It is also possible that Rufina of Smyrna, a large city on the Aegean coast of Asia (now the modern city of Izmir), was a proselyte.[106] She calls herself *Ioudaia*, the Greek term for a Jewish woman that translators have routinely rendered "Jew." Oddly, however, *Ioudaia* and its male form *Ioudaios* rarely occur in Greek inscriptions considered Jewish on other grounds or in Latin inscriptions using the equivalent *Iudea/Iudeus*. In the small percentage of individuals described by this term in the singular (as opposed to cases where the plural designates the community or synagogue), an astonishingly high number show some indications of not having been born Jewish. In fact, I have suggested that this is precisely the connotation of the term in some instances.[107]

Whether this is what Rufina intends by this self-designation is by no means certain. The inscription she commissions establishes a burial site for her slaves and freedpersons (though not for herself or any members of her family, suggesting there was another burial site for them), and sets forth penalties for anyone who violates the tomb by an unauthorized burial. These penalties are common enough in burial inscriptions from Asia Minor, whether Jewish, Christian, or pagan. Often they take the form of fines payable to local temple treasuries, or in the case of Jewish inscriptions, to the local community or synagogue of the Jews. Ambiguous curses invoking various deities are not infrequent, including some distinctively Jewish and Christian forms. Lest anyone think to evade punishment by effacing the terms of the penalty on the inscription itself, many carry warnings that a copy of the inscription has been deposited in public archives.

The penalties prescribed by Rufina are unusual principally in that they call for a split fine, part to be paid to the sacred treasury[108] and part to the Jewish people. To modern sensibilities, something seems incongruous about a Jewish synagogue president authorizing a fine to be paid to a pagan treasury. This may, of course, be our problem and not theirs: temples, after all, also functioned as banks in antiquity. Since the purpose of the fine is to encourage the prospective recipient to bring any violators to justice, the choice of a temple treasury may reflect Rufina's perception that this was a better guarantor of the sanctity of her household's burial site than the Jews alone. But only a handful of other Jewish inscriptions from western Asia Minor employ this split fine, suggesting that something more may be at issue. One of those is the inscription for a young woman who is either a convert to Judaism or the daughter of a Jewish woman and a non-Jewish man.[109]

Perhaps Rufina's slaves and freedpersons are not all Jewish. Maybe Rufina herself has strong ties to the non-Jewish community, by virtue of having been born there.

The circumstantial evidence can be pushed a little further. In this inscription, Rufina mentions neither father nor husband. She may have been unmarried, but she must have had a biological father. Perhaps he is just dead. If she was herself a freedwoman, we would not expect a father's name, but we might expect the name of her patron. One scholar has suggested that Roman Jews refused to include their freed status (or their slave status) on their own burial inscriptions, since this would be tantamount to acknowledging a master other than God.[110] The nature of this inscription, a burial dedication for slaves and freedpersons, renders such an explanation less likely. Of course other inscriptions by Jewish women contain no reference to a father or husband, such as that of Juliana of Naro. But one possible explanation is that Rufina adopted Judaism later in life and severed her connections with her male relatives as a consequence, a situation which can be adduced for numerous other converts in Greco-Roman antiquity.

It may simply turn out that the Greco-Roman environment, especially in Asia Minor, was particularly conducive to women's leadership in public life, particularly in those offices where civic responsibility and religion intersected. But we should be sensitive to the possibility that women's leadership was particularly likely in Jewish synagogues with relatively high numbers of proselytes (both male and female) for whom the participation of women in public life, including religious *collegia*, was familiar and acceptable.

Evidence for women's charismatic leadership in Greco-Roman Jewish communities is virtually nil. Of course, we know little about charismatic Jewish men outside of Judea, but in any case there are no identifiable, historical female counterparts to figures such as John the Baptist or such messianic candidates as Jesus of Nazareth, Theudas, or Simon ben Kosiba (also known as Bar Kochba). There is only the tantalizing and elusive tale of the three daughters of Job in the pseudonymous *Testament of Job*, whose inheritance consisted of three mysterious bands, which caused the women wearing them to utter ecstatic angelic prophecies. As I have indicated, whether the text reflects the actual behavior of some Jewish women we can only surmise: whether such behavior would have had leadership ramifications is even further beyond our grasp.

It is fruitful to apply Mary Douglas's cultural theory to the divergence between Jewish women's lives as depicted in rabbinic sources and as evidenced for Diaspora communities. Assessing the location of the rabbis on Douglas's grid is complicated by the fact that rabbinic sources do not provide or only randomly provide the kinds of data on which Douglas's method relies: who marries whom, how marriages are arranged, rituals and activities surrounding birth, death, and other crucial *rites de passages*, the scope of social relations, and so forth. Further, even when rabbinic sources do touch on subjects such as marriage and weddings or death and funerals, we cannot tell how much they portray reality and how much they merely offer a utopian vision. All of the communities from the Greco-Roman world present us with this difficulty to varying degrees, but it is particularly acute for rabbinic writings. Nevertheless, I would argue that the ideals attributed to the rabbis are still significant, especially to the extent that they are consonant with a particular cluster of attitudes as outlined by Douglas.

As a whole, the rabbis seem to exhibit many characteristics of strong group. They are intensely concerned with delineating boundaries not only between Jews and non-Jews, but between themselves and insufficiently observant Jews, such as the *am ha'aretz*. Their attitudes toward admission to the group are precisely what we would expect with strong group. They discourage proselytism and conversion.[111] Their concern to define inclusion in Israel matrilineally may conceivably reflect a desire to tighten the boundaries by switching from the less verifiable patrilineal principle.[112] While it would be absurd to suggest that the rabbis had no contact with those outside their group boundaries, they clearly wish to minimize such contact and to present such interaction as irregular. To the extent that we can surmise what the daily lives of the rabbis were like, we may say that they lived, ate, worked, worshipped, and perhaps even played with the same communities, yet another test of the strength of group.

The question of rabbinic grid seems somewhat less clear. Two prominent characteristics of high grid, hierarchy and ascribed status, do not seem terribly strong among the rabbis. For example, birth does not make one a sage, although few sages come from the *am ha'aretz*. Only years of study of Torah enable one to achieve the status of a sage. But the rabbis do display other characteristics of strong grid, especially when we utilize Douglas's description of grid as strong classification and strong regulation. In many ways, in fact, the rabbis are a textbook case of an intensely complex classification system, much of it implicit and taken for granted in the arcane speech of the Mishnah and the Talmud. If a "web of prescriptions" cast over all of life constitutes prima facie evidence of high grid, there can then be no doubt that the rabbis exhibit both high group and high grid.

By comparison, Jews living in the cities and towns of the Diaspora, particularly in portions of Asia Minor and probably in Egypt prior to the revolts of 115–117 C.E., display weaker group, which is not to say weak group. For example, there is considerable evidence of their comfort and acceptance in Gentile culture, albeit to varying degrees.[113] The location and prominence of the synagogue at Sardis, especially from the second to the fourth centuries C.E., can be read no other way, and whatever central role the synagogue there played in the life of Sardian Jews, from its very location alone, we would know that Jews at Sardis had to have substantial contact with their non-Jewish neighbors. Those vexing persons called *theosebeis*, usually translated as "Godfearers," may also suggest far looser boundaries between Jews and non-Jews than seems to have characterized rabbinic life, whether we consider these people partial proselytes or righteous Gentiles who supported Jewish synagogues, or even just friendly neighbors. "Godfearers" are epigraphically attested in the courtyard of the Sardis synagogue, and now in the recently published inscription from Aphrodisias, not all that far from Sardis itself. Among the many other examples that could be adduced to support this point are the participation of Jews in civic life in the city of Venosa in southern Italy in the late imperial period, the continuing attraction of the synagogues in Antioch for pagans and Christians alike, which we learn particularly from Chrysostom's irate sermons in the fourth century C.E., and even the evidence for intermarriage between Jews and non-Jews, both pagans and Christians.[114]

Thus we may suggest first that those Jews in the Greco-Roman period whose social experience was characterized by relatively low grid were more likely to accept the active participation of women in public life. This seems to be particularly true for those Jews whose sense of group was looser than that of the rabbis, who did not draw quite such hard and fast lines between themselves and Gentiles, whose social intercourse with non-Jews extended beyond the mere necessities of economic survival, and who read Gentile literature and attended events in the theater or the athletic arena. For slaves, of course, all this needs to be qualified: Jewish slaves owned by Gentiles would have had a great deal of intercourse with Gentiles, of all sorts, but not necessarily by choice, and for them, these questions require additional analysis.

Second, Jewish women who lived in communities like those implied in rabbinic literature are likely to have experienced the insulation and isolation characteristic of strong group with low grid. For such persons, Douglas suggests that the only satisfaction available comes from fulfilling one's station in life, which seems to reflect the place of women in rabbinic thought, if not in rabbinic reality. This also suggests that the wives and daughters of rabbis might not have wholly shared the cosmology of their husbands and fathers, and probably not their rituals and beliefs, at least not to the extent that the tradition would have us believe. Insulated and isolated persons frequently display attributes that are often insultingly attributed to women, such as the fondness for superstition and magic and lack of critical judgment which are quite familiar from rabbinic portrayals of women.

By now, it might seem obvious that the intense concern with the menstrual purity of women on the part of the rabbis makes the most sense for men experiencing high group and high grid. Douglas proposes that although attitudes toward sexuality ought to be profoundly affected by group and grid, in all cases, "the effects of the division of labor between males and females, and the extent to which valued property is inherited on lines laid by marriage alliances"[115] skews things considerably. For different reasons, women may be sources for ritual defilement and shame not only where group and grid are both strong, but also where they are both weak and where the emphasis on individualism might lead us to predict no such regulations. For Douglas, the key is the separation between marriage, procreation, and the inheritance of valued property, so that only where the two are severed, as for example in the promulgation of asceticism among the Therapeutics or among many early Christians, do we find strong gender equality and would we expect little interest in menstrual and sexual purity regulations.

Nevertheless, I would still suggest that the weakening of both group and grid among many Jews in the Diaspora was likely to be accompanied by a relaxation of menstrual purity laws and lessened interest in compliance, especially to the extent that such regulations are used to reinforce social distinctions between Jews and non-Jews. Further, we would expect, if we were fortunate enough to have such evidence, that the arguments put forth by Diaspora Jews in support of the menstrual purity laws would have differed considerably from those offered by the rabbis, such as the punishment of Eve.[116]

The segregation of Jewish women in ancient synagogues has been widely assumed by modern scholars, mostly by reading back later practice into antiquity. In fact, as we have seen, no extant evidence confirms that Jewish women sat apart from Jewish men in synagogues, although the ascetic Therapeutics did sit separately for their communal gatherings. Douglas's work allows us to see that issues of gender and other hierarchical discrimination in seating are integrally related to social location. We would most expect persons who experience high grid to organize space hierarchically, in the likeness of their social life, whereas those experiencing decreased grid should display much less interest in physical representations of social hierarchy. In the case of the rabbis, it seems eminently plausible that they would have supported the physical segregation of women from men: whether they would also have endorsed other forms of stratification in the synagogue would be interesting to know.[117] Conversely, we would expect the seating patterns of Diaspora synagogues to reflect the differing social structures and concerns of those communities, as regards both divisions between Jews and non-Jews, and those between women and men. We might also make similar arguments for the allocation of household space and the attempt to confine at least free adult women to the interior of complex houses.

In the case of Jewish women's education, Douglas's model allows us to see that more is at stake here than just different practices. Rather, the debate about the education of women reflects the perspectives and concerns of the different quadrants. One of the crucial means of enforcing insulation is limiting access to knowledge. Douglas argues in fact that by controlling information which might undermine authority, insulation sustains the boundary system as effectively as high group does.[118] I would suggest that these concerns underlie rabbinic and other ancient discussions about the education of women, and further explain the kinds of arguments used to justify restricting women's access to education, particularly the use of arguments from nature, such as the claim that women have inferior minds. Similar arguments are also used to deprive male slaves and serfs of sophisticated knowledge.

The discrepancy between rabbinic attitudes toward women's cultic leadership and the demonstrable presence of women synagogue officers may also reflect these disparate locations (both theoretical and real). Further, rabbinic perception of a strong connection between religious knowledge and public religious leadership may have intensified their inability to see women in such roles. Conversely, in Diaspora communities where such connections were weaker, women who held religious office would have seemed far less anomalous.

Finally, we may consider the application of Douglas's model to Philo's Therapeutics, who present an example of an egalitarian community closer in certain respects to the Cynics or the Epicureans than to the Qumran Essenes or the Jesus movement. Hardly a protest or a renewal movement, and lacking any apparent millenarian consciousness, the Therapeutics drew their members from wealthy, educated men whose prior experience was likely to have been that of strong group and grid, and from wealthy, educated women whose prior location was probably moderate group and strong grid. Its long-term existence depended primarily on its ability to continue to attract relatively small numbers of affluent persons. The Therapeutics display certain traits of strong group in their withdrawal from the rest of society and their communal meals, worship, and festivals. But the real degree of group they experienced was mitigated by the fact that the Therapeutics actually spent most of their time alone in their cells.

The Therapeutics espoused a high degree of bodily control and a moral cosmology consonant with strong grid. The level of communal organization was modest: from Philo we learn only of a president and of choral leaders patterned after Moses and Miriam, who led the male and female choirs, respectively. Gender and seniority played moderate roles in seating and service for meals, suggesting overall that the level of grid experienced by the Therapeutics in their community was moderately weakened in comparison to their earlier social existence.

The egalitarian impulse among the Therapeutics was legitimized not by an eschatological asceticism, which would have negated women's traditional roles as wives and mothers, but rather by a philosophical asceticism, adopted freely by a small number of individuals. This form of asceticism, which the Therapeutics do not seem to have advocated for the populace at large, did not so much change the rules for all women as it simply redefined Therapeutic women as men. Such a motif was not altogether absent among Christians whose promulgation of eschatologically based asceticism would license alternatives for women in numerous early Christian communities. Here, however, it seems a blanket characteristic, at least in Philo's presentation.

# — 10 —

## Autonomy, Prophecy, and Gender in Early Christianity

When the pagan critic Celsus denigrated Christianity in the second century C.E. as a religion of women, children, and slaves, he articulated a connection between women and early Christianity that has only recently received serious and critical attention.[1] As we have already seen, the association of women and religion was rampant in Greco-Roman antiquity, particularly the association of women with the wrong sorts of religion, such as "superstition."[2] By his equation of Christianity with women, children, and slaves, Celsus may have meant merely to confirm what elite, educated men already knew: obnoxious new religions are the provenance only of the intellectually and socially inferior. To take his observation as a useful historical datum would then be to miss the polemical effect of such calumny. Yet it is interesting to consider Origen's refutation, less than a century later, of this particular accusation from Celsus. He does not deny that Christianity appeals to "'the foolish, dishonourable and stupid, and only slaves, women and little children.'" He merely asserts, "Not only does the gospel call these that it may make them better, but it also calls people much superior to them."[3] Origen's reply might thus be taken to concede that many Christians did fall under Celsus's unflattering rubric.

Was Celsus accurate? Were women, in fact, disproportionately represented among the earliest communities of Christians? From Adolph von Harnack at the turn of the century to recent feminists, some scholars have found the relative visibility of women in the New Testament and other early Christian texts indicative of a quantitative and qualitative difference in the roles of women in early Christian communities.[4] Others have argued that women were no more prominent in early Christian circles than they were in many other Greco-Roman religious environments, particularly in the various so-called mystery religions.[5]

Yet material that seems on its face to point to the significant presence of women, such as stories in the canonical gospels, may in fact be deceptive. The two-part work consisting of the Gospel According to Luke and the Acts of the Apostles (Luke-Acts) contains numerous stories about women in the early

128

Jesus movement, including many unique to their author, commonly referred to as Luke. Many have taken these stories as reasonable representations of the activity of women in the Lukan community, if not as indicative of even greater activity than Luke reveals. Yet Mary Rose D'Angelo has demonstrated that many of Luke's stories about women are tailored to suit Luke's several agendas, including his desire to demonstrate the respectability of the early Christian movement and its acceptance within aristocratic Greco-Roman society.[6]

Intriguingly, Luke's portrait of those attracted to the Jesus movement contradicts Celsus's calumny in a rather precise fashion. Luke's Christians are anything but foolish and dishonorable or slaves and little children. When they are women, they are frequently free, aristocratic, affluent, and respectable, if not also demonstrably educated and intelligent women. Seven times in Acts, Luke describes those who flocked to the preachings of Christian missionaries as men and women: in two of those instances, he emphasizes that they were of high social status (17:4, 17:12). On the surface, this pairing of men and women could be mistaken for a kind of gender inclusivity on Luke's part, particularly when the emphasis is placed on the specific mention of women. As D'Angelo shows, however, Luke's consistent pairing of women with men reflects his pervasive concern to demonstrate that Christian women are properly associated with and subordinated to men in accordance with Greco-Roman norms.[7] Almost certainly, Luke's insistence on this should be seen as a response to a Greco-Roman critique of Christianity, which was in fact grounded in the social reality of those Christian communities that denied or minimized the significance of marriage and women's subordination to husbands and fathers. When Luke insists that Christians are men and women, he may in fact be emphasizing not the women, but the men. It is unlikely, if not impossible that Luke's portrait is here a direct response to Celsus, whose polemic against Christianity is usually dated to the last quarter of the second century C.E. Recent scholarship, however, has begun to support a mid second-century date for Luke-Acts, circa 125 to 140, making it quite plausible that Luke is responding to the same pagan stereotype about Christians that Celsus so eloquently articulates.[8]

The difficulty of extracting reliable information about early Christian women extends beyond the tendentiousness of specific authors. Consider in detail the treatment of resurrection appearances to women. Though the accounts of the canonical gospels differ in many respects, all four concur that women were the first to discover the empty tomb of Jesus and thus to learn of his resurrection.[9] In the Gospel of John, not only is a woman, Mary of Magdala, the first to discover the empty tomb, she is also the first to see and speak with the risen Christ, who instructs her to go and tell the other disciples.[10] This tradition is found also in the long ending of the Gospel of Mark.[11]

But the version of the resurrection appearances given by Paul in 1 Corinthians 15:6–8 contains no mention of one to Mary of Magdala, nor any reference to the tradition that women were the first to discover an empty tomb. Since Paul's letter antedates the written form of the gospels by at least two decades, we might be tempted to argue that his account is the earlier. But by a principle known as the criterion of dissimilarity, we might argue that the traditions of women as the earliest witnesses to the resurrection and of Mary of Magdala as the first to see the risen Christ are so problematic that they would surely have been omitted had they not been firmly believed to be true by tradents, or transmitters, of the gospel traditions. Since the purpose of Paul's recitation of resurrection appearances here is the legitimation of authority, especially his own, the absence of Mary of Magdala from the list of those who saw the risen Jesus would constitute early evidence of the attempt to exclude women from claims to the authority that such appearances confer. Although many scholars believe Paul has here utilized earlier lists of appearances, perhaps joining two separate traditions (Cephas and the Twelve; James and all the apostles), the absence of women is still striking.[12] It would seem easy to conclude that Paul's omission ironically points both to women's inclusion in the resurrection appearance traditions and to their implications for women's authority.

Vexingly, there is an alternative, more negative interpretation of the history of the empty tomb tradition, which must be separated from the history of the appearance traditions. F. W. Beare argued that the empty tomb stories do not derive from the earliest level of traditions about Jesus, but arise later, out of response to beliefs in the bodily resurrection.[13] Citing precisely Paul's discussion of the nature of the resurrection in 1 Corinthians 15:12–57, Beare points out that Paul does not expect the dead to rise uncorrupted from their tombs, but rather to rise with a mysterious spiritual body. Since Paul sees Jesus as the prototype of the resurrected, we must conclude that he did not think of the resurrection of Jesus in this manner, and is unlikely to have known the empty tomb traditions, which portray precisely such an uncorrupted resurrection.

When Christians eventually crystallized beliefs about the bodily resurrection of Jesus, they told stories[14] about the empty tomb designed to confirm the physical resurrection, even though, as Beare points out, the absence of the body most readily suggests that someone has removed it, not that the person has risen from the dead. Beare did not consider the significance of the fact that women are "credited" with the discovery of the empty tomb. But why? Positing women as the original parties at the empty tomb would have provided an adequate answer to the question of why this information had not surfaced earlier, by claiming unofficial and secret witnesses and by blaming women, indirectly at least, for a failure of nerve that led to the "suppression" of this important information. To those who might have asked, "Why doesn't the Apostle Paul mention these stories?" or "Why haven't we heard them before?" the answer would have been: because women discovered the vacant tomb and either they did not tell anyone (Mark 16:8) or they did, but no one believed them (Luke 24:11), and no one then retold the story until now.[15]

This reconstruction undercuts the theory that these stories are told because they reflect some underlying truth too difficult to repress. Instead, it suggests that the stories of the empty tomb reflect subtle early Christian denigration of women. Great caution, then, must be exercised in the attempt to extract any reliable historical data about women from earliest Christian sources, especially literary works such as the canonical gospels and Acts. Ironically, this is true because early Christian communities, especially after the death of Jesus, experienced considerable conflict over the appropriate roles of women, and tended to retroject their positions about this conflict back onto the stories they told about the women who encountered the earthly Jesus. The attempt to reconstruct early Christian women and early women's Christianity is thus fraught with difficulties, but it is not altogether hopeless. With care, we can make some important observations about both.

In the first few Christian centuries, many, perhaps even most, Christians, were not born such. Christian predilection for both sexual asceticism and martyrdom clearly contributed to this state of affairs, by increasing the likelihood that Christians would not bear and raise children. Although the initial followers of Jesus were Jews, earliest Christian communities rapidly began to draw primarily from the pagan population. For many women, being a Christian was a conscious decision.

Were women any more likely than men to make this decision? Was the profile of women who became Christians meaningfully different from that of men? Were women any more likely, at least before the fourth century, to have become Christians (of whatever stripe) than to have been attracted to other religions in the Greco-Roman world, including Judaism? Were women's motivations for becoming Christians significantly different than those of men? Were the consequences of becoming Christian different for women than for men? Bearing in mind the class distinctions implicit in Celsus's characterization, is it appropriate to speak of women generally here? What role was played, if any, by ancient social class distinctions?

Two of these questions are particularly difficult to answer. Because we lack the sources for an accurate demographic survey of any early Christian community, it is impossible to know whether more women than men became Christians, or whether women were more likely to choose Christianity than to choose Judaism, or Isiac initiation, or the mysteries of Dionysos, or anything else. Still, we may fruitfully examine the sources we do have to see what pattern of gender distribution emerges.

It would be ideal to begin at the beginning, with the Jesus movement in Roman Galilee and Judea in the fourth decade of the first century C.E. As we have just seen, though, subsequent Christian debates about women's roles and authority distort not only the broad portrait of women in the early movement around Jesus, but the specific details as well. Feminist scholarship has demonstrated irrefutably that women constituted a significant presence in the Jesus movement, despite all the attempts by both ancient writers and subsequent transmitters and translators to obscure that presence.[16] Yet the exact nature of women's involvement in the movement is debatable.

In a study that became a classic almost instantaneously, *In Memory of Her:*

*A Feminist Theological Reconstruction of Christian Origins,* Elisabeth Schüssler Fiorenza argues that women not only constitute the paradigm of the faithful disciple in the earliest layers of the gospel traditions, but also that women were the first Gentiles to become followers of Jesus and were intimately involved in the development and articulation of a theology which made possible an extension of the Jesus movement to non-Jews.[17] She bases this claim on her analysis of several stories about Jesus and non-Jewish women. The Gospel of Mark 7:24–30 presents an encounter between Jesus and a woman described as "Greek, a Syrophoenician by birth," in which the woman begs Jesus to cast out a demon that has possessed her small daughter. When Jesus demurs with the saying, "Let the children first be fed, for it is not right to take the children's bread and throw it to the dogs," the woman counters his argument immediately: "Yes, Lord; yet even the dogs under the table eat the children's crumbs." In turn, Jesus acknowledges the force of her response and grants her request. Returning home, the woman finds her child healed. The story is absent in the Gospel of Luke, but is preserved also in the Gospel of Matthew 15:21–28, which identifies the woman as a Canaanite and has Jesus accede to the woman's request, praising her not for the character of her response, but for the magnitude of her faith.

With the exception of minor additions, Schüssler Fiorenza does not believe the story in Mark is the product of the author. Rather, she considers it a pre-Markan tradition whose function is to counter sayings of Jesus that seem to oppose a Gentile mission with stories of Jesus which affirm such a mission. Noting also the encounter between Jesus and a Samaritan woman in the Gospel of John 4:7–42 in which many Samaritans come to believe in Jesus because of the woman's testimony, Schüssler Fiorenza interprets the identification of women and non-Jews as a reflection of historical verisimilitude.

As we saw for the empty tomb traditions, an alternative and more negative interpretation is possible here. First, in the encounter between Jesus and the Syrophoenician woman in Mark, there is no indication that the woman becomes a follower of Jesus. Unquestionably, she recognizes his power as a healer and seeks him out on that account, but the tale here says nothing more. Only in the Matthean version does the woman acknowledge Jesus as Son of David and does Jesus acknowledge the woman's faith. Both these elements are consistent with Matthean redactional concerns[18]

Second, it is highly plausible that the core of the Markan story is the saying attributed to Jesus: "It is not right to take the children's bread and throw it to the dogs." The entire narrative framework may be a construct designed both to explain the circumstances under which Jesus might have said such a thing and to counter the negative implications of this saying for a Gentile mission. The Matthean version attempts to kill two birds with one stone by incorporating not only this saying, but also the equally problematic "I was sent only to the lost sheep of the house of Israel."[19]

Since the saying about children's bread suggests the rejection of Gentiles, early tradents may have asked under what circumstances Jesus would have said such a thing. Answer: he said it to a Gentile, who sought his help. For Schüssler Fiorenza, the choice of a woman as the Gentile petitioner alludes to the historical role of women in the mission to the Gentiles. But since the story portrays Jesus as rebuffing the Gentile, at least initially, it seems equally possible that it was less threatening and problematic for Jesus to respond this way to a woman than to a Gentile man. Additionally, if this story intends to stress the marginality of those whom Jesus included, the portrait of the petitioner intensifies the message that Jesus accepted all who sincerely sought him out, even the dregs of the dregs. The one who here seeks Jesus is not only a Gentile, not only a woman. She is also presented without the respectability of a husband, and as the mother not of a son, but of a daughter, a condition sometimes so lamented in the ancient world that people who only had daughters were considered childless.

As is the case with so much testimony about women in antiquity, we cannot resolve this question with any certainty. It is important, then, to understand what may be at stake in differing interpretations of ancient documents. For Schüssler Fiorenza, women's instrumentality in the extension of the Jesus movement to Gentiles has only positive implications: it provides evidence of women's early leadership in the Jesus movement that ultimately serves to legitimize women's leadership in contemporary Christian communities. Since I am less concerned to seek the legitimation of contemporary women's lives in the precedents of antiquity, for me, women's role in the mission to the Gentiles is at best neutral. From a Jewish feminist perspective, I suppose it could even be construed negatively, given the degree to which a mission to the Gentiles involved not only separation from Judaism, but ultimately the denigration of that tradition. The caveat here, as elsewhere, is that in the absence of definitive evidence, we tend to construe the ancient sources in ways that are consonant with our own concerns, particularly when there are significant contemporary implications for our readings, acknowledged or not.

Remarkably few women in the early Jesus movement appear to conform to the most socially acceptable categories of virgin daughter, respectable wife, and mother of legitimate children. Frequently, they are anomalous not merely by virtue of their gender, but also by additional marginal traits, often specific to women. If the women who followed Jesus and who were members of the earliest communities after his death had living husbands, virtually nothing in the gospel traditions attests to this. In Luke 8:3 the benefactor Joanna is identified as the wife of Chuza, Herod's steward, but this detail accords with Luke's program to portray the Christians as financially and socially respectable too conveniently to have much claim to historicity. According to the Gospel of John 19:25, one of the three women present at the death of Jesus was Mary, the wife of Clopas, here differentiated from the mother of Jesus (who is not named as Mary in John) and Mary of Magdala.

A few women are explicitly designated as widows in stories about Jesus.

Mark and Luke both include the parable of the widow whose donation of two copper pennies to the treasury was a greater gift than the large donations of the wealthy.[20] Two other narratives about widows are unique to the Gospel of Luke. One concerns Anna the aged prophet, whom the author would have us believe spent something like fifty or sixty years of her life as a widow, living in the temple of Jerusalem, fasting and praying.[21] The second portrays an unnamed widow in the city of Nain, whose only son Jesus raises from the dead.[22] The exceedingly close resemblance between this story and one about Elijah in 1 Kings 17:17–24 makes it unwise to use the story as evidence for women in the community around Jesus.

This interest in widows, visible also in Acts, appears uniquely Lukan, at least in comparison to the other synoptic gospels.[23] If we may correctly date Luke-Acts to the second century, this emphasis on widows may reflect the development of orders of widows known from other second-century texts, especially 1 Timothy, as we consider shortly. If so, we should be extremely cautious drawing conclusions about first-century Christian women on the basis of these passages.

The widowhood of other women may be inferred, though by no means demonstrated, from the fact that they are described as the mothers of grown men, without any explicit mention of husbands. These include Mary, the mother of James and Joses[24] and Simon's mother-in-law.[25] Curiously, Simon (Peter)'s wife is unidentified. The husbands of a few women are alluded to, yet absent from the drama itself: the unnamed mother of the sons of Zebedee[26] and of course Mary, the mother of Jesus, whose husband is conspicuously but not surprisingly absent from the gospel narratives.

Other women are described neither as mothers, nor as wives, nor as widows: about their status we can say little except that the absence of such information is itself interesting. Among these women are the patrons mentioned in Luke 8:3, Mary of Magdala and Susanna, Salome in Mark 15:40–41, and the sisters Mary and Martha in Luke 10:38–42 and the Gospel of John 11:1–44. Although Luke and John contain significantly different stories about the sisters, both associated Martha with the public realm and Mary with the domestic. It is worth noting that in Matthew's treatment of Mark 15:40–41, the name of Salome is omitted and her description as one who followed Jesus from Galilee to Jerusalem and ministered to him is instead assigned to a nameless plurality of women.

One story about Jesus portrays him healing a woman who had been hemorrhaging for twelve years. On the reasonable assumption that such a flow of blood was gynecological in nature, the woman would have found herself in a state of ritual impurity, both according to the purity laws of Leviticus 15:25–30 and to subsequent rabbinic regulation.

Interestingly, women are virtually never possessed by demons in the gospel traditions, with two exceptions. Luke claims that Jesus had driven seven demons out of Mary of Magdala, but only Luke reports this story.[27] Quite possibly this constitutes an intentional attempt to mute the traditions of authority associated with Mary. The only other female possessed by a demon in the gospels is the daughter of the Gentile woman just discussed.

When we have stripped away the redactional layers of early gospel traditions about women in the Jesus movement, very little remains that bears on the questions I have posed. The letters of Paul and his early disciples (excluding 1 and 2 Timothy and Titus) offer a better glimpse into the demographic composition of some first-century Christian communities. From these letters and Acts, Wayne Meeks compiles a list of just under eighty persons, of whom about one-fifth are women.[28] Since it is by now a well-demonstrated historical principle that women are systematically underrepresented in virtually all historical sources, we should by no means infer that this represents the proportion of women to men in these churches accurately. Further, since such sources tend to exhibit disproportionate interest in community leaders, we cannot even assume that these women provide a representative sampling.

Although Meeks's demographic data are, by his own admission, "fragmentary, random and often unclear,"[29] they are also quite suggestive. The churches so represented are urban communities, whose members seem to be clustered in the middle ranges of the Greco-Roman social scale. Of those whose occupation can be deduced, most seem to be artisans and small-scale traders. Some were slaves, some were of free birth, others perhaps had been born slaves but were subsequently freed. Some were heads of households; some were slave owners. Some seem to have had significant financial resources of their own; others seem to have been relatively poor.

If this portrait does not exactly bear out Celsus's subsequent description, it nevertheless suggests to Meeks an underlying pattern of status. By virtue of their birth, or their citizenship status, or their inclusion (or exclusion) from the various Roman *ordines*, many new Christians ranked low in the Greco-Roman prestige system. Often, however, the actual wealth of these persons exceeded their social prestige as measured according to these other categories. The dissonance, or discomfort, generated by this "status inconsistency" would have been significant.[30]

Meeks detects evidence of divergent status indicators for the women of this sample as well, several of whom seem to be of relatively low social prestige, yet are comparatively wealthy, such as Lydia in Acts 16 or Phoebe, whom Paul salutes in Romans 16:1 as the deacon of the *ecclesia* at Cenchreae, the port of Corinth. Though he does not pursue the issue fully, Meeks recognizes that the very independence of these women may have created a sense of dissonance in a culture where autonomous women were usually, though perhaps not always, anomalous.[31] Women like Lydia would appear to have attained such independence prior to joining a Christian community, whereas for others, their independence is only attested by their roles within the new community. This distinction becomes significant when we consider the possible motivations for women joining Christian churches.

The social status of most of these women seems similar to what Meeks observes for the men. Lydia of Thyatira resides in Philippi, where she deals in purple cloth, an expensive commodity in antiquity. On the basis of their names, which divulged considerable social data in the Greco-Roman world, Meeks proposes that two other women of Philippi, Euodia and Syntyche, were similarly resident foreign merchants. Junia, whom Paul calls prominent among the apostles in Romans 16:7,[32] may have been a freedwoman of the *gens* Junia, although Meeks correctly points out that this is not necessarily the case.[33]

What is particularly intriguing about this list is the absence of much information about those components that so affected the status of women: their derivative identity as daughters, wives, mothers, and even sisters of men. None of the women are identified by the names of their fathers (or mothers): filiation is also rare for the men in Meeks's sample.[34] Two women are identified by the names of their sons. In Romans 16:13, Paul salutes one Rufus and his otherwise unnamed mother, whom Paul calls "my mother also."[35] This apparently honorary designation may allude to this woman's role as a patron and supporter of Paul.[36] In Acts 12:12 the house of Mary, the mother of John also called Mark, is large enough to hold a gathering of "many" Christians, which suggests to Meeks that the family, if not Mary herself, had significant financial resources.[37]

Only one woman, Priscilla, is explicitly said to be married. Her name atypically precedes that of her husband Aquila in Acts 18:18 and 18:25.[38] Two other women mentioned in Paul's list of greetings in what is now Romans 16 are usually taken to be married. Meeks considers Julia to be the wife of Philologus (16:15) and Junia to be the wife of Andronicus (16:7).[39] In addition, Apphia, the second addressee of Paul's letter to Philemon, is usually taken to be Philemon's wife.[40]

Yet none of this is certain. In Paul's letters, Prisca (as he calls her) is clearly linked to Aquila: in 1 Corinthians 16:18 he conveys to the Corinthians the greetings of Prisca and Aquila and the church in their house. In Romans 16:2 he sends his own greetings to Aquila and Prisca. But Paul never explicitly identifies them as a married couple. It is only Luke who designates her as Aquila's wife. This additional detail is consistent with Luke's desire to demonstrate the respectability of Christians in general, and perhaps also to refute accusations that Christian women have excessive autonomy and lack the proper deference to male authority.[41]

Neither Julia, Junia, nor Apphia is explicitly said to be married or to be the wife of her putative husband. Apphia is called only "the sister," a term whose significance eludes us. "Sister" is a frequent designation for women members of Christian communities, as "brother" often designates men, but it may also, as D'Angelo points out, carry connotations of missionary partnership.[42] In Romans 16:15, Paul greets a woman he identifies only as the sister of Nereus, a relationship that is usually taken at face value because of the possessive pronoun, but here, too, as D'Angelo shows, it may allude to a missionary pairing that guarantees nothing about the biological or legal relationships between the two. In any case, the letter to Philemon has in fact three addressees: Philemon, Apphia, and Archippus (1:1). How they were related to one another, if at all, is simply not apparent in the letter.

Julia is similarly taken to be the wife of Philologus by virtue of the proximity of their names: "Greet Philologus, Julia, Nereus and his sister, and Olympas." Nothing, however, compels us to read the text this way, and conceivably we have here again two missionary teams, or perhaps even a larger grouping of five, analogous to the list of five men that immediately precedes it in 16:14.

What can we extract from this? In a culture that routinely identified free persons by the names of their fathers, the absence of filiation for both women and men (with a few exceptions) is intriguing. Obviously, all these persons once had biological fathers, but if they were slaves or freedpersons, they had no legal fathers, only owners or patrons, to whom they stood in a hierarchical relationship analogous, but by no means identical, to that of a freeborn legitimate child to its father. Further, in a culture where free men rarely entered into licit marriage much before they were thirty, many adults would not have had living fathers, least of all women who were themselves now the mothers of grown children. Additionally, many persons who joined the Christian movement did so over the objections of their families, both natal and marital, leaving them effectively fatherless, if not motherless as well. Thus the absence of filiation in our early Christian lists does not demonstrate much significant distinction between the experiences of Christian women and those of Christian men.

The absence of explicit evidence that these women were married requires a similarly nuanced inquiry. Under the prevailing Greco-Roman legal systems, licit marriage was the prerogative only of free persons, since it was closely linked with the transmission of property, particularly land, which only free persons could own.[43] Slaves could and did enter into sexual relationships and partnerships they intended to be permanent, but the children born to a female slave were the property of her owner, and slave families could be and often were callously separated at the whims of their owners. Since legally slaves had neither fathers nor husbands, if any of the women in our small sampling were slaves, we would expect to find them identified only by their owners, who could have been either men or women.

For Jews, the situation was somewhat different. Although the legal status of Jews in the various cities and regions of the Greco-Roman world is a complex matter beyond the scope of this book, in general, Jews were allowed to live by their ancestral laws (however those were understood), and surely that extended to the realm of marriage. This would hardly have protected Jewish slaves whose owners wished to separate spouses, parents, and children, but it would have meant that, at least from an internal Jewish perspective, all Jews could contract licit marriage, which had certain significant and enforceable implications within Jewish communities. It may not be insignificant then that the best attested marriage in our list is that between two Jews, Priscilla and

Aquila, although the suspect nature of Luke's portrayal makes even this somewhat dubious.

For many of the women and men in our Pauline sample, then, the absence of paternal identification or evidence for spouses may be related to their status as slaves and freedpersons. For the women explicitly identified as householders, such as Nympha of Laodicea (Colossians 4:3), Lydia of Thyatira, residing at Philippi, and Phoebe of Cenchreae, the failure to mention spouses may be more telling, and may point either to their status as widows or to divorce, whether instigated before or after their entrance into a Christian community.

Along similar lines, it would be rash to argue that because only two women are explicitly identified as mothers, very few of these Christians were parents. Yet the same circumstances that would have affected their ability to contract marriage would have affected their ability to have licit children, and over illicit children they often would have had little control, and perhaps even little contact.

All these factors prevailed by virtue of the cultural contexts out of which Christians came. The cultural constructs of the community into which they had moved would have further complicated these issues.

The Jesus movement began as a protest and renewal movement wholly within Judaism, one of many responses to the extraordinary conditions of political, social, economic, and religious crisis experienced by Jews living in Roman Palestine in the first centuries B.C.E./C.E.[44] It preached the imminent end of the current world order and the advent of the Reign of God, where the last would be first, and the first would be last. Consonant with these intense eschatological beliefs, the Jesus movement advocated a radical interim ethic that had far-reaching ramifications for social roles, including those associated with gender distinctions. We can see clear traces of this in sayings attributed to Jesus in the synoptic gospels of Mark, Matthew, and Luke, as well as in the Gospel of Thomas. The disciples of Jesus are repeatedly admonished there to renounce their families, occupations, and residence to follow Jesus.[45]

Material common to the Gospels of Matthew and Luke, but absent in the Gospel of Mark, points both to an early collection of sayings of Jesus and to a community that compiled and utilized this collection, which scholars call Q, from the German for *Quelle* (source). Several Q sayings have Jesus warn his followers that they must expect to sever their family ties, or at the very least to subordinate those ties to their loyalty and love for Jesus.[46] "Whoever comes to me and does not hate father and mother, wife and children, brothers and sisters, yes and even life itself, cannot be my disciple."[47] The Gospel of Matthew 19:10–12 attributes to Jesus a saying that legitimized sexual asceticism for the sake of the coming reign.

Such teachings were reinforced in a letter that Paul wrote to the church at Corinth in the fifties. Prompted by a question from the Corinthians that is difficult to reconstruct through his answer, Paul advised the Corinthians that celibacy was preferable to sexuality. For those unable to resist the strength of sexual desire, Paul conceded that one could have licit sexuality through marriage, though in light of our discussion about Greco-Roman marriage, one wonders exactly what he meant by the term, which he does use.[48]

Just why sexual asceticism was so prominent an aspect of earliest Christianity continues to engender scholarly debate. Some have seen its origins as pious imitation of Jesus himself, whose chastity must be largely inferred from the absence of any explicit evidence to the contrary in the canonical gospels.[49] Aline Rousselle has elegantly demonstrated the strong ascetic tendencies inherent in the general cultural environment of the Greco-Roman world[50] and their influence and effect on subsequent Christian asceticism. Still, I continue to see the specifics of Christian asceticism as a logical extension of earliest Christian belief in the imminent end of the universe as they knew it, which rendered normal social relationships, particularly marriage and childbearing, more or less irrelevant.[51]

For women, this constellation of intense convictions that the end of the world was at hand and that marriage, childbearing, and the transmission of property from one generation to the next were consequently no longer of any concern had major ramifications. As we saw in our opening discussion of Mary Douglas's cultural theory and women's religions, it is precisely when traditional divisions of labor according to sex and concerns for the transmission of property are invalidated that women stand to achieve significant parity with men. The negation of sexuality, marriage, and childbearing brought with it the possibility of expanded roles for women within the Jesus movement, including substantial participation in the public life of Christian communities. It also effectively freed at least some women from the control of husbands and fathers. Women who rejected marriage effectively refused to submit to the authority of a husband in the first place; women whose new beliefs impelled them to separate from both natal and marital families were now potentially beyond the control of their families and husbands. While the meager hard evidence, then, does not enable us to demonstrate significant differences between the circumstances of Christian women and those of Christian men, these observations suggest that the underlying social realities were meaningfully different for women than for men.

What motivated women to join the Jesus movement and subsequent early Christian communities, and what were the consequences of such decisions? To inquire about motivation is far more problematic than to inquire about consequences. Intertwined with our own perspectives, the very concept of motivation is thorny. In the second century, Celsus would contend that converts to Christianity were motivated by nothing impressive: superstition perhaps, or the inability to know any better. Christians, on the other hand, found questions of motivation essentially uninteresting, though early Christian popular writing such as the various apocryphal Acts of the Apostles or the later lives of holy persons suggests that the circumstances of conversion held considerable appeal for Christian audiences. For modern scholars, particularly Christian theologians whether feminist or otherwise, the question of motivation is troubling at a more explicit level. To inquire about human motivation seems to threaten the validity of particular theological explanations. To propose, as I have done in some of my earlier work, that some women became Christians because they were childless, did not wish to marry and bear children, or because the community they joined advocated beliefs that minimized the importance of children could be taken as an implicit challenge to the beliefs these women espoused. Nevertheless, the fact remains that in the first several centuries of what we now designate as "our" era, most people did not join the Christian movement. To then ignore factors of human motivation would be to jettison the inquiry entirely.

With the sole exception of Paul's enigmatic description of his call in Galatians 1, supplemented perhaps by his account of a heavenly ascent in 2 Corinthians 12:2–4, we have no identifiable firsthand accounts of those who joined the movement in the first century. Our only access to the motivation of such persons is by inference. For subsequent centuries, the situation improves somewhat, although we still have no firsthand conversion accounts known to be authored by women. Although we cannot definitely demonstrate the self-conscious motives of early adherents, the possible connections between the consequences of becoming Christian and the various levels of motivation warrant a detailed analysis of the consequences for women who made such choices.

What happened to women who joined Christian communities? Much depended, obviously, on the community they joined. Members of the church at Corinth in the mid first century C.E., whose practice of celibacy Paul commended, found themselves in a vastly different environment from members of the church presupposed by the author of 1 Timothy, probably in the mid second century C.E., who claimed that women would be redeemed from the sin of Eve through childbearing.[52]

While it is difficult to know just what women actually believed, the systems of belief that undergird the early gospels and the seven undisputed letters of Paul would have afforded women a flexible system in which departure from the normative standards for women was legitimized while the standards themselves were not wholly obliterated. The Jesus movement and the early Pauline communities supported the renunication of natal and marital ties, but left just enough room for those who were able to join without breaking such ties altogether.

Within the reconstructed Q sayings source, the female figure of Wisdom is prominent. Some scholars have proposed that the Q community would have advocated a theology particularly conducive to women by virtue of this more nuanced perception of the gender of the divine.[53] Amy-Jill Levine warns, however, that the earliest levels of Q traditions may not have supported the full inclusion of women into the mendicant lifestyle reconstructed for that community, which in turn warrants caution for any conclusions about the significance of a Q Wisdom Christology.[54]

What was the nature of the communities that women joined? The communities that coalesced first around the person and then around the figure of Jesus of Nazareth were characterized by strong group identity coupled with a rejection of status distinctions, of hierarchical structure, of ritual purity, and of social conformity, to name but a few of their features.[55] In Douglas's cultural model, they were quintessentially egalitarian (strong group, weak grid).

These communities afforded individuals substantial personal autonomy. In other low grid communities, this is frequently manifest in the form of choices of clothing, food, associates, and residence and the ability to dispose of time and goods at one's own discretion. In the synoptic gospels, not only are the disciples of Jesus repeatedly admonished to renounce their families, occupations, and residence to follow him, as we noted earlier,[56] but they are also are depicted as forming new associations, often with such social undesirables as the taxpayers and sinners of Matthew 9:10–11.

In the Acts of the Apostles, the emergent community of Christians comprises a new voluntary association that in many ways replaces the family and community its members have left. The core members of this new community are depicted as having taken up residence, at least temporarily, in a communal room in Jerusalem. Among them, at least initially, are the "eleven" (Judas having died a fittingly gory death), "the women," Mary the mother of Jesus, and Jesus's brothers.[57] All are portrayed as having abandoned their prior lives and occupations, choosing new uses for their time. The amount of time members spend together, their concern for group identity and boundaries, and other concerns typify the strong degree of group that characterizes early Christian communities.

Members of the new movement simultaneously rejected the classification systems typical of high grid. This includes the hierarchy of priesthoods (Jewish and otherwise), the hierarchies of Roman social class structure, and the hierarchies of gender. The baptismal formula that Paul quotes in Galatians 3:28 reflects this rejection: "As many of you as were baptized into Christ have clothed yourself with Christ. There is no longer Jew or Greek, there is no longer slave or free, there is no longer male and female; for all of you are one in Christ Jesus."

After the death of Jesus, women who joined a Christian community found themselves within a restructured version of the Greco-Roman household, both literally (to some extent) and figuratively. The letters of Paul demonstrate that early Christian communities were centered in the homes of individuals whose houses could accommodate such a gathering, although it is hard for us to say just how many persons gathered and just how large a house would have been needed. The familial terminology Christians adopted for members of the community is relatively egalitarian: sister or brother is the label of choice. There are no fathers in this terminology, and there may be no mothers as well, depending on exactly what Paul meant in Romans 16:13, when he called the mother of Rufus "mine also."[58]

The adoption of the familial metaphor almost certainly facilitated women's ability to reject their traditional roles. Earlier, we considered Riet van Bremen's interpretation of titles such as "mother of the city" as a means of legitimizing the roles of women benefactors in Greco-Roman cities. By rendering the public realm an extension of the domestic, women were able to exercise power and public functions without explicitly challenging the division of society into public and domestic realms, and the association of women with the domestic.[59] With the location of early Christian churches within private households and with the use of sibling terminology, early Christian communities seem to have brought the public into the domestic sphere,[60] and in the process to have further supported women's activity. As anthropologist Michelle Rosaldo proposes, this blurring of the distinction between the public and the domestic has enormous ramifications for gender roles and the values attached to them.[61]

What kinds of rituals did women encounter when they joined early Christian communities? As Meeks points out, we know relatively little about earliest Christian rituals. If the earliest followers of Jesus, in his lifetime, engaged in any ritual activity specific to the community, we cannot reconstruct it. They may have practiced baptism, but only the interpretation and not the act itself would have been unique to their community. Shortly after the death of Jesus, though, his followers appear to engage in some specific rituals while, at least according to Luke's portrayal in Acts, continuing to attend local synagogues and to worship at the temple in Jerusalem as appropriate.[62] We know that baptism was the rite of entry for both women and men, and that it was understood as the enactment of a ritual death and rebirth into a holistic community, which for Paul could be described symbolically as the body of Christ.

We also know that early Christians celebrated a ritual meal, the Lord's Supper, attested explicitly only in 1 Corinthians 11:17–34 and 10:14–22. This communal ritual meal commemorated the death of Jesus. As Meeks emphasizes, such celebrations were a well-known feature of Greco-Roman burial societies, which guaranteed all members a respectable funeral, and regularly gathered for commemorative banquets.[63] Customarily, the hosts of a Greco-Roman banquet provided better food and wine for some guests than for others, expressing and reinforcing social stratification. In his letter to the Corinthians, Paul berates them for using this ritual as an opportunity to express Greco-Roman status distinctions inappropriate for a community that constitutes, in Paul's eyes, the body of Christ.

Conceivably, women's presence and participation in the Lord's Supper were a source of conflict. In classical Greek culture, respectable women did not dine in public. Rather, Greek male citizens employed professional courtesans (called *hetairae*) and female musicians to amuse them at dinner parties. By the first century C.E., however, it had become acceptable for aristocratic Roman women to dine with their husbands, both at home and in public. In other quarters of Greco-Roman society, though, some taint may still have attached to women who dined at banquets.

In Acts 6:1, Luke describes a conflict between those he characterizes as "Hellenists" and "Hebrews," with the latter complaining that their widows were neglected in the daily distribution. Schüssler Fiorenza suggests that this passage reflects disagreement over women's presence at communal meals. At issue may have been women's right to serve, or perhaps to receive their fair share of the distribution. She proposes that the "Hebrews" may have found women's presence at public meals more disconcerting than the "Hellenists."[64] On all these issues, though, the text is unclear. Further, in light of our earlier discussion of Luke's redactional concerns, it seems wise to remain skeptical that this passage is historically reliable.[65]

Very early in the formation of Christian communities, the inclusion of Gentiles raised the question of continued observance of certain aspects of the Mosaic law. According to both Acts and Paul, the circumcision of male converts and the observance of kosher food laws quickly become points of contention. From an anthropological view, this makes eminent sense, since such practices are fine symbolic and actual manifestations of group boundaries.[66] The issue of circumcision would have affected women only indirectly, in the event they gave birth to sons, and the general ascetic tone of these communities lessened the likelihood of early Christian women becoming mothers. The rejection of kosher laws would have more impact on women's experience, since to the extent that women were involved in the preparation of food and the maintenance of cooking equipment and dishes, they would have borne the major responsibility for the enforcement and observance of such requirements.

Frustratingly, early Christian sources make no mention one way or another of the observance of menstrual purity regulations. We would expect Christians to reject these along with circumcision and food laws, since such regulations functioned as signifiers of purity and therefore of social boundaries, which Christians were in the process of redefining and broadening.[67]

Why, then, are they absent? Are our sources too delicate to mention such intimate matters? Assuming they had observed them to begin with, did Jewish women who joined a Christian community continue to observe menstrual and childbirth purity regulations? Were Gentile women instructed to observe such regulations? In part, the absence of any such discussion may be related again to the ethic of celibacy, which would have rendered the laws of menstrual purity largely irrelevant, since they principally governed the times when a man could have sexual relations with his wife. Celibate Christian women, whether Jewish or Gentile by birth, would have had little need to worry about such matters. On the other hand, what about those married persons for whom Paul advocates regular sexual satisfaction to avoid the dangers of illicit sexual pleasure, *porneia*? It would be hard to imagine that Paul instructed Corinthian Christian wives to observe the laws of menstrual purity, given his stance on these other issues, but the texts are regrettably silent.

If we cannot say much about what early Christian rituals were like, we can say something about what they were not like, particularly with regard to women. That is, there is no evidence that early Christian rituals addressed the specific life circumstances of women or that any early Christian rituals were gender-specific. This comes as no surprise, given the intense early Christian emphasis on sexual asceticism and its associated relative disinterest in women's roles as daughters, wives, and mothers. The putative monotheism of earliest Christianity and its absence of female divine figures (Wisdom notwithstanding) are likely to have been further factors in the absence of gender-specific religious rituals.

Released from many of the responsibilities and obligations associated with the roles of daughters, wives, and mothers, some early Christian women may have had both the time and the opportunity to become literate and to study. Two stories unique to the author of Luke-Acts hint at such possibilities. In Acts 18:26, the missioner Priscilla appears as a woman knowledgeable enough to correct the insufficient teachings of the eloquent Alexandrian Apollos. Luke 10:38–42 recounts Jesus's praise for a woman named Mary, who chooses learning at the feet of Jesus over her household responsibilities, in the face of objections and resentment from her sister Martha, who is left to do all the work herself. The story of Priscilla may imply some degree of literacy on her part (which conceivably she could have acquired before she joined the Jesus movement); the story of Mary and Martha has Jesus give explicit approval to women who reject traditional roles for discipleship.[68] "Mary has chosen the better part, which will not be taken away from her" (10:42). Both suggest that Luke's community would have found such women consonant with their own experiences.

Women's increased devotion to study and spiritual pursuits need not have been tied directly to any increased literacy, though. The earliest members of the Jesus movement relied extensively on oral tradition for the transmission of the teachings of Jesus and stories about his life, and even their knowledge and use of Jewish scripture may have been predominantly oral. Thus women would have had to acquire little if any degree of literacy to be active participants in the formulation, transmission, and interpretation of traditions about Jesus. The same, of course, would be true for men who joined the Jesus movement, many of whom are unlikely to have had any meaningful degree of literacy. But to the extent that women in the first century C.E., Jewish or not, were on the whole less likely than men to have been literate, the emphasis on oral tradition would have had a greater impact on women's opportunities to play a significant part in the emergence of Christian traditions.

Conversely, when access to authority and prestige in the community is related to the ability to read and interpret written texts, women can effectively be confined by their relative illiteracy and dependence on literate men to tell them both the contents and the meaning of authoritative texts. An increasing emphasis on written texts, including those understood to be Scripture, would ultimately have negative repercussions for women.

Along similar lines, the propensity of many early Christian communities for ecstatic experience and prophecy was integrally related to women's opportunities for public roles and for communal authority and prestige. Those few members of the earliest Jesus movement who had actually been disciples in Jesus's own lifetime were able to legitimize their claims to authority by virtue of that real acquaintance: this is the criterion set forth in Acts 1:21–22 for the selection of a twelfth person to replace Judas. But for many others, and perhaps even for those who did know Jesus in his lifetime, visions of the risen Christ and knowledge of teachings imparted secretly to the visionary became of particular importance, as did the personal experiences of spirit possession evidenced by speaking in tongues and prophecy.[69]

The very nature of such experiences meant that, at least in principle, anyone could have them. God (or Christ, or the Holy Spirit) could choose the recipient without regard to social class, literacy, or gender. So it is not at all surprising to find that one of the major religious experiences of early Christian women was that of ecstatic prophecy. From an anthropological perspective, this centrality of personal religious experience, often ecstatic, that dominated early Christianity is characteristic both of weak group and strong grid (the combination which, according to Douglas, describes most women's lives) and of strong group with low grid (the conditions that prevailed in many early Christian churches). The resistance that women's prophecy encountered in some circles in the form of attempts to regulate and verify such experience may be understood as pressure for increased classification and social control, particularly of women.

The earliest example we have of this is the situation at Corinth in the mid first century C.E. that Paul addresses particularly in the letter now known as 1 Corinthians.[70] From their readings of Paul's letters, many scholars believe that a group of *spirituals* (persons characterized, perhaps sarcastically, as the "strong" in 1 Cor. 4:10 and 2 Cor. 13:9) had emerged in Corinth who believed that in becoming Christians they had attained perfection and had recognized the fundamental separation of the body and the spirit (hence the label spirituals). The spirituals may have been divided into two factions: those who denied the body in order to live a spiritual life and those who claimed that their spiritual freedom made all bodily activity irrelevant and therefore permissible.

Regardless of whether all or only some Corinthians were spirituals, and whether there were then two camps of spirituals, denying the body effectively severed the identification of women with the body, fertility, childbirth, lactation, and child raising, leaving them free to assume new roles and statuses within the Corinthian community. Whether the Corinthian position was a function of intense eschatological expectation, a consequence of strong body-spirit dualism, or some combination of the two, the net effect for women was similar. As Wire explores in great detail, we may conclude from 1 Corinthians 7:1–40 that many Christian women had renounced sexuality.[71] The passage in 1 Corinthians 11:2–16, in particular, demonstrates that women at Corinth were praying and prophesying within the public context of Christian community and almost certainly deriving from those experiences a status, authority, and prestige that Paul found extraordinarily offensive and problematic. In these verses, Paul attempts to regulate the behavior of women prophets in the assembly, insisting that women should cover their heads when they pray or prophesy in these gatherings.

The argument of 1 Corinthians 11:2–16 has puzzled and frustrated commentators for centuries.[72] To begin with, Paul proffers an ordered hierarchical classification in which the woman's head is man; the man's head is Christ; Christ's head is God. The next portion of Paul's argument here is tortuous and opaque and commentators have agonized over it. Paul argues from the order of creation in Genesis 2:18–24, in which God creates first the male human, then the female from the rib of the male. Then he asserts that women should have *exousia* (which usually means power or authority) on their heads because of the angels. Perhaps the Corinthians understood the reference instantly, but modern interpreters do not. Conceivably, Paul alludes here to the sons of God in Genesis 6:1–4 who lusted after the daughters of men and fathered grotesque giants on them. If these arguments are not sufficient, Paul then throws in the appeal to nature and ends with a blunt attempt to assert his own authority: "If anyone is disposed to be contentious, we recognize no other practice, nor do the churches of God" (11:16).

Much scholarly discussion has been devoted to what Paul had in mind here—whether he was discussing head coverings or hairstyles, refuting pagan practices whereby men covered their heads when they sacrificed, or intentionally and symbolically attempting to exercise control and authority over Corinthian women prophets. Few have considered the ways in which Douglas's theory illuminates the significance of arguments about head coverings for women in ritual contexts.[73]

Let us leave aside whether it was common practice in late antiquity for "respectable" women (pagan and Jewish) to cover their heads when they went out in public, which it probably was.[74] Instead, let us remember that hair amply illustrates the use of the human body as a metaphor for the social body, offering a particularly effective medium for the expression of social constraints. As Douglas points out, the distance between the social body and the human body reflects the constraints of group and grid.[75] Hairstyles convey significant messages about degrees of social constraints.[76] While Douglas is careful to insist that the interpretation of hairstyles be undertaken relative to a particular culture, tightly bound styles seem to reflect tight social control; wild, uncontrolled hair is expressive of relaxed social control and protest. Elaborate, formal hairstyles may be typical of high grid, perhaps with high group: they are certainly what we see in the elite classes in Rome and Greece.[77]

It seems quite likely that here (as at many other points in his correspondence with the Corinthians, which are beyond the scope of this study) Paul opposes the extremes of low grid that the Corinthians exhibit, now in the form of uncontrolled hair. From an anthropological perspective, whether this passage pertains to head coverings or hairstyles is probably irrelevant: control and release are really what is at issue. Significantly, Paul has no interest in curbing women's ritual activity—their praying and prophesying are givens. The issue is the expression of control and subordination and the imposition of grid, of classification and hierarchy, here according to gender.

For Paul, what seems to be particularly offensive is that women, by praying with their heads uncovered, apparently do what men do. Conceivably, what upsets Paul is that the Corinthians have it all backward; the men are covering their heads and the women are not: the men have long hair and the women do not.[78] But the thrust of Paul's argument is the differentiation between what men do and what women do. I do not think this is a debate about the form such distinctions should take: I think Paul wishes here first to impose distinctions and second to dictate the form of those distinctions. (That this may be seen at odds with Galatians 3:28 is beside the point here.) For Paul, Corinthian practice is simply unacceptable, and the justifications he provides (hierarchial body metaphors, appeals to nature, and so forth) are predictable in their consonance with Douglas. Despite the difficulty we have deciphering Paul's logic, all these arguments, I propose, are characteristic of the higher classification that Paul both experiences and wishes to promulgate.

Paul's insistence on women covering their heads when praying and prophesying and his attempts to control the ecstatic religious behavior of the Corinthians in general are not unrelated. The Corinthians were exceedingly prone to various sorts of religious enthusiasm: they prophesy, they speak in tongues, they perform miracles. At least as Paul describes it, the entire attitude of the Corinthians toward religious behavior was what Douglas calls antiritual. The Corinthians fail to perform the Lord's Supper properly, and they are rather careless about other spiritual experiences. They do not worry about the details of head covering or hairstyle when they prophesy and pray; they are not too fussy about interpreting their ecstatic experiences for the benefit of the community, and so forth.

For Douglas, social constraints and religious ecstasy are integrally related.[79] The Corinthians' social situation was characterized by precisely the role confusion, intimacy, and informality that Douglas associates with low grid and that is reflected in a preference for ecstatic religious behavior. Though it was not,

Douglas's description of the phenomenon of effervescence might have been taken directly from 1 Corinthians.

> Emotions run high, . . . the favored patterns of religious worship include trance or glossolalia, trembling, shaking and other expressions of incoherence and dissociation . . . The movement is seen to be universal in potential membership.[80]

The Corinthians seem to have had a rather positive attitude toward trance and spirit possession (as long as that spirit was the Holy Spirit), an attitude Paul shares in general. Douglas suggests that communities that view trance positively and as a source of power and guidance for the community as a whole will be loosely structured, with group boundaries unimportant and social categories undefined, or where distant control exists but impersonal rules are strong. That is, they are likely to be found along a diagonal running from strong grid with weak group (insulation and isolation) to weak grid with strong group (egalitarian communities).[81] In other words, the Corinthians' predilection for trance reflects the social experience of the Corinthian church, which is loosely structured, has weaker group boundaries than Paul would like, and displays little in the way of classification and hierarchy. This, ironically, is one of the areas that Paul finds most difficult to deal with, because his own authority rests on the validity of spiritual experience, yet he is acutely aware of the dangers this poses to his ability to impose his authority on the Corinthians.

Paul's attempt to structure and control the religious ecstasy of the Corinthians by regulating the behavior of women suggests, of course, that one key feature of Corinthian enthusiasm and low grid was its lack of gender differentiation. It is hard to argue from Paul's letter that women were any more disposed to such activity than were men, but it seems clear that low grid at Corinth expressed itself in ways that gave women opportunities which Paul found problematic.

Wire offers the further suggestion that the conflict over head covering is closely related to the revised constructions of public and domestic space that prevailed among the Corinthians. Within the privacy of their own homes, respectable women would not have covered their heads. Taking with utmost seriousness the view of the *ecclesia* as the household, Corinthian women saw no reason to cover their heads within it. "[Corinthian women prophets] retain the house-church as their own space and signify that they are no longer determined by shame through sexual subordination but are determined by honor through the spirit as persons who have put on Christ, God's image not male and female. . . ."[82]

At Corinth, autonomy for women was manifest particularly in women's roles as prophets, that is, as conduits between the divine and the community, and was integrally related to the constellation of low grid, comparatively strong group, and sexual asceticism, whether rooted in a dualistic denial of the body or in eschatological rejection of sexuality and childbearing. Paul's distress over the autonomy of Corinthian women, particularly Corinthian women prophets, and over an attendant minimization of gender discrimination is clear enough from his otherwise unclear outpouring in 1 Corinthians 11:2–16. The conclusions we can draw from this passage may be limited, but they are quite significant: Corinthian Christian women prayed and prophesied in public, and Paul recognized their right to do so, despite his attempts to bring them under some control.

No analysis of Christian women at Corinth can refrain from some comment on 1 Corinthians 14:33b–36, those (in)famous verses that attempt to impose more serious and clear-cut restrictions on women's public participation in the Christian assembly. These verses are sufficiently difficult that no English translation can be neutral: no translation can avoid at least a preliminary interpretation of what it is women are not supposed to do and where they are not supposed to do it. According to the newest Revised Standard translation (NRSV), the text reads,

> As in all the churches of the saints, women should be silent in the churches. For they are not permitted to speak, but should be subordinate, as even the law says. If there is anything they desire to know, let them ask their husbands at home. For it is shameful for a woman to speak in church.[83]

Ancient commentators already recognized that the passage was in conflict with 1 Corinthians 11:2–16. In antiquity, one prayed and prophesied aloud, not silently, so how could women pray and prophesy in the assembly if they were forbidden to speak there? Some ancient exegetes solved the problem by arguing that 1 Corinthians 14:33b–36 applied only to married women, not to virgins. Others wrestled with the meaning of the term *to speak*, a tactic also taken by some modern commentators.

Several ancient manuscripts place these verses after 14:40, rather than in their current location. This suggests that even in antiquity, some people perceived that the verses break the flow of Paul's argument about prophecy where they are currently found. On the basis of this, the contradiction with 11:2–16, and other evidence, many modern scholars favor a simpler, though more radical solution. Paul did not write these verses: rather, someone else interpolated them into the text. From the location of the interpolation in Paul's regulation of prophecy, it seems most plausible that the interpolator intended to silence not just women in general, but women prophets in particular. Although we are unlikely ever to know the identity of the interpolator, we see in Chapter 11 that Montanist women prophets of the late second century C.E. are particularly likely to have been the targets of repression. Whether or not 1 Corinthians 14:33b–36 retrojects Montanist controversies into the first century, the high degree of probability that these verses are not Paul's own makes it unnecessary to discuss them further in any reconstruction of the experiences of Christian women at Corinth in the first century C.E.

Regrettably, none of this discussion yields a conclusive response to our earlier question: namely why Corinthian women became members of the church to which Paul writes his several letters. We have seen that in doing so, they attained a level of autonomy and prestige that we doubt they experienced prior to joining the community, and we have seen that their autonomy and prestige were closely linked to a severing of the connections between sexuality, childbearing, and the transmission of property. Yet we do not know enough about individual women at Corinth to draw firm connections between consequences and motivation. We cannot establish that Corinthian women were drawn to the Christian community *because* it promulgated a denial of women's traditional roles, but our inability to do so has more to do with the scarcity of the evidence than with the plausibility of such connections.

A constellation of second-century Christian texts enables us to construct a fuller and more complex portrait of the experiences of early Christian women. By the mid second century, the letters of Paul circulated widely among some Christian communities, where they were perceived as both enigmatic and ambiguous. The text of 1 Corinthians in particular provided support both for Christians who advocated asceticism and celibacy and for those who advocated marriage and childbearing as the appropriate norms for Christian behavior. These different norms had significant ramifications for women's autonomy and authority in early Christian churches.

Nowhere do we see this more clearly than in two letters addressed to Timothy in the name of Paul and in a work called the *Acts of Thecla*, sometimes also known as the *Acts of Paul and Thecla*.[84] The letters to Timothy advocate marriage and childbearing and call for the subordination of women to men, prohibiting women from teaching or having authority over men. The *Acts of Thecla* advocates abstinence and chastity, denies the validity of childbearing, and ends with a scene in which the apostle Paul commissions Thecla to go and teach the word of God. Both then attempt to make explicit and unambiguous what was previously much less clear, and both claim Paul as their authority.

The classic New Testament expression of misogynism, 1 Timothy 2:11–16 forms the basis of most later Christian restrictions of women, together with 1 Corinthians 14:33b–36. The author of Timothy writes,

> I desire then that women should dress themselves modestly and decently in suitable clothing, not with their hair braided or with gold, pearls or expensive clothes but with good deeds, as is proper for women who profess reverence for God. Let a woman learn in silence with full submission. I permit no woman to teach or to have authority over a man; she is to keep silent. For Adam was formed first, then Eve; and Adam was not deceived, but the woman was deceived and became a transgressor. Yet she will be saved through childbearing, provided they continue in faith and love and holiness, with modesty.[85]

In 1 Timothy, the proper sphere for Christian women is carefully delineated. Good Christian women keep their mouths shut, exercise authority only over their households and children and never over men, and generally confine themselves to the private, domestic sphere. When and if they become released from their household obligations by virtue of widowhood, they are not to avail themselves of the inherent opportunities for freedom, but are to continue to confine themselves to private prayer.[86] The text of 1 Timothy clearly evidences precisely the opposite behavior on the part of some Christian women and a compulsive concern to keep Christian communities in conformity with perceived Greco-Roman norms of ordered and orderly households.

Dennis MacDonald argues cogently that 1 and 2 Timothy and the *Acts of Thecla* represent different interpretations of the Pauline tradition (particularly 1 Corinthians) and different responses to the alternatives for women offered by earliest Christianity.[87] Although in its current written form, the *Acts of Thecla* may indeed be later than 1 and 2 Timothy, MacDonald argues convincingly that 1 and 2 Timothy allude to precisely the kinds of stories told in *Thecla* and were composed to wrest Pauline authority away from those who used Paul to legitimize the authority of women teachers and baptizers.[88]

The *Acts of Thecla* recounts the experiences and adventures of an upperclass pagan woman, Thecla of Iconium, a city in west-central Asia Minor. When Paul comes to Iconium preaching Christ and the denial of marriage and sexuality, Thecla is spellbound. Glued to her window for three days and nights, she listens intently to his teachings. After she renounces her engagement to Thamyris, a prominent citizen, her outraged fiancé arranges for Paul to be imprisoned. Undaunted, Thecla bribes her way into his jail cell, where she is discovered kissing Paul's chains, "united with him in loving affection."[89] Hauled before the authorities, Thecla endures the irate testimony of her horrified mother, who implores the governor to "Burn the lawless one! Burn her who is no bride in the midst of the theater in order that all the women who have been taught by this man may be afraid."[90] But a miraculous rainstorm quenches the fire of her intended pyre, and Thecla is allowed to go free. Reunited eventually with Paul, she and the apostle travel to Antioch, where a prominent citizen attempts to rape her. When she rebuffs him, she finds herself once again before the authorities and sentenced to die, this time battling beasts in the public arena.

This next portion of the story has particularly intriguing implications for women's relationships with one another in Greco-Roman antiquity. Fearing for her purity, Thecla convinces the governor to give her over into the custody of a wealthy woman named Tryphaena, who is once even called queen in the text. Not only does Tryphaena shield and succor the hapless Thecla, but Thecla prays for the eternal life of Tryphaena's own deceased daughter, Falconilla, whose salvation Tryphaena has seen in a dream. Not only human women, but female animals protect our heroine. A lioness licks her feet in a preliminary procession. When Thecla ultimately goes to fight the beasts, a lioness intervenes on her behalf, ripping a bear to shreds and battling with a male lion to the death so that Thecla might be safe. When the lioness fails to save her, the

women of Iconium intervene, throwing down such a profusion of flower petals
that the rampaging animals virtually fall asleep.

In a symbolically charged scene, the governor then agrees to tie Thecla
between fearsome bulls, who are prodded "from underneath with red-hot
irons at the appropriate spot that being the more enraged they might kill
her."[91] But a miraculous flame envelopes Thecla and burns the rope without
burning her, and Tryphaena faints dead away. At this, the governor concedes
that something more powerful is at work, and Thecla is again free.

In the midst of all this gruesome torture, however, one of two scenes significant for the subsequent history of the story take place. In between the
beasts and the bulls, fearing she is about to die unbaptized, Thecla throws herself into a pool of hungry seals and declares herself baptized.[92] The other takes
place at the end of the tale, when Paul commissions Thecla to go and teach the
word of the Lord. After forgiving her mother, Thecla goes off to the region of
Seleucia, where she enlightens many before her death.[93]

From the North African church father Tertullian, we know that by the end
of the second century C.E. the story of Thecla was used to legitimize women
baptizing, which Tertullian vehemently opposed. He attempted to discredit
the *Acts of Thecla* by attributing their authorship to a presbyter in Asia Minor
who had, so Tertullian asserts, admitted to the forgery.[94] MacDonald suggests
that the story of Thecla or ones very similar were already circulating when
2 Timothy was written, and that 2 Timothy contains strong allusions to such
tales.[95]

The connections between the epistles to Timothy and *Thecla* are apparent
at several points. The text of 2 Timothy asserts that Paul has run into severe
opposition in "Asia" (without any further geographic specificity), especially
from men named Phygelus and Hermogenes. He has, however, received support from Onesiphorus and his household and has been deserted by a man
named Demas, who has left him to go to Thessalonica.[96] Those who oppose
"Paul" teach "godless" things, specifically that the resurrection has already
come.[97]

In the *Acts of Thecla*, set in Iconium in Asia,[98] Paul is opposed by two men
named Demas and Hermogenes, who inform Thamyris, Thecla's fiancé, about
Paul's errors:

> we [Demas and Hermogenes] will teach you concerning the resurrection which
> he [Paul] says is to come: that it has already taken place in the children whom
> we have and that we are risen again because we have full knowledge of the true
> God.[99]

In contrast to his opponents, Paul teaches not only that the resurrection is yet
to come, but that asceticism and the consequent lack of offspring are the necessary prerequisites for salvation. He is also befriended by a man named Onesiphorus, and his wife and children.

The text of 1 Timothy also assigns false ascetic teaching to the opponents of "Paul." In a retrojected prophecy, the author reports that according to the Holy Spirit,

> in later times some will renounce the faith by paying attention to deceitful spirits and teachings of demons [*daimones*], through the pretensions of liars whose consciences are seared with a hot iron. They forbid marriage and demand abstinence from foods, which God created to be received with thanksgiving by those who believe and know the truth.[100]

Interestingly, Thecla's Paul not only denies marriage, but seems to adhere to dietary restrictions, as evidenced in a scene in the tombs outside the city, where Paul and Onesiphorus and his family eat a meal of bread, water, and vegetables.[101]

Of course, parallels and similarities do not of themselves indicate the nature of the relationships between two texts or two stories. There is a strong allusion in 1 Timothy to stories like that of Thecla as a provocation, an allusion that was obscured in the RSV translation of verse 4:7: "Have nothing to do with godless and silly myths."[102] As MacDonald rightly observes, the translation "silly" myths conceals the true meaning of *graōdeis* and offers a subtle modern commentary. For *graōdēs* more accurately means "like an old woman," and *graōdeis mythos* are more accurately "old women's myths/tales." The NRSV translation of "profane myths and old wives' tales" is closer to the Greek. But the point of this depiction, as MacDonald realizes, is not simply that these are "silly" tales, the ancient equivalent of foolish "old wives' tales," but rather that they really are stories told by women and for good reason.[103] MacDonald and others, myself included, have long argued that the story of Thecla and the many others like it in the various apocryphal Acts of the Apostles are women's stories: at the very least, told and read by women, and perhaps even (although this is much more problematic) written by women.[104]

In 2 Timothy 3:6 the author confirms the connection of the false and offensive doctrines with women, while asserting that women accept these teachings for lack of any critical discernment:

> For among them are those who make their way into households and captivate silly [literally, little] women overwhelmed by their sins and swayed by all kinds of desires, who are always being instructed and can never arrive at a knowledge of the truth.

MacDonald concludes that the Pastoral epistles on the one hand and the Thecla traditions on the other represent two vastly different interpretations of Paul's teachings, particularly as set forth in his correspondence with Christians at Corinth.[105] There, as we have seen, while reluctantly conceding the legitimacy of marriage, Paul argued for the priority of celibacy and asceticism over marriage and parenting, a preference that had significant implications for women's participation and authority.

By the early second century, two distinctive interpretations of Christian life emerged that are reflected in the texts at hand. Those who wrote, read, and promulgated the Pastorals favored marriage, social conformity, hierarchy, and structure, and bitterly opposed any leadership roles on the part of women. Those who told stories like that of Thecla and ultimately committed them to writing and circulated them favored asceticism, rejected social conventions, denied the value of contemporary hierarchy, and believed that women could baptize and teach just as men. Both sought legitimation and conceivably even derived their positions from the same writings of Paul, even though the battle was hardly only an exegetical one. Both camps were equally willing to fabricate in defense of their positions, hoping to clarify once and for all what had proved incredibly ambiguous in Paul's letters to the Corinthians.

MacDonald believes that those who advocated the rights of women to teach and even baptize were quite capable of formulating stories that supported their cause. The author of the *Acts of Thecla*, perhaps even the anonymous presbyter villified by Tertullian (routinely assumed to be a man, but not inconceivably a woman)[106] may not have been above composing the tale of Thecla to prove that Paul intended women to teach and even baptize.[107] The specific story of Paul and Thecla is almost certainly a fabrication. But there must have been women just like Thecla, who did deny marriage and childbearing, authority and hierarchy, who taught and baptized, who were accepted and revered by many, and who fully saw themselves within the tradition of Paul and his troublesome Corinthians. It is conceivable that behind the tradition of Thecla lies a historical woman, who ultimately came to be linked with the figure of Paul, but we have no way to determine this.

Tales like that of Thecla permeate the collections of stories told about the various apostles, known as the Apocryphal Acts, which achieved great popularity in the second and third centuries C.E. In the *Acts of Andrew,* Maximilla renounces marital relations, and manages to live chastely within the household of her husband despite his vigorous protests. In the *Acts of John,* Drusiana persuades her husband Andronicus to join in her celibacy. She nevertheless finds herself pursued by a perverse rapist who even attempts to defile her entombed corpse. In the *Acts of Thomas,* Mygdonia, wife of a king named Charisius, encounters stiff resistance to her renunciation of sexuality. These and many other stories detail women's desire for celibacy and the rejection of traditional roles assigned to women. Not infrequently they culminate in martyrdom of the apostle, whose death is brought about by irate powerful husbands and fiancés. The Apocryphal Acts do depict men who also adopt the continent life, but the implications and consequences for such men are vastly different. The resistence they encounter from irate wives and mothers is virtually nonexistent, and although the *Acts of Thecla* does contain accusations that Paul deters young men from marriage, it is always the conversion of women which prompts the death of apostles. If women objected to the celibacy of their male Christian relatives, they had much less political and social clout with which to express their opposition.

These multiple popular accounts of the conversion of women to ascetic forms of Christianity, retrojected from the second and third centuries back to the first, are not ever explicit about the motivations of these women, but we should not find this surprising. They could hardly have functioned as effective inducement and reinforcement for other conversions had they explicitly suggested that women became ascetic Christians precisely and only to reject traditional roles and obtain autonomy not otherwise accessible.

Yet a closer look at the portrait of the women in these stories reveals an intriguing pattern. Many appear to be of marginal or transitional status, particularly with respect to the classifications for women in the Greco-Roman world. Some aristocratic women are betrothed virgins, like Thecla: others are newly married, without children.[108] None are explicitly identified as mothers.[109] In the *Acts of Peter*, Agrippina, Nicaria, Euphemia, and Doris are all the concubines of Agrippa. The explicit self-understanding of those real women whose experiences are cloaked in the portraits of our fictional Christians may be beyond our reach. But the persistent themes of the denial of sexuality and childbearing and the rejection of women's traditional roles as dutiful daughters, wives, and mothers strongly suggests to me that ascetic forms of early Christianity were most appealing to women who either could not, or would not, accept the traditional standards of worth for women in their culture, and who found in ascetic Christian communities alternative standards and alternative roles that were infinitely more rewarding.[110]

In her ground-breaking feminist revisioning of early Christianity, Elisabeth Schüssler Fiorenza has critiqued this view. Consonant with the work of Meeks, she proposes that in the first few centuries, many Christian communities were appealing to women for their ability to address the specific status inconsistencies that certain women experienced. In particular, she suggests that many women sought access to the Greco-Roman prestige system by serving as patrons and benefactors of numerous Greco-Roman voluntary associations, yet found that their financial largess did not bring them the expected prestige and satisfaction they sought. In the Christian ecclesial household, by contrast, they found wholly different standards of worth that recognized their roles as patrons without denigrating them as women.[111]

I find it conceivable that membership in a Christian community did resolve certain social inconsistencies for both women and men. Ironically, though, I find this essentially the same kind of argument that I proposed in my earlier work: namely that early Christianity was appealing for its ability to provide people with standards of worth more consonant with their perceptions of the cosmos in which they lived. The details may differ here, but we are still essentially operating with models of deprivation and compensation. As I indicated earlier, such models often encounter resistance from scholars of early

Christianity for their negative theological implications. They seem to imply that if these are the reasons why people became Christians, then the ultimate theological truth of Christianity is called into question. Our divergent interpretations of the sources are then as much a function of our concerns about these issues (or the lack thereof) as they are about the sources themselves.

In the end, I am still convinced that not only were the consequences of becoming Christian, particularly an ascetic Christian, different for women than for men, but also that the complex reasons why many women chose to become Christian are likely to have been different than the reasons why many men did so. It makes eminent sense to speak of women's Christianity (or probably Christianities) in the first few centuries and to explore the ways in which women's experience of being Christian differed from those of men, without in any way advocating, therefore, that all women experienced Christianity as fundamentally the same.

Finally, from our analysis of the reconstructed experiences of Christian women in the first and second centuries, I propose within a few decades of the death of Jesus, early Christian communities may be mapped at two fairly different locations on Mary Douglas's grid-group diagram. The first-century Corinthian women prophets, with their comparatively strong group and minimal grid, exemplify the first location, a location that we see again for those communities reflected in the legends of women's conversions in the various apocryphal Acts of the Apostles.

The second location is spread across the strong grid half of the diagram, with some Christians (particularly women, children, and slaves) experiencing the constraints of relatively weak group and strong grid while others (particularly the male leadership) experienced comparatively strong group with strong grid. The concerns of the Pastoral Epistles strongly suggest that these letters emanate from communities fitting this description, whose authors themselves experience strong group and grid. The experience of Christian women in these differing communities varied in predictable ways. In relatively egalitarian low grid communities, women experienced increased autonomy, wider scope of public roles, increased access to education and information, decreased emphasis on childbearing and marriage, and so forth. Conversely, Christian communities spread across the high grid axis insisted that women (and men) conform much more to traditional gendered Greco-Roman norms and consistently sought to confine and constrain women. We might plausibly argue that this pattern continues through many centuries of Christianity, but at the very least, we now see that it continued through the remainder of the centuries under discussion here.

# — 11 —

## Heresy as Women's Religion:
## Women's Religion as Heresy

For Jerome, at least, the connections between women and heresy were already established in 2 Timothy[1] and are amply manifested in a long list of historical instances:

> It was with the help of the harlot Helena that Simon Magus founded his sect. Bands of women accompanied Nicolas of Antioch that deviser of all uncleanness. Marcion sent a woman before him to Rome to prepare men's minds to fall into his snares. Apelles possessed in Philumena an associate in his false doctrines. Montanus, that mouthpiece of an unclean spirit, used two rich and highborn ladies Prisca and Maximilla first to bribe and then to pervert many churches. Leaving ancient history, I will pass to times nearer to our own. Arius intent on leading the world astray began by misleading the Emperor's sister. The resources of Lucilla helped Donatus to defile with his polluting baptism many unhappy persons throughout Africa. In Spain, the blind woman Agape led the blind man Elpidius into the ditch. He was followed by Priscillian, an enthusiastic votary of Zoroaster and a magian before he became a bishop. A woman named Galla seconded his efforts and left a gadabout sister to perpetuate a second heresy of a kindred form.[2]

Jerome, it seems, was of the opinion that behind every heretical man was an heretical woman. It is in fact the case that among the many early Christian movements ultimately deemed heretical by their catholic opponents are a number in which women were apparently prominent, both as followers and as leaders. Jerome notwithstanding, women may not have been prominent in the majority of so-called heresies, but most movements we know to have been characterized by the prominence of women were ultimately judged heretical.

Among the more familiar of these was a movement first known as the New Prophecy, which arose in the middle of the second century in a village in Phrygia (modern west central Turkey), not far from the ancient major city of Hierapolis. The conflicting testimony of several ancient authors makes reliable reconstruction difficult, but some approximation of the origins, history, and nature of the movement is possible.[3]

157

The beginnings of the movement are uniformly attributed to a man named Montanus, who began to have strange ecstatic experiences and to prophesy.[4] According to an anonymous source cited by Eusebius, Montanus was a recent convert to Christianity. He was accompanied by two women prophets named Maximilla and Priscilla whom he had brought into the movement. Quoting extensively from a refutation of Montanus written by a man named Apollonius, Eusebius reports that the two women left their husbands to join Montanus.[5] Apollonius also claims that Montanus taught the annulment of marriage, established new fasts, appointed people to collect funds, organized the receiving of gifts under the name of offerings, and paid wages to his preachers.

Judging from the reactions it provoked, the New Prophecy garnered considerable support. Eusebius's anonymous source reports that he and other opponents of the movement lectured for many days in the town of Ancyra in Galatia, refuting the claims of the New Prophecy: that such a defense was deemed necessary attests to the strength of the movement.[6] In a confrontation between Maximilla and two "eminent men and bishops," Julian and Zotikos, the opponents of the New Prophecy were thwarted when supporters of another Montanist prophet named Themiso "muzzled [their mouths] and did not allow the false spirit which deceived the people to be refuted by them."[7] The author claims that the "faithful" in Asia met many times and in many places to examine the prophecies of the new movement, which they ultimately declared inauthentic. That many others did not share that judgment is reflected in his report that the New Prophecy was thrown out of the church and excluded from fellowship.[8] They were not, however, persuaded to change their minds!

While Eusebius's sources emphasize the central role of Montanus, an earlier heresiologist, Hippolytus, who lived in the second/third centuries C.E., stresses the role of Priscilla and Maximilla. Members of this Phrygian heresy, he claims, have been deluded by the two women. Though acknowledging Montanus as a prophet prior to Priscilla and Maximilla, they attribute their practices and beliefs to the prophecies and instructions of the two women, and magnify them above the apostles, and even, perhaps, above Christ.[9] According to Hippolytus, the Phrygians engaged in new fasts, feasts, meals of dry food (without water, oil, or other liquids) and of radishes. Numerous sources testify to books written by members of the New Prophecy, including Priscilla and Maximilla, who presumably wrote down their prophecies. Apollonius complains that Themiso dared to compose a "catholic" or general epistle in imitation of Paul.[10]

Exactly what members of the New Prophecy believed, particularly in contradistinction to other Christians in western Asia Minor, is a subject of dispute among scholars. Sayings attributed to Maximilla suggest an apocalyptic dimension to the movement. Her specific predictions of wars and cataclysms were ridiculed by opponents of the movement as evidence of false prophecy when they failed to come true.[11] A saying attributed to either Priscilla or a prophetess named Quintilla (who first receives mention in the fourth-century heresiologist, Epiphanius) appears to predict the descent of the heavenly Jerusalem at a site in western Asia Minor, which has led many scholars to characterize the movement as highly eschatological. Douglas Powell, however, finds the saying both in form and in content somewhat at odds with others attributed to Montanus, Maximilla, and Priscilla, and suggests that it represents the vision of a later prophetess, perhaps named Quintilla. Powell also notes the grammatical ambiguity of the saying. The Greek usually translated "Here Jerusalem *will descend* from heaven" can just as easily sustain the translation of "Here Jerusalem *descends* from heaven," transforming a future expectation into what Powell calls a "partly realized" eschatology.[12]

It is conceivable that unfulfilled predictions created difficulties for the movement after the deaths of its original leaders. According to the anonymous source in Eusebius, Montanus and Maximilla are both said to have committed suicide, though at different times. Nothing is reported on the death of Priscilla.[13] Powell suggests that a prophecy attributed to Maximilla "After me, there will be no more prophecy, but [only] the *synteleia* [the end, the completion, the fulfillment]" would have posed enormous problems for subsequent prophetic pretenders. Indeed, from numerous sources it appears that no prophetic leader arose immediately to replace the original trio of Maximilla, Priscilla, and Montanus. Powell proposes that "An obvious solution . . . was to produce the only *synteleia* possible, a spiritual Jerusalem descending wherever Montanists gathered together,"[14] and argues that this is precisely what we find in the saying attributed uncertainly to Priscilla or Quintilla.

The fate of the movement in Asia Minor is difficult to reconstruct, although it clearly persisted in some form. The account of Apollonius, as transmitted by Eusebius, may report the emergence of subsequent Montanist prophets and leaders.[15] Within a relatively short time the movement had spread at least to North Africa, for we know of it there both from the writings of Tertullian and an important text for women's religion, the *Martyrdom of Saints Perpetua and Felicitas*.

The preface of the *Martyrdom of Saints Perpetua and Felicitas* offers an effective apology for a point of view consonant with that of the New Prophecy. The writer demonstrates the divine basis of contemporary prophecy, quoting Acts 2:17–18, itself a paraphrase of Joel 2:28: "in the last days, God declares, I will pour out my Spirit upon all flesh and their sons and daughters shall prophesy. . . ."[16] The phrase "new prophecy" is specifically invoked: "So, too, we hold in honor and acknowledge not only new prophecies but also new visions as well, according to the promise."[17]

The preface explicitly acknowledges that the writer has incorporated the prison diary of a young woman named Vibia Perpetua, who was martyred on the emperor's birthday along with a slave named Felicitas and three other Christians: Saturninus, Secundulus, and Revocatus. The diary of Perpetua may thus represent the earliest *known* writing of a Christian woman.

According to the preface, Perpetua was the daughter of a respectable family, and both her parents were still alive. A recent convert to Christianity, she had two living brothers, one of whom had also become a Christian. From the diary, we learn of another brother, who died in childhood from cancer. Although the preface relates that Perpetua was newly married and still nursing her infant son, neither the preface nor the diary ever identifies her husband.

Perpetua's diary offers us fascinating insights into the conflicts between Perpetua and her family, particularly her father, over her decision to die as a Christian. During repeated visits to the prison, her distraught father implores her to renounce her Christianity and spare him and all her relatives the great pain of her loss. Perpetua reports that she grieves for him, but refuses to relent. Her own conflict centers on her distress for her infant son, which is finally relieved when the child suddenly loses interest in breast milk, leaving Perpetua free to concentrate all her energies on her impending martyrdom. Perpetua experiences no physical discomfort from this abrupt end to her nursing, and attributes both circumstances to divine providence.[18]

Apparently a gifted visionary, Perpetua asks for and receives several visions while awaiting death. In one, she and Saturus ascend to heaven on a bronze ladder, at whose foot an enormous dragon lies, and at whose sides weapons of all sorts are attached. Once safely in heaven, Perpetua encounters a gray-haired man in shepherd's clothing milking sheep in an immense garden. Welcoming Perpetua as his child, he offers her some of his milking. When she consumes it, all present say "amen," and Perpetua awakes with a sweet taste in her mouth.[19]

In a second vision, Perpetua sees her dead brother Dinocrates suffering the torments of the unfortunate dead, according to pervasive Greco-Roman beliefs: he is hot, thirsty, dirty, pale, and still bears the marks of the facial cancer that killed him. In front of the dead child she sees a pool of water, but the refreshment the dead seek is beyond his small reach. Day and night Perpetua then prays for her brother, until finally she has another vision, in which she sees him now clean and refreshed, a scar where the wound had been, drinking his fill from the pool of water and playing happily.[20]

Perpetua's final vision comes on the night before her death. Led by the deacon Pomponius to fight the beasts, she is astonished to find in the arena no beasts, but only an enormous Egyptian. Attendants come out to prepare her for the fight, and suddenly Perpetua realizes, as they strip off her clothes, that she has become a man. Miraculously successful against the Egyptian, Perpetua receives a green branch with golden apples from a larger-than-life trainer, who greets her once again as a woman. Perpetua concludes from this vision that it is not beasts she will combat but the devil himself, yet she will be victorious. On this note, the text of her diary concludes.[21]

To Perpetua's visions, the author appends a vision of Saturus in which he and Perpetua are implored to intervene in a dispute between a bishop named Optatus and a presbyter and teacher named Aspasius.[22] The remainder of the text details the deaths of the martyrs, who endure all with the expected faith and calm of the righteous.

The *Martyrdom of Perpetua and Felicitas* confirms both the centrality of prophetic vision in the New Prophecy and the prominence of women. In this account, the visions of Perpetua are given pride of place over those of Saturus, and the deaths of Perpetua and Felicitas predominate over those of their male companions. The absence of any mention of Perpetua's husband or of the father of Felicitas's baby may confirm the connections of the New Prophecy with sexual renunciation. Interestingly, none of the other martyrs is said to have been married.

Other explanations are possible. We may have here simply another instance of the family tensions that we know accompanied conversion to Christianity for many people. More compelling, though controversial, is the suggestion of Mary Lefkowitz that the relationship between Perpetua and her father typifies at least emotionally what can be demonstrated for incestuous fathers and daughters in the twentieth century.[23] If Lefkowitz is correct, we may even wonder whether there ever was a husband and whether the true father of the baby was Perpetua's own father. In addition to the evidence adduced by Lefkowitz, such an extreme reading is tentatively supported by the father's excessive concern for the child and his ultimate refusal to give the baby back to Perpetua, events that trigger the instantaneous weaning. If this reading is correct, then Perpetua sheds little light on the ascetic orientation of the New Prophecy in North Africa.

We might also wonder whether Perpetua's husband was opposed to her conversion and somehow instrumental in her fate. Within the Apocryphal Acts, numerous tales like the story of Thecla portray aristocratic ascetic Christian women as victims of torture and persecution instigated by their pagan husbands who bitterly resent their wives' denial of marital relations.[24] But if Perpetua's situation was similar, neither she nor her redactor chose to divulge this.

More information is forthcoming from Tertullian, a convert to Christianity who lived most of his life in North Africa and was a member of the New Prophecy around the turn of the second/third centuries C.E. His writings confirm some of the assertions in later sources, including Eusebius and Epiphanius, while apparently contradicting others. Some of his treatises expound and defend the movement's penchant for fasting, including total abstinence from food, modified fasts of dry foods (without meat, oil, or liquid), and fasts of vegetables.[25] The sympathetic and ingenious scriptural support he provides for these fasts and related rituals substantiates the evidence in anti-Montanist writers, while allowing us to see how members of the movement understood such practices.

Tertullian's treatises on monogamy, chastity, and modesty, all written during his adherence to Montanism, present a somewhat different portrait of New

Prophecy practices and attitudes regarding sexuality and marriage.[26] Eusebius's sources accuse Montanus of preaching the annulment of marriages, although Powell astutely points out that the only evidence offered by such sources is the separation of Priscilla and Maximilla from their husbands.[27] Although Tertullian admires celibacy, most of his treatises concerning sexuality focus on monogamy, and oppose not simply polygamy (which doesn't really seem to be the issue, although he addresses it), but remarriage after death of the spouse or divorce. If Tertullian accurately reflects the New Prophecy in North Africa at the turn of the third century, it was not a movement slavishly dedicated to celibacy and the renunciation of marriage, although sexual asceticism would certainly have been consonant with Tertullian's beliefs.

Neither Eusebius's sources nor Hippolytus provides any indication that women served in any leadership capacity other than that of prophet. Of course they do not testify to any other form of leadership in the New Prophecy for men, either. It is only a later report in Epiphanius, to which we return, that associates Montanist-type movements with women priests, presbyters, and bishops.[28] Yet both Tertullian and the *Martyrdom of Saints Perpetua and Felicitas* demonstrate that women's leadership and authority were issues among North African members of the New Prophecy. Tertullian's acceptance of women prophets is implicit in his description of a member of the New Prophecy who dutifully subjected the visions she experienced in church to the scrutiny of the community.[29] However, Tertullian is strongly concerned with the control of prophecy in this passage. Elsewhere, he is adamant that women may not serve in any ministerial capacities, and as we have seen, he lambasts those who use the *Acts of Thecla* as legitimation for such practices, acknowledging in the process that some churches were ordaining women, and were using the *Acts of Thecla* in this way.[30]

Many of Tertullian's writings are pervasively and almost viciously misogynist, so that it is difficult to assess the basis of his opposition to women's leadership. Some scholars believe that Tertullian was probably not yet a member of the New Prophecy when he wrote the treatises containing these arguments, so that their divergence from Montanist practice need not be a serious problem.[31] But nowhere in Tertullian's writings can we identify any evidence that women members of the New Prophecy in North Africa baptized or served as priests, presbyters, or bishops.

Tertullian's treatise against flight in the face of persecution[32] and his strong advocacy of martyrdom also contradict accusations by (later) church writers that Montanism and martyrdom did not go hand in hand. There is no evidence that the earliest Montanists were martyred, however, while there are those rather nasty tales that Montanus and perhaps Maximilla died a suicide.[33] By Tertullian's time, Christians in North Africa were much concerned with the appropriate responses to persecution. Was it permissible to avoid martyrdom by flight? Could one renounce the faith to save one's life, and then receive a second baptism? Persecutions in the third century, especially those of Decius in

250 C.E., would render these issues acute, but Tertullian is evidence that the problem was already substantial at the turn of the century. He argued against flight and against second baptism. If Tertullian was the editor of the *Martyrdom of Saints Perpetua and Felicitas*, as some have alleged,[34] it was not necessarily because he sought to advocate women's leadership, but because he saw a need to defend the movement against accusations that it dodged martyrdom by glorifying some New Prophecy martyrs.

The fate of the New Prophecy in the third century is obscure. The refutation of the Montanists composed in the fourth century by the now much maligned Epiphanius, Bishop of Salamis, drew heavily on a source that many scholars date to the second or third century. According to this unknown author, whom Heine believes wrote from Phrygia itself, the Montanists forbade marriage and prescribed abstinence from certain foods.[35] Lest anyone think that they did so to demonstrate their superior virtue, or some such thing, the author assures his readers that they do so because they consider abominable certain things which come from God.[36] He is obviously aware that much of what the Montanists teach resembles catholic positions and agrees that the "holy church" also glorifies virginity and celibacy and values chastity and widowhood, but it accepts sanctified marriage, condemning only *porneia*, *moicheia* (adultery), and *aselgeia* (impurity). The Montanists absolutely deny the possibility of second marriages.[37]

Epiphanius also includes descriptions of movements apparently extant in his own time. He reports that enclaves of Montanists live in Cappadocia, Galatia, and Phrygia, in Cilicia, and even in Constantinople, under the name Cataphrygians: he also reports that they go under the names Quintillians, Priscillians, and Pepuzians (after two prophets, Quintilla and Priscilla, and the village where the Montanists expected Jerusalem to descend to earth).[38]

The testimony of Epiphanius that interests us the most is his refutation not of the Montanists but of the related "heretics" he called Quintillians, Pepuzians, Artotyritai, and Priscillians. According to Epiphanius, not only did the Cataphrygians (a title under which he subsumes all these groups) revere Priscilla and Quintilla as prophets and founders of the movement, but women among them continued to prophesy and served as bishops, presbyters, and so forth.[39] They based their beliefs on fascinating interpretations of Scripture to which we shall return.[40]

A few references to the ritual activities of the Cataphrygians may be found in Epiphanius's testimony. Seven virgins carrying lamps and dressed in white often entered their assemblies to prophesy. They were given to ecstasy and weeping and to the use of bread and cheese, presumably in a Eucharistic ritual, hence their name *Artotyritai* (bread and cheese eaters).[41]

Scholars have offered a number of explanations for the specific practices and symbols associated with the Cataphrygians and Montanists, especially as reported by Epiphanius. Some have looked to the indigenous cult of Cybele, which does have some undeniable elements in common with these accounts of

Montanism. In Jerome's denunciation of the Montanists, who approached his disciple Marcella in 385 C.E., he calls Montanus "mutilated and emasculate."[42] Some have taken this to mean that Jerome believed Montanus to have been a Gallus, one of those priests of Cybele who underwent self-castration in ecstatic devotion to the goddess. The weeping of Cataphrygian women is compared to the mourning of women for Cybele's consort Attis.[43] The attire of Montanist prophets, which, according to one source in Eusebius included makeup and flamboyant dress, is remarkably reminiscent of descriptions of Cybele's priests, the *galli*, who were famous for their self-castration and other forms of bodily affliction while in a state of ecstatic possession.[44] The vision of Christ in the form of a woman attributed to a Montanist prophetess is explained as the natural transformation of Cybelean converts to Christianity, who were used to imaging the divine in the form of a woman.

But other scholars have denied any significant connections with the worship of Cybele, while acknowledging the centrality of the goddess in this part of the ancient world. Instead, they have suggested that the entire constellation of Montanist symbols may be explained as exegesis of Revelation. The vision of Christ as a woman in brilliant clothing is explained as similar to the appearance of Christ as the Church in the form of a woman "clothed with the sun" in Revelation 12:1.[45] The seven virgins carrying lamps are seen as indebted to the five wise virgins of Matthew 25:4, who carried oil with their lamps, but numbering seven to accord with the seven spirits before the throne and the seven candlesticks of Revelation 1.[46] An even more creative though highly improbable interpretation sought to locate the origins of Montanist practices and symbols in Judaism, or Jewish Christianity, seeing references to mourning and dancing on Yom Kippur in the weeping of Montanist women and connecting the repasts of radishes with the horseradish of the Jewish Passover.[47]

My own sympathies are with those who would revive discussion of connections with the worship of Cybele, not so much because Christian tradition is insufficient to account for the elements of Montanism but because I find it unbelievable that Christians living in Phrygia and neighboring regions through the fourth century or so would have been able to avoid the cultural contacts with Cybelean worship. The symbols of Revelation need to be discussed in this light; there is no question that the worship of Cybele far antedates Christianity in Asia Minor and may account for the imagery of the church as female in Christian communities living in this region. To view the connections with Revelation and the connections with Cybele as mutually exclusive alternatives is to ignore the cultural realities of western Asia Minor in late antiquity and a possibly fruitful avenue of inquiry.

Powell offers an intriguing suggestion that the origins of rituals utilizing bread and cheese may come from one of the visionary experiences of Perpetua described earlier. Entering heaven, Perpetua encountered a kindly gray-haired man milking sheep who gives her something to consume. Translated literally, the text is somewhat strange, and seems to imply that the milk was instantaneously transformed into cheese. "He called me, and from the cheese which he was milking he gave me a mouthful, as it were—and I accepted it with cupped hands and chewed it."[48] After Perpetua consumes this substance, all those present say, "amen." Powell correctly observes that no testimony to the use of cheese in Montanist rituals antedates *Perpetua*. Rather, we learn of it from Epiphanius in the fourth century. Powell therefore proposes that the use of cheese in Cataphrygian rites may be a deliberate emulation of the Perpetua vision, whose ritual nature is suggested by the affirmative "amen" of the onlookers.[49]

Given our primary interest in the New Prophecy and related movements as evidence for women's religion, there remains one ancient source worth consideration: a debate between a Montanist and an "Orthodox" Christian dated mid to late fourth century.[50] Although the work may well be a fiction composed by an orthodox Christian to refute the Montanist position, it is conceivable that the work represents an actual debate. Eusebius's anonymous source makes it clear that such conversations took place and were committed to writing,[51] and we certainly know that public debates were commonplace in the Greco-Roman period. The relative evenhandedness that characterizes the work buttresses the possibility it represents a real discussion.

In one key passage the Montanist asks his opponent why the Orthodox turn away from Maximilla and Priscilla and say that it is forbidden for women to prophesy, giving numerous scriptural examples of women who prophesied. The Montanist also points out that when Paul requires women to pray and prophesy with their heads covered, this of course implies that women may pray and prophesy. The Orthodox representative does not dispute women's right to prophesy; their position, rather, is that women may not speak in church or have authority over men, "as in the books written in their [Priscilla and Maximilla's] names."[52] Unable to adduce an acceptable letter in the name of Paul that speaks directly to women's authorship of books, the Orthodox speaker claims that writing books in their own names is the equivalent of women praying or prophesying with uncovered heads. To prove this, he offers the example of Mary the mother of Jesus, who could have written in her own name, but did not, because that would mean having authority over men.

When the Montanist asks whether Mary prophesied with her head covered, the Orthodox advocate claims that she had the evangelist for her head covering, since the Gospel was not written in her name. To this, the Montanist replies, "Don't give me allegory for dogma!" After this the discussion wanders for a while. The Montanist finally returns to the central problem, asking whether the real reason the Orthodox reject Priscilla and Maximilla is because they wrote books. The Orthodox disputant concedes that it is because of this, and also because they were false prophetesses.[53]

For all the difficulties we encounter in reconstructing the history of the New Prophecy, it is clear that women played significant leadership roles in the movement from its inception in the second century on, which was a source of contention between Montanists and their opponents. Leadership was by no means restricted to women. On the contrary, both women and men appear as Montanist leaders, because leadership and authority are conveyed especially by prophetic experience, which is equally accessible to women and men.

Before offering some analysis of these movements, we should look briefly at Epiphanius's testimony about the heretical beliefs and behavior of some Christian women in Arabia, whose ideas derived from heretical Thracians.[54] The connections between Phrygia and Thrace in antiquity are long-standing, and it is not inconceivable that these various "heresies" were in fact related in some way. Perhaps Epiphanius erroneously perceived these women, called Collyridians, to be distinct from the Cataphrygians, when they were in fact simply a group of Montanists.[55]

This second heresy Epiphanius attributes exclusively to women, who are, he asserts, easily mistaken, fallible, and poor in intelligence. The special error of these women is their worship of Mary, the mother of Jesus.

> For some women prepare a certain kind of little cake with four indentations, cover it with a fine linen veil on a solemn day of the year, and on certain days they set forth bread and offer it in the name of Mary. They all partake of the bread. . . .[56]

Earlier, he claims, "They attempt to undertake a deed that is irreverent and blasphemous beyond measure—in her name they function as priests for women."[57] His argument against the Collyridians is in equal parts a refutation of the possibility that women might legitimately function as Christian priests and a denial of the divinity of Mary.

If some truth underlies Epiphanius's testimony, in the fourth century some Christian women living in Arabia but of Thracian descent or with Thracian connections worshipped Mary, with women functioning in a priestly capacity. Unlike the Cataphrygians, whose rites and beliefs are traced to the visions of historical persons, the origins of Marian worship are not reported; nor can we surmise anything about those women beyond what they actually do, and even about that we know very little.

On the other hand, the parallels with women's worship of goddesses, particularly the Great Mother in Phrygia and other regions of Asia Minor, spring immediately to mind. Epiphanius connected these women and their worship of Mary with the women berated by Jeremiah for their worship of the Queen of Heaven, to whom they baked cakes and whose favor they sought in bringing fertility to themselves, their cattle, and their crops.[58] Stephen Benko believes that the worship of Mary was closely connected with the worship of the ancient Great Mother and that the women in Arabia might well have been recent Christians, who simply carried over their worship of the Mother, which was associated with Thrace as well. Pointing out the role of women as priestesses in the cult of Cybele and in worship of the Great Mother in other forms, he suggests that it might have been exceedingly logical for these women to go on performing such rites.[59] Valerie Abrahamsen has also suggested that women who played cultic leadership roles in pagan contexts would have expected to play such roles in their newfound Christianity as well.[60]

But the question that particularly concerns us is what percentage of members of the New Prophecy and subsequent Montanist churches were women, and whether the presence of women leaders necessarily correlates with the presence of women among the rank and file. Regrettably, literary sources offer little information about the actual makeup of Montanist churches. In fact, we might suggest that the movement was not unusually female in membership, since this would have been precisely the kind of detail heresiologists would have highlighted in order to further discredit the movement.

Epigraphical testimony is somewhat more intriguing in this regard. There are few surviving inscriptions that are explicitly Montanist.[61] Elsa Gibson argues for the Montanist identity of a small group of inscriptions from the museum at Uşak on the grounds that one of them commemorates a female presbyter (with the ambiguous name Ammion)[62] and that only the Montanists had female presbyters. Gibson also considers the probability that inscriptions dating from the third century C.E. in Phrygia containing the phrase "Christians for Christians" are Montanist, but finds the evidence frustratingly inconclusive.[63]

Nevertheless, it is worth discussing what these inscriptions tell us about the social location of some third-century Christians, particularly about the visibility of women. Like most ancient epitaphs, they are deceptively simple. A woman named Marcia commemorates her brother Loukios (Lucius), whom she calls *adelphoteknos* (he who loved his sibling) and his wife, Tatia. The bishop (*episkopos*) Diogas who commemorates the presbyter Ammion is himself buried by his wife, whose name is incomplete on the monument she established as the family tomb. Another bishop, Artemidoros, is also buried by a man named Deiogas, who does not call himself *episkopos*, and who may or may not be the same as the man of the other two inscriptions.[64] The names of these people are unremarkable for the area and the time. Several are married, including the bishop called Diogas. Of the eight persons mentioned whose gender is known, four are women, all of whom are known by their own names, although one name is only partly visible.

Judging from their funeral monuments, Gibson characterizes the dedicators of the "Christians for Christians" inscriptions as affluent and exuberant, partaking fully and unabashedly of the Greco-Roman cultural milieu of imperial Asia Minor.[65] Their names confirm their status as indigenous, free people: their ability to afford large, elaborate burial monuments suggests their economic status, and perhaps their intellectual and educational attainments as well, although this is less clear.[66]

If these inscriptions are Montanist, they fail to support the thesis of pervasive sexual asceticism. Marriage and childbearing is unquestionably the norm for these people. Husbands, wives, children, and grandchildren abound in these inscriptions. Large families are not uncommon.[67] If the reports of Montanist asceticism were more likely to be reliable, we might argue that the abundant evidence for marriage and family weakens the likelihood that these inscriptions are Montanist, but since those reports are questionable, this is not the case.

Women are visible and even prominent in these inscriptions, often commissioning inscriptions to their dead husbands. Conceivably, this reflects their tendency to outlive their husbands, reasonable enough in a society where fifteen-year-old girls often married men close to twice their age,[68] even when maternal mortality was a routine fact of life. This may seem a simple enough explanation of their presence, until we remember that at Rome the largest single category of burial inscriptions is that of husbands to wives. In contradistinction to many Greek and Latin epitaphs from elsewhere, the inscriptions from Asia Minor are usually family affairs: when someone dies, everyone in the immediate family (at least descendants) seems to get mentioned in the burial inscription.

Interestingly, however, the frequent mention of women in inscriptions from Phrygia and other regions of Greco-Roman Asia Minor is hardly unique to Christian epitaphs. In the absence of better controls, it is difficult to suggest that these inscriptions demonstrate the particular characteristics of Christians in general, let alone a specific Christian sect.

The question remains then to what extent Montanism constitutes women's religion. Apart from the ambiguity of the epigraphical evidence, the literary sources confirm the visibility of women in the promulgation of the story of Perpetua and Felicitas as exemplary; in the instance of the woman prophet whose visions Tertullian incorporates into his treatise *On the Soul* and in the development of a movement in which women are priests, bishops, and so forth, where those roles are legitimized through fascinating scriptural exegesis.[69] Nothing suggests that these movements are ever wholly women's movements; in fact, what may make them most interesting is not that their membership was limited to women, or primarily women, but rather that both men and women found compelling a movement in which women were not relegated to the background, but were significant, even central players.

It is significant that such a movement flourished in western Asia Minor, where the visibility of women can be documented in numerous other Greco-Roman contexts.[70] How this correlates with the ancient tradition of Mother Goddess worship in Anatolia remains to be investigated, as do the factors behind its success in North Africa where there is, for example, important evidence for women synagogue donors, although perhaps dating to a later period.

To comprehend the significance of the New Prophecy and other Montanist phenomena for women's religions, we should remember that the New Prophecy apparently emerged in the late second century C.E. as a renewal and revival movement within early Christianity analogous in some ways to the emergence of Christianity more than a century earlier. Just as the Jesus movement generated alternatives for women through its eschatological urgency and associated asceticism, so the New Prophecy and subsequent movements advocated beliefs and practices that similarly licensed the public participation of women in the life of the community. Although Powell demonstrates that the evidence for Montanist asceticism is much thinner than has been thought, the stories of Perpetua, Felicitas, Maximilla, and Priscilla, even perhaps Quintilla, reinforce our impression that autonomy for women is closely linked in early Christianity with separation from men, usually accomplished through sexual chastity. Sexual asceticism constitutes release from men's authority and also generates attempts to reinstitute such authority in one form or another. We are not surprised to find no tales of currently married women prophets here.[71]

The Montanist understanding of prophecy was a major component in the legitimation of such roles for women. Heine proposes that in Rome the prophetic claims of the New Prophecy and Montanism engendered substantively different opposition from that encountered in Asia Minor, opposition which I suggest had significantly different repercussions for women.[72] In Asia, the key issue was simply whether the Montanist prophets were true or false prophets, for which there were ample tests: conformity with biblical descriptions of prophets, conformity with apostolic prediction, and so forth.

In Rome, Heine argues, the whole notion of postapostolic prophecy was contested: the Roman church contended that there could be no prophets after the apostles. This conflict has important ramifications for questions about the formation of Christian scriptural canon, and, like many other issues, has subtle gender implications as well.

Heine is certainly correct that while Eusebius's Asia Minor sources (the Anonymous and Apollonius) worry about the tests of prophets, Hippolytus of Rome faults the Montanists for claiming to learn more from their prophets than from the law, the (biblical) prophets, and the Gospels (which seems to be shorthand for what he considers Scripture). Heine argues that the process of canonization created a severe dilemma for Montanists, for canon implied that no further divine revelation or prophecy could be forthcoming. He notes that the list of received works now known as the Muratorian canon denies a book called *The Shepherd of Hermas* precisely because it was recent and prophetic. Interestingly, Hermas receives revelations from an elder woman in shining garments holding a book in her hands.[73] Heine also points out that the Montanists were accused of writing "Scripture."

The point here is that in the face of this debate, Montanists began to use the Paraclete passages in the Gospel of John to support their contention that Christians could continue to expect prophetic revelation. Heine proposes that only when Christians in Rome opposed the very notion of continuing prophecy did the Montanists offer this particular reading of the Paraclete passages.

The arguments about women who write in their own names then emerges more clearly as part and parcel of the whole debate. This suggests yet again that Asia Minor, with its view of continuing prophecy, was a more hospitable climate for women, whereas orthodox Christianity in Rome was considerably more hostile. It would be interesting to know whether the implications for women were in some way already a factor. It would be appealing to argue that the gender implications of prophecy were less offensive in Asia Minor because of the long traditions associated with Cybele and with women's generally strong public status. In Rome, by contrast, license for women was yet another reason to deny prophecy. But such an interpretation would require careful consideration of the important evidence that elite pagan women had considerable public presence and authority in late republican and imperial Rome.

Equally at issue here are clear political ramifications, which Elaine Pagels has explored in depth.[74] The denial of prophecy guarantees a particular apostolic succession while the advocacy of continuing prophecy forever threatens a stable political order, and it is hard to believe that even in Asia Minor this was not an issue for those esteemed men and bishops. Pagels illuminates the role of arguments about beliefs in the bodily resurrection of Jesus in this regard. Bodily resurrection restricted resurrection appearances to a limited few, during a limited period of time, narrowing future claims to ecclesiastical authority to those few who could establish chains of transmitted authority between themselves and the recipients of those appearances. Denial of the bodily resurrection paved the way for unlimited continuing appearances, with attendant political ramifications for women as well as for men.

Plotted on the grid/group matrix formulated by Mary Douglas, the New Prophecy almost certainly qualifies as strong group with weak grid. As we have seen, the extant sources shed little light on women who joined the New Prophecy in Asia Minor other than Priscilla and Maximilla, but if our assessment of the movement as strong group and weak grid is correct, we would not be surprised to find a substantial number of women joining the movement and finding legitimation of expanded options for women in it.

When we meet with the New Prophecy in North Africa in the writings of Tertullian, we see some subtle differences that cohere with Douglas's theory. The North African New Prophecy lacks any emphasis on radical eschatology[75] and tempers asceticism, particularly sexual asceticism. Tertullian's writings, even those generally agreed to come from his Montanist phase, exhibit strong interest in male control over women and in the regulation of marriage. Here the movement, or at least Tertullian's representation, displays a shift diagonally upward toward stronger group and grid, which, particularly in the negation of asceticism and eschatology, bodes less well for women's options. Indeed, it is Tertullian who is explicit about forbidding women to baptize, teach, and have authority over men: even when he describes a woman prophet, he emphasizes the precautions taken to assure that her prophecy does not violate any of these restrictions.[76]

The picture of the New Prophecy gleaned from the *Martyrdom of Saints Perpetua and Felicitas* suggests a movement closer in social location and corresponding cosmology to that of the early Phrygian movement. As I explore in detail in Chapter 12, here we have evidence for a woman exercising the authority of a martyr-confessor: here there is at least an implicit asceticism in Perpetua's presumed separation from her husband.

In the attacks on the Montanists and associated heterodoxies by their so-called orthodox opponents, we also see the issues that characterize the intra-Christian debate between groups such as those at Corinth and the community of 1 Timothy: permissible sex, permissible food, appropriate forms of death, appropriate forms of worship, and gender discrimination, among others. Not surprisingly, the Phrygian movements propound attitudes and beliefs that typify their surmised strong group and weak grid, including their willingness to assign positions of prominence and authority to women, legitimized through the experience of charismatic, ecstatic prophecy and possession; their orthodox opponents line up predictably at high group and grid, relegating women in the process to the confines of low group and high grid.

We may see one final example of these conflicts in a fourth-century text that illustrates the so-called orthodox imposition of the same sorts of constraints. Falsely attributed to the authority of the apostles, *The Constitutions of the Holy Apostles*[77] contains regulations for Christian worship and for the appropriate behavior of widows, among others, that reveal the constellation of concerns typical of strong group and strong grid, in marked distinction from what we have seen for Christian communities characterized by their asceticism and sense of an imminent eschaton.

In the name of the apostles and elders, Christian worship is regulated here according to classic distinctions of ascribed status that Douglas associates with high grid. Age and gender are primary factors in determining who sits where during the worship service; for women, marriage and motherhood provide additional means of discrimination. Young men sit together, if there is room— otherwise they stand while older men take the available seats together. Younger women sit together, again if there is room, or stand behind seated older women. Priority of seating for women goes to older women, virgins, and widows, then to younger women. Married women with children are assigned space by themselves. Parents are expected to control young children who attend the services. Separate entrances are marked for women and for men, and proper seating is enforced by porters and deaconesses, who stand at the doors. The deacon is charged with rebuking those who sit in the wrong place and with moving them to the right place.

The text displays strong concern for control over behavior in church services. There is to be no whispering, no talking, no sleeping in church! Of course, we suspect that such regulations testify to the opposite: people do in fact whisper, talk, and doze off during dull sermons and readings, but the point here is the strong degree of concern for order and hierarchy that the text manifests, however successful it was in imposing such order.

Worshippers' participation in several rituals is similarly ordered by hierarchical rankings. The Eucharist is received "every rank by itself . . . in order."

Men give the Lord's kiss to men, women to women. Newcomers (catechumens and penitents) are excluded from the later portions of the service, including the Lord's kiss and the Eucharist. Women are to receive the Eucharist with covered heads, and the text manifests great concern for ensuring that outsiders do not inadvertently wander into the service.

Unquestionably the person or persons who composed the *Constitutions* and the community that received it were heavily and directly influenced by their reading of 1 Timothy. The admonition to pray for the whole world and for universal peace are reminiscent of 1 Timothy 2:1–2. Earlier portions of the work, devoted to regulations for widows and deaconesses, are even clearer in their reliance on 1 Timothy.

In the *Constitutions,* we see once more the relationship between strong grid and restrictions on women. Of particular interest in the *Constitutions* is what appears to be a deliberate omission of references to women prophets. When the author(s) paraphrases 1 Timothy on the prohibition against women teaching, he[78] claims that women may not teach in the church, but may only pray and hear those that teach. Although 1 Corinthians is explicit on the point that women pray and prophesy in church, the *Constitutions* omits any such mention.[79] In general, prophecy is of little interest in the *Constitutions,*[80] and where it is discussed, it is minimized. Those who prophesy are not superior to apostles, and those who receive the gift of prophecy should be humble. This accords with Douglas's theory, for communities experiencing strong classification (grid) with strong group are the least likely candidates for ecstatic prophecy. For Douglas, this is largely because ecstatic prophecy itself expresses weakened constraints, particularly of grid. We may also observe that ecstatic prophecy, which is believed to come directly from God and is therefore outside the control of the community, threatens to undermine the hierarchical transmission of authority from one generation of leaders to another presumed by the *Constitutions* (and western orthodox Christianity until the Reformation).[81]

The *Constitutions* does not simply interpret 1 Timothy (and to a lesser extent 1 Corinthians); it also introduces issues that those writings do not address, such as women performing baptism. In this regard, it stands in the tradition of debate reflected elsewhere in the *Acts of Thecla* and in Tertullian, among other sources, and, as we observed earlier, it testifies that the practice was still very much current when the *Constitutions* was composed.

The ideal community envisioned by the author(s) is clearly characterized by strong group and strong grid. It is apparent, however, that the author sees the appropriate place for women at a social location where they will experience all the constraints of strong classification, with less inclusion in the community than men. The classification of women according to their sociosexual status is one illustration of this.

For example, the *Constitutions* categorizes grown women in three groups: virgins, married women, and widows. Now of course, these three groups correspond to the real-life experiences of most women in antiquity—upon sexual maturity they may be categorized as virgins until they marry, whereupon they become married women, usually with children. Since the age discrepancy between husbands and wives at first marriage frequently meant that women would survive their first husbands (barring death in childbearing), many women were destined to become widows, at least for some time.[82] It is the Christian transformation of those social roles into religious categories that is particularly interesting.

We have already seen that early Christian renunciation of sexuality and childbearing enabled Christian women to achieve a significant degree of autonomy and to function in positions of community responsibility and authority that were difficult for other women to attain. Of course it is precisely virgins and widows, especially those who choose to perpetuate those roles for life (or for the remainder of their lives) who by virtue of such renunciation may achieve autonomy and exercise authority. In some early churches, orders of virgins and widows may emerge in recognition of the privileged status of such women. As we saw in the case of 1 Timothy, the development and articulation of orders of virgins and widows represents a clear attempt to regulate and control those women whose denial of traditional women's roles and responsibilities is likely to give them access to power and authority that the authors of 1 Timothy, the *Constitutions*, and numerous other works found dangerous, threatening, and offensive. Their response is to attempt to impose rigid structures of hierarchy and classification that negate the implications of virginity and chastity for ancient women.

So, in the *Constitutions*, not only are women enjoined from teaching or baptizing, but widows are explicitly told to be meek, quiet, passive, obedient, and so forth. They are to remain at home, singing and praying, fasting, spinning and weaving for the poor, and reading (which suggests, incidentally, either that they are educated or that they have educated slaves to read for them!). The image the *Constitutions* loves for widows is that of the fixed altar, which stays in one place, as widows are exhorted to do.[83]

# — *12* —

# *Women's Leadership and Offices in Christian Communities*

In the Greco-Roman period, virtually the only context for the discussion of women's religious leadership is women's Christian ministry. Aside from rabbinic discussions about why women cannot discharge cultic obligations for others,[1] no ancient Jewish sources debate whether women can hold synagogue office. Perhaps not surprisingly, the subject never comes up in pagan literature either, where women's right and ability to hold religious office is never in dispute.

Only among Christians is women's religious leadership an issue. Only Christians both attempt, sometimes successfully, to exclude women from religious office and community authority and argue about it. In fact, the whole issue of women's leadership roles in early Christianity cannot be divorced from the larger framework of conflict over women's roles and alternatives that we explored in Chapter 11.

Feminist scholarship of the last several decades has demonstrated that women clearly exercised leadership, in the modern sense, within the earliest movement around Jesus of Nazareth; evidence for that leadership has been largely and intentionally suppressed.[2] Women such as Mary of Magdala were key members of the circle around Jesus of Nazareth, and they subsequently played instrumental roles in the growth and development of the early Christian movement.

Financial patronage for the Jesus movement came from both women and men. Like their male counterparts, the women who funded the Jesus movement almost certainly did so within the context of the ancient benefactor system. They were of independent financial means sufficient to support a considerable number of itinerant Jewish charismatics, and they are likely to have expected prestige, respect, and honor in return, although presumably not in the form of statues and decrees. Schüssler Fiorenza and D'Angelo both argue, though, that the reference to women patrons in Luke 8:2-3 is intended there as a subtle denigration of these women, assigning to them *merely* the role of financial supporters.[3]

In the letters of Paul, we find ample evidence that women continued their

174

roles of patronage and leadership in the early missionary phases of the Christian movement following the death of Jesus. We see a multiplicity of roles played and functions performed by members of the early churches, assigned, so Paul believes, according to the gifts each one possesses from the Holy Spirit. In his so-called first letter to the church at Corinth, Paul even ranks these appointments: "And God has appointed in the church first apostles, second prophets, third teachers, then workers of miracles, then healers, helpers, administrators, speakers in various kinds of tongues."[4]

Paul acknowledges women prophets[5] and at least one woman to whom he accords the title apostle, Junia.[6] But as we have seen, he found women prophets problematic,[7] and he may also have had similar responses to women teachers. At the very least, women prophets (and presumably male prophets as well) troubled Paul because they derived from their ecstatic, prophetic experiences status, authority, and prestige that threatened his own. Paul must have found himself in an extraordinary bind, for his own claims to leadership and authority were located not in the traditional apostolic claims to knowledge of the earthly Jesus, but in a revelation from the risen Jesus and in the continuing experience of precisely such prophetic ecstasy.

The controversial passage 1 Corinthians 14:34–36b might seem to suggest that something more than the authority of prophets is at stake. It has been read as evidence that in other Pauline churches in the early decades of the Christian movement, women were already excluded from positions of cultic leadership. At Corinth, however, women were participating in public in a way which threatened that practice.

If this passage is authentic, which, as we have seen, is questionable, or if it is an early interpolation, we may then take it as unambiguous evidence for the presence of women leaders in the Pauline churches, which probably extended to whatever cultic acts early Christians performed.[8] But if MacDonald is correct that 1 Corinthians 14:33b–36 was inserted into the letter in the late second century, probably as a rebuke against Montanist women prophets, the passage loses its utility as historical source for the first-century Corinthian church.[9] Such leadership was vigorously defended in some circles and vehemently opposed in others, as we have seen in our earlier discussion of 1 and 2 Timothy and the *Acts of Thecla*.

Within the New Testament itself, there are several other examples of women leaders in early Christian communities. One is the woman known as Prisca or Priscilla who, together with her (husband) Aquila, was a key member of Paul's circle.[10] Some clue to her importance may be extracted from the fact that both the author of Acts and Paul himself sometimes refer to the couple as Priscilla and Aquila, rather than vice versa.[11] They are credited in Acts with correcting the insufficient teachings of the Alexandrian Apollos, and depicted as instrumental in the early Christian missionary movement.[12] Priscilla and Aquila may have been active in the missionizing effort even before they met Paul, although Acts tends to link them with the apostle.[13]

Two other intriguing examples have received much less scholarly attention than Mary of Magdala or Prisca. One definitely and one probably comes from that region of the ancient world where women's leadership was particularly successful and controversial, western Asia Minor. The author of the book of Revelation admonishes the church at Thyatira for following a woman whose name the author obscures behind the hostile pseudonym of Jezebel.[14] Thyatira was also the hometown of a seller of purple goods named Lydia, who according to Acts met Paul in the city of Antioch and may have established a church in her house.[15] The author accuses "Jezebel" (who calls herself a prophet) of teaching Christians to practice sexual immorality and to eat food sacrificed to pagan deities.[16]

As Adela Collins observes, we have seen this cluster of concerns before in Paul's correspondence with the Corinthians.[17] She suggests that in Thyatira, as elsewhere, women prophets were associated with controversial teachings which have substantial implications for women's history. Among other things, Collins points out that eating food sacrificed to idols is as much a social question as anything else; it allows Christians to partake of meals with their non-Christian neighbors whereas abstaining circumscribes the social parameters of their lives. For Mary Douglas, strict rules regarding permissible food and sex are typical of strong classification, and often associated with strict control of women as well. We may surmise that for the author of Revelation, more is at stake than doctrinal differences. It is interesting, however, that the author does not quarrel with the right of women to prophesy, nor adduce 1 Timothy or even the authority of Paul against the right of women to teach, although the verb used to describe Jezebel's activity is *didaskein* (to teach). This suggests that the author is only offended by the content of Jezebel's prophecy and teaching, and not the mere fact that she engages in either. It may also imply that the author does not know a Pauline tradition against women teaching (perhaps because it has not yet been composed!) or does not accept such a tradition.[18]

Our second example, also associated with the Johannine tradition, is even more problematic than the prophetic woman leader at Thyatira. The salutation of the letter known as 2 John begins with the phrase, "The elder to the elect lady and her children." The letter itself is quite brief and concerns itself principally with a warning about false teachers, who should not be received into the recipient's house.

At the turn of the second/third centuries C.E., Clement of Alexandria wrote the following description of 2 John:

> The second epistle of John, which was written to virgins, is (the) most simple. It was written in truth to a (certain) Babylonian woman, by the name of Eclecta, which signifies, moreover, the election of the holy church.[19]

It would seem that Clement thought the addressee female, and that the woman's name symbolized the election of the church. Most modern scholars, however, have denied that this letter is addressed to a specific woman who was the head of a house church, probably someplace in Asia Minor.[20] Instead, they have interpreted the term *Kyria*, lady, as a metaphor for a church itself. In support of this argument, they have pointed to known instances where the church (itself feminine in the Greek—*hē ecclēsia*) is symbolized by a female figure[21] and to the closing of the letter that reads, "the children of your elect sister greet you," which they interpret as a reference to a sister church. They have also argued that the term *eclectē* (elect or chosen) is more likely to be applied to a Christian community than to a person.

As a few scholars have recognized, none of these arguments suffices to resolve the matter.[22] Romans 16:13 applies the epithet *eclectos* to one of Paul's co-workers, Rufus. Conceivably, the recipient of the letter was named Kyria, which is well attested in this period, like Martha in Aramaic and Domna in Latin. The opening of the letter would then conform closely to the greeting of 3 John that reads, "The elder to the beloved Gaius."

One argument for considering the letter addressed to a church rather than a person is the apparent use of the term *tekna* (children or offspring) for members of the community. In verse 1, the elder professes to love all the lady's children in the truth, which is somewhat ambiguous, but in verse 4 he writes that he rejoiced to find some of her children following the truth, which seems to refer to members of the group. Since the children are metaphoric, scholars have concluded the lady must be as well. But the lady could be quite real, even though the term *children* refers not to her biological offspring, but to the community she leads, probably in her house (verse 10).

We will probably never know for certain the true addressee of 2 John. But behind all of the scholarly debate lies an implicit and unaddressed problem that only Elisabeth Schüssler Fiorenza has broached, namely the dilemma posed by a letter explicitly written to a woman as the head of a church in Asia.[23] If we do not begin with the assumption that such a letter is improbable, we are more likely to concede that such a reading is probable. Finally, we might remember if we are correct that the author of Revelation was incensed at the doctrines of a woman prophet in Thyatira, but not at her leadership per se, it should not be surprising to find another woman leading a church, this time one the author of Revelation might have found eminently more acceptable, to whom an epistle warning of false teachings (perhaps like those of "Jezebel") is addressed.

Evidence for women's leadership and its attendant conflict comes also with the flourishing of Montanism, or the New Prophecy, in villages, towns, and cities of Asia Minor in close proximity not only to Thyatira, but to Thecla's imputed hometown of Iconium. We have previously examined the testimony that Priscilla and Maximilla shared equally in the leadership of the group, or even surpassed Montanus in their authority and prestige.[24] Since these reports come from opponents of the movement, it is conceivable that the attribution of great authority to the two women is intended less as historical fact than as denigration of the women and their followers. When, for example, Hippolytus attributes innovations of fasts and feasts, including meals of radishes, to the teachings of Priscilla and Maximilla, it is hard to know whether this reflects their authority within the community or Hippolytus's insult (or conceivably even both!).[25]

The leadership of all three, though, Priscilla, Maximilla, and Montanus, was prophetic and charismatic: it was derived from their prophetic experiences and from the communal acceptance of their utterances as divinely inspired. The same may be said for the handful of others explicitly named in the anti-Montanist writers: the prophets Themiso and Alexander, Quintilla, and possibly another unnamed female prophet. Whether it was associated with any specific office seems unlikely by the very nature of the movement.

It is difficult to determine what cultic offices existed among the Montanists. We have examined Epiphanius' report that groups he associated with Priscilla and the otherwise unknown female prophet named Quintilla had women bishops, presbyters, and so forth.[26] Whether Epiphanius's description applies to the second century is almost impossible to tell: some scholars think that Quintilla was a later prophet and that this passage cannot be safely read back into the second century.[27] On the basis of Epiphanius, Elsa Gibson has argued for the Montanist identification of the third-century C.E. epitaph of a woman presbyter from the general area in Asia Minor where Montanism flourished.[28] It reads simply:

Diogas the bishop to Ammion (feminine) the presbyter/elder, in memory.

Gibson suggests that were it not for Epiphanius, we would not have been able to recognize this as Montanist, an observation in itself methodologically interesting. Presumably she would otherwise have thought Ammion to be male (since the name can be either and male presbyters pose fewer methodological problems), considered Ammion to be an "old" woman, or concluded that there were sometimes women presbyters among the "orthodox." Gibson is not surprised that the monument contains no "communion paten in a wreath above a table," for she asserts that administration of the Eucharist was not one of the duties of a female presbyter.[29]

References to Quintilla may shed some light on leadership within Montanist communities. If Quintilla was in fact a later prophet, perhaps even well after the deaths of Maximilla, Priscilla, and Montanus, we could conclude that charismatic authority was still the preferred mode. This seems to be substantiated by the attacks on Themiso and Alexander, who seem to have lived later. At the very least, we suspect that female leadership continued within Montanist churches in Asia Minor for some time.

Women's leadership in the New Prophecy in North Africa is more problematic. Tertullian, our major source for the movement there, clearly opposed women baptizing and teaching, although he acknowledged their prophetic gifts, provided, of course, that they submitted their prophetic experiences to the proper procedures for verification.[30] Whether other members of the New Prophecy shared Tertullian's aversion to women leaders is less clear.

We may draw some tentative conclusions about women's leadership in the

New Prophecy from the *Martyrdom* of *Saints Perpetua and Felicitas*, where we see Perpetua as a leader of the Christian community in several respects. She is first of all a visionary, who not only receives visions by divine volition, but has the ability to request them.[31] Such an ability enables her to learn the future and therefore to guide the community, at least those imprisoned with her.

Perpetua's leadership extends to the earthly, historical realm as well. The redactor of the *Martyrdom* appends to her diary the story that it was because of Perpetua's intervention and persuasion that the imprisoned Christians were treated with dignity on two occasions.[32] First, Perpetua convinces the Roman military tribune to improve their conditions while awaiting death, arguing that the Christians will offer a spectacle more befitting the occasion of the emperor's birthday if they are in better physical condition. Acceding to her requests, he allows her brothers and others to visit the prisoners and dine with them. Second, on the day they are to die, Perpetua holds the tribune to an earlier promise that the Christians would not have to wear the dress of pagan priests and priestesses. Twice, then, Perpetua is portrayed as the effective speaker on behalf of the group, a group that included numerous men, including the noble Saturus, whose death is also glorified in the *Martyrdom*.

Further, in a vision attributed to Saturus, he and Perpetua are asked to intervene in a quarrel in heaven between the bishop Optatus and the presbyter and teacher Aspasius.[33] It seems that Saturus, Perpetua, and the other martyrs will only be able to rest if the two are reconciled. Optatus and Aspasius even throw themselves at the feet of the martyrs, who express amazement at this seeming reversal of status: "Are you not our bishop and are you not our presbyter? How can you fall at our feet?" Perpetua then speaks with them in Greek (the *Martyrdom* is extant in Latin), but the text is ambiguous as to the outcome.

At least one scholar has suggested that beneath the story of Perpetua we may detect traces of a fascinating debate about leadership and authority, especially as it pertained to women. Frederick Klawiter proposes that it was precisely the value the New Prophecy placed on martyrdom that accounted for its willingness to accord women ministerial status.[34]

Klawiter points out that through the second century, Christian martyrs were presumed to have great powers while awaiting death, including the "power of the keys," that is, the power to forgive Christians who had denied their faith under pressure and therefore fallen from grace. Since this particular power was also the prerogative of bishops/presbyters, "anyone who exercised such power was thereby demonstrating a ministerial power."[35] He adduces the testimony of Hippolytus that by 190 C.E. "a male confessor [a person awaiting martyrdom by virtue of having publicly confessed to being Christian] released from prison automatically had the status of a presbyter in the Roman church."[36] While the evidence is only implicit, Klawiter also suggests that released confessors in the New Prophecy attained similar formal rank.

In particular, he points to the visions of Perpetua concerning her younger brother Dinocrates, who had died of cancer at age seven. Seeing Dinocrates at first dirty and thirsty, with his cancerous face, she prays that he might be pardoned on her behalf. In a subsequent vision, she sees the child clean, healed, and refreshed, and realizes that he has been saved through her intercession.[37] From this story, Klawiter concludes that Perpetua, as a confessor, held the power of the keys, and therefore ministerial rank.

Klawiter further argues that the assignment of ministerial rank to confessors created a serious problem precisely by virtue of its implications for women. Perpetua and Felicitas were hardly unique in their status as martyrs: the Roman government rarely saw reason to spare Christian women from torture and death.[38] But if confession and imprisonment brought ministerial status by their very nature, surely that extended to women. Klawiter suggests that the *Letter of the Churches of Lyons and Vienne*[39] (which describes the martyrdom of a woman named Blandina and her three male companions, and which Klawiter attributes to Irenaeus) intentionally draws careful distinctions between confessors and martyrs

> to undercut the authority of released confessors who claimed the title of martyr and the power of binding and loosing intrinsic to such status. Irenaeus' point must be that only imprisoned confessor-martyrs have the power of the keys; once a confessor is released, he or she can no longer claim the title of martyr or the power inherent in such a title.[40]

Klawiter believes that one of Irenaeus's concerns must have been the implications for women's ministry, which the New Prophecy took much more seriously. His provocative essay concludes with the suggestion that women's ministerial status in the New Prophecy derived precisely from the priestly power they obtained as confessors and retained when released.

Klawiter's analysis is susceptible to criticism at several points. In order to explain why the New Prophecy allowed women to retain the power of the keys upon release whereas "orthodoxy" did not, Klawiter must assume a prior disposition toward women's ministry. Even so, it is conceivable that the distinction drawn between confessor and martyr *is* designed to combat the authority conferred on women through the experience of imprisonment. It also seems to me that the evidence for women's ministry as Klawiter conceives it is somewhat circular. The early evidence, after all, is mostly drawn indirectly from Perpetua herself and from the late testimony of Epiphanius, which may or may not reflect second- and third-century practices. But it is difficult to use Perpetua to explain practices that must be inferred from her *Martyrdom*. Klawiter is, I think, too certain that women held ministerial rank, as he understands it, in the New Prophecy. But unquestionably, they exercised great leadership and authority, and that itself is significant.

When we attempt to peer behind the veil of male Christian polemic against women's leadership to ascertain the extent to which women did in fact hold specific religious offices, we find the evidence meager, and the interpretive dilemmas substantial. A number of Christian women are known to have held titles of ecclesiastical office, including those of deacon, presbyter, and teacher.

Our efforts to understand precisely what these titles meant have been hampered by the tendency to interpret the evidence in conformity with ancient orthodoxy, downplaying the significance of women's roles and assuming that any evidence for women officers other than deacons *must* come from heterodox communities.

## Baptizing by Women

The debate over women's leadership in early Christian communities became particularly acute around the question of women's authority to baptize. The *Acts of Thecla* are our earliest explicit evidence for women baptizing although some suspect that the practice was much earlier.

The administration of baptism by women was opposed by numerous Christian writers through the fourth century, testimony to its persistence in at least some communities. Among its most prominent opponents was Tertullian, who as we have seen denounced the *Acts of Thecla* precisely because it was used to support the practice.[41] He imputes such activity to (unspecified) heretics, claiming that women among them "are bold enough to teach, to dispute, to enact exorcisms, to undertake cures, it may be even to baptize."[42] Precisely at whom these charges are leveled is uncertain. The extant *Acts of Thecla* explicitly describe her teaching and recount her self-baptism: it is difficult to say what else might be inferred from the final phrase "and she enlightened many." Conceivably, Tertullian even has in mind the New Prophecy itself before he was drawn to it, but this is only speculation. In *The Veiling of Virgins*, Tertullian understands the prohibitions of 1 Corinthians 14:33b–36 and 1 Timothy 2:11–12 to cover not only speech and teaching, but also baptizing, offering, and any masculine function, let alone sacerdotal functions.[43]

The early third-century *Didascalia Apostolorum*, or *Teachings of the Apostles*, also proscribes baptism by women.[44] In the fourth-century *Constitutions of the Holy Apostles*,[45] which drew extensively on the *Didascalia* and are presented as the instructions of the twelve apostles, women are admonished not to baptize, for no small danger threatens any woman who does so.[46] The arguments adduced are not surprising, relying heavily on the order of creation, the teachings attributed to Paul, and the failure of Jesus to authorize women to baptize, especially Mary. The *Constitutions* explicitly connect baptism and the priesthood.[47]

## The Deaconate

One office that women are known to have held with relatively little ancient debate is the deaconate. The term is derived from the Greek *diakoneō*, meaning to serve or minister. The noun *diakonos* was already used to designate an attendant or official in a religious context before the rise of Christianity.[48]

The earliest Christian reference to a deacon occurs in Romans 16:1, where

Paul acknowledges his debt to Phoebe, called deacon of the church at Cenchreae.[49] He asks the recipients of the letter to extend to Phoebe whatever assistance she may require, for she has been the *prostatis* of many, including himself. *Prostatis* (*prostatēs* in the masculine) carries a range of meanings, from patron or benefactor to a senior Macedonian civic officer.[50] In the Greco-Roman period, it frequently designates the patron of a religious association and clearly connotes status, prestige, and authority.

As feminist commentators have pointed out, modern English translations of these verses have obscured the probable significance of Phoebe.[51] The RSV, for example, translated the Greek phrase *ousan [kai] diakonon* as "a deaconess," although *diakonon* is simply the accusative form of the masculine,[52] while the subtle addition of the indefinite article implies that Phoebe is one of several, minimizing her importance. The NRSV, however, has taken feminist critique seriously, now calling Phoebe a deacon and noting that an alternative translation would be "minister." The significance of *prostatis* has been similarly obscured. According to the RSV, Phoebe "has been a helper to many, and myself as well"; the Jerusalem Bible reads, "she has looked after a great many people, myself included." The NRSV reads, significantly, "she has been a benefactor of many and of myself as well."

It is obvious that many modern translators have demoted Phoebe based on their own assumptions about gender and status in late antiquity and early Christianity. Regrettably, it is not so obvious that a formal office of deacon was established by the mid first century, and if so, what it entailed. The author of 1 Timothy knows the offices of bishop, deacon, widow, and elder, and his prescriptions for deacons envision both men and women in that office.[53] But 1 Timothy concerns itself only with the qualifications of deacons and not with the precise nature of their responsibilities. There is also a brief allusion to women deacons in Pliny, who acknowledges torturing two *ancillae*, who were called *ministrae*.[54] As we have considered, *ministra* can designate pagan cult officers:[55] it is usually assumed to represent here the Latin equivalent for *diakonoi*, but it offers no information on their duties. The likelihood that these women were slaves, or of modest status, is sometimes used to argue for the inferiority of the office of deacon, but such arguments may be blunted when we remember the emphasis early Christians put on the reversal of social status in the Kingdom and the denial of worldly rank.

The *Constitutions of the Holy Apostles* describes some of the duties of deaconesses, who are to be present whenever a woman speaks with a male deacon or with a bishop.[56] Traditional scholarship has usually seen the origins of women's deaconate as a necessary concession to ancient norms of propriety and decorum. Women were "allowed" to hold this office only because there were some cultic functions men could not perform with respect to women without raising the specter of impropriety. Women deacons, therefore, are seen as analogous to the stereotype of women nurses in the offices of modern male gynecologists: they are there primarily to vouch that nothing untoward took place, not to do anything of "real" consequence.

The epigraphical evidence does not wholly support this perception. Consider a fourth-century C.E. inscription from Jerusalem, the epitaph of a woman named Sophia, who calls herself a deacon, using the masculine form with the feminine article, a servant (*doulē*), and bride (*nymphē*) of Christ.[57] Most importantly, she calls herself the second Phoebe. Greg Horsley points out that this is an allusion not only to the designation of Phoebe as a deacon but to her position as *prostatis*. The phrase "the second Phoebe" may well have been a title conferred on Sophia by virtue of her benefactions. Titles of the "New Homer" or the "New Dionysos" are known to have been conferred on benefactors in acknowledgment of their gifts, and eventually became a virtual part of the person's name.[58]

Women called by the masculine form are also known from Patrai in Greece, from Melos,[59] and from Bulduk and Archelais in Asia Minor. The deacon Domna was the daughter of Theophilos, called elder, probably meaning a presbyter rather than a designation of age, suggesting perhaps a linkage of family and cultic office.[60] Deaconesses are known from Stobi in Macedonia, from Delphi, from Mt. Hymettos,[61] and from numerous places in Asia Minor.[62] It is doubtful that we can extract any clear patterns of date or geography in the choice of the masculine versus the feminine, or even links to movements deemed heretical by subsequent orthodoxy, although it is noteworthy that so many of these inscriptions come from Asia.[63] While the inscription of Sophia links her directly with the Phoebe of Romans, the sixth-century C.E. epitaph of Maria of Archelais explicitly endows her with the characteristics of the virtuous widow in 1 Timothy: she reared children, practiced hospitality, washed the feet of the saints, and distributed bread to the afflicted. It is difficult to say whether the association of the deaconate with the duties of widows is intentional or coincidental.[64]

## Women Exercising Priestly Functions

There is considerable evidence that in the early Christian centuries, priestly functions (especially blessing the Eucharist) were performed by persons with the title presbyter, or elder. The formal establishment of a Christian priesthood is exceedingly late. Christian writers of the Greco-Roman period use verbs like *hierourgein* or *leitourgein* to indicate priestly activities, but virtually all uses of the noun forms refer to Israelite priests or to pagans. Unfortunately, English translations sometimes obscure this distinction.[65] All evidence for Christian women with the title presbyter (elder) is therefore potentially testimony to women performing priestly functions.

Women presbyters are attested in both literature and inscriptions. As we have noted, Epiphanius asserted that among the heretical followers of Priscilla and Quintilla, women were bishops and presbyters and the rest.[66] What other offices he intended by this we cannot say.

Less well known than the testimony of Epiphanius is a letter sent in 494 by Gelasius, Pope of Rome, to three churches in southern Italy and Sicily.[67] He writes:

> Nevertheless, we have heard to our annoyance that divine affairs have come to
> such a low state that women are encouraged to officiate at the sacred altars and
> to take part in all matters delegated only to the offices of the male sex, to which
> they do not belong.[68]

Otranto argues that this can hardly refer to women's deaconate, (which would not have been terribly controversial) but rather to their exercise of actual priestly functions, which they perform with official encouragement. He concludes that in the late fifth century, women in a large area of southern Italy were ordained to priestly offices by their bishops. A similar situation may have prevailed in France, for only seventeen years later three bishops of Gaul wrote to two priests named Lovocatus and Catinernus criticizing them for allowing women to hold the chalice of the Eucharist and for distributing it during a service.[69]

In support of his position, Gelasius invoked the canons of various church councils that prohibited women from liturgical office,[70] including (although he does not specify) Canon 11 of the fourth-century Council of Laodicea (in Asia Minor), which reads: "It is not allowed for those called *presbytidas* to be appointed to preside in the church."[71] Otranto calls our attention to the interpretation of an early medieval bishop, Atto of Vercelli, who was questioned by a monk named Ambrose on the meaning of the terms *presbytera* and *diacona*. Atto interpreted the deaconate of Phoebe as evidence that women were originally ordained in the early churches, but subsequently excluded from priestly office by the later Canons:

> For just as those women called *presbyterae* had assumed the duty of preaching,
> ordering and instructing, in the same way clearly the deaconesses had assumed
> the duty of ministering and baptizing, a practice which today is not at all in
> use.[72]

Otranto points out that Atto had good reason for denying the more traditional interpretation of *presbytera* as the wife of a male presbyter, for he was a strong supporter of the antiquity of the celibate male priesthood. But we should hardly conclude that Atto has fabricated his interpretation solely to avoid the implication that males exercising priestly functions were once married. Instead, we may see Atto as evidence that even in the medieval period the tradition of women exercising priestly function in what was considered *orthodox* Christianity had not been altogether forgotten or suppressed.

In further support of his thesis, Otranto adduces an inscription found in a catacomb of Tropea in Bruttium (modern Calabria), one of the regions addressed by Gelasius, memorializing a woman named Leta, called *presbitera*. Not surprisingly, Ferrua considered it the epitaph of the wife of a presbyter[73] and dated it to the mid fifth century. But Otranto, surveying the known epitaphs of male Christian presbyters, points out that the wives of such men are not called *presbytera*, making it quite plausible that Leta was precisely one of those women whose priestly activities offended Gelasius.

Otranto collected several other inscriptions that may testify to women presbyters,[74] to which we may add the presbyter Ammion[75] and a few others. A mummy in the Louvre carries an inscription for Artemidora, daughter of Mikkalos and Paniskiane, who is designated *pres`b(yteras)*.[76] Greg Horsley disagrees with the editors' conclusion that the term here denotes age, since the other abbreviation in the inscription is a so-called *nomen sacrum* for Lord, suggesting that this abbreviation, too, has special Christian significance. Two other women with such an abbreviation are known from Sicily and Thera.[77] In the family tombstone from Melos, which contained reference to the female deacon we noted, all the children have church titles. Of the three designated *presbyteroi* one, Asklepis, may be female. The two remaining daughters are *partheneusasa*: they have committed themselves to virginity. Another inscription commemorates an elder named Kale who died at age fifty: its date and provenance are unknown.[78]

Regrettably, none of the known inscriptions other than Ammion's is unambiguously that of a Christian woman holding the office of presbyter. But the problem may be as much our perspective as anything else. If the pattern we have seen for pagan and Jewish women holds for Christian women as well, we should expect to find at least a few women holding cultic office here and there. Unfortunately, the typical scholarly response to such evidence has been to read in light of Epiphanius and consider it heretical, to employ the same dubious hermeneutical principles used on the Jewish inscriptions to explain away the data by denigrating the office of deacon because women held it, or by interpreting the abbreviation of presbyter as a simple designation of age rather than office.

The best explicit evidence for women exercising priestly functions is that of Epiphanius, who details two different heresies in which women function as priests. The first he associates with groups he calls by the umbrella label of Cataphrygian, after the region in which Montanism developed (Phrygia): he also knows them as Pepuzians (for the village where Montanist prophecies expected the descent of Jerusalem); Artotyritai (for the bread and cheese they apparently used in eucharistic rites); and Quintillians and Priscillians, after two Montanist prophets, Quintilla and Priscilla.[79] What distinguishes these from other Cataphrygians is not clear, although it may be the practice of ordaining women to the clergy (the focus of Epiphanius's description) and the attribution of founder to Priscilla and Quintilla, not Montanus.

These groups, comprising both women and men, traced their beginnings to the teachings and prophecies of Quintilla and Priscilla and accepted the authority of both the "old" and "new" Testaments. But they clearly read Scripture differently from Epiphanius. Apparently, they agreed that priority was significant in Genesis, but rather than focusing on the order of creation in

Genesis 2:21–25, they attributed special status to Eve because she was the first to eat of the fruit of the tree of knowledge (Gen. 3:6). We would not be surprised if they gave more weight to the creation account in Genesis 1:26–28. They cited Miriam the prophet and the four daughters of Philip as precedent for women clergy, making a linkage between prophecy and other Christian offices, which is perhaps not surprising given the centrality of prophecy in the Montanist movement(s). Further, they read Galatians 3:28 ("in Christ there is no male and female") as proof that women could hold all Christian positions of authority, perhaps the only evidence for this reading of Galatians in antiquity. Epiphanius does not say explicitly that they ordained women as priests, but this may be implied in his statement that women among them are bishops and presbyters, as we have seen. Priestly function is also implicit in his assertion they perform rites using bread and cheese, rites that Epiphanius can only consider heretical variants on the Eucharist, and therefore involving the exercise of priestly function.

To refute their exegesis of Genesis, Epiphanius counters with Genesis 3:16: "Your orientation will be toward your husband and he will rule over you," and with Paul's argument that "Adam was not deceived but Eve was first." He also offers up that odd inversion of obvious biological fact in 1 Corinthians 11:8 that man is not from woman, but woman from man, and 1 Timothy 2:12, with a significantly different reading from our text. Where 1 Timothy 2:12 prohibits women from teaching (*didaskein*), Epiphanius's version prohibits women from speaking (*lalein*). Conceivably, he has simply confused 1 Timothy 2:12 with 1 Corinthians 14:35, which does use the verb "to speak."[80] Perhaps, however, something more is at stake. These Cataphrygians are accused of deceiving people by pretending to experience ecstasy and grief, but interestingly, they seem on the whole doctrinally bland except for their use of cheese together with bread in their Eucharist, and of course, their ordination of women.

Epiphanius is absolutely explicit that women functioned as priests in a heresy he attributes to women now living in Arabia who originally came from Thrace, a region of northern Greece in close proximity to western Asia Minor, which long had had cultural ties with Phrygia.[81] In the name of the virgin Mary, these women prepare small cakes with four indentations that they cover with a fine linen veil and offer in a eucharistic service in which women function as priests. Presumably this means they distributed the cake to those assembled in the same manner as "orthodox" male officiants distributed the Eucharist to their churches.

Epiphanius finds this objectionable on two fronts: it inappropriately exalts Mary, and women should not function as priests. In approximately equal parts he refutes both the argument that women may serve as priests and the suggestion that Mary is due worship.

He begins by an *ad feminam* argument. Such beliefs and practices are absurd on their face, for they are taught by women, who are themselves easily mistaken, fallible, and poor in intelligence. Ever since Eve, women have been deceived easily by the devil.[82]

Should this prove insufficient, Epiphanius proceeds to point out that even Eve in her fallen state did not presume to exercise priestly office, nor did any woman ever offer sacrifices to God. A long list of exemplary male priests is offered, ending with the observation that "never did a woman exercise priesthood" in the Old Testament.[83] Moving to the New Testament, Epiphanius offers the argument from "tradition." If God had wanted women to be priests, or to hold any church offices for that matter, he would have surely made Mary the mother of Jesus such, but he did not. Mary rather than John, the son of a priest, could have baptized Jesus, but she did not, thereby indicating that God did not want women to baptize. Linking the apostolate with the priesthood, Epiphanius points out that God could have appointed women apostles, but did not: only men were appointed to perform what he calls "the priesthood of the gospel," and James, the brother of the Lord, was appointed the first bishop of Jerusalem. The succession of bishops and presbyters came from the twelve male apostles, and there were no women among them.

Epiphanius concedes that women prophesied in the earliest church, but insists that prophecy is not priesthood, an association which the Cataphrygians may well have made, if not also the cake-offering Collyridians. He also concedes that there is a clerical order of women deacons, but insists again that this order has no priestly component. Women deacons are necessary only to preserve the modesty of the female sex. Interestingly, he views the male deaconate in a similar way, as providing assistance, but having no priestly function. Epiphanius denies any ecclesiastical function or authority to widows, pointing out that the author of 1 Timothy, whom he unquestioningly took to be Paul, did not appoint them to be female presbyters or priests.[84]

Opposition to women exercising priesthood in principle, if not in actuality, also comes from John Chrysostom, who asserted women's exclusion from the priesthood but acknowledged that women effectively exercise power over the office nonetheless. "Since they can effect nothing of themselves, they do all through the agency of others; and they have become invested with so much power that they can appoint or eject priests at their will. . . ."[85] The *Constitutions of the Holy Apostles,* while similarly denying that women may exercise priesthood, is one of the few Christian sources to allude to pagan practices as part of the argument, asserting that pagans ignorantly ordain women as priests to female deities.[86] Whether this alludes to persons who argued from pagan precedent is unknown, though not inconceivable.

## Women Teachers

In discussions of Paul, Timothy, and Thecla, we have seen some of the debate over the legitimacy of women teachers in Christian communities. Those churches that accepted 1 Timothy denied women's authority to teach and cited 2:12 again and again as the definitive Scripture on the subject.

There never seems to have been much dispute over women's authority to teach other women. If anything, Christians seemed to have worried about the possible impropriety of men teaching women in private households. Clement of Alexandria thought that the apostles took along their wives so that they could preach to women within their quarters.[87] What really bothered many (male) Christians was the possibility that women could teach men and have authority over them.

Christian scripture made life difficult for those who read 1 Timothy 2:12 in the narrowest sense, for the book of Acts depicts the missionary Priscilla correcting the inadequate teachings of Apollos and instructing him in the proper doctrine.[88] As a result, at least some early Christian exegetes felt compelled to temper their interpretations of 1 Timothy accordingly. John Chrysostom, for example, concluded that since Priscilla clearly taught Apollos, 1 Timothy could not ban all teaching of men by women. Rather, the story of Priscilla must indicate the circumstances under which women could, in fact, teach men. He argued that 1 Timothy only restricts women from teaching men in public, and only when their husbands were also believers. Women married to non-Christians must be able to teach them, albeit in private, for how else would they be able to bring their husbands into the salvation of Christ if not by instructing them, as Paul asserts in 1 Corinthians 7:16.[89] Chrysostom also interpreted 1 Corinthians 14:34–35 as a prohibition against women's chatter, but not against their public speech. Why different male exegetes interpreted these passages as they did is clearly complex. In Chrysostom's case it is probably not incidental that he had powerful women friends, notably in the person of Olympias.[90]

Not surprisingly, we have scant literary evidence for Christian women acknowledged as teachers and virtually no evidence from nonliterary sources. A fragmentary papyrus of a Christian letter from the fourth century twice mentions a Kyria who is called *didaskalos*.[91] The modern editor found the notion of a woman teacher sufficiently problematic to suggest that the letter emanates from a non-orthodox community.

How women themselves viewed these issues may possibly be illuminated by an obscure letter largely overlooked by contemporary scholarship. Among the corpus of letters considered falsely attributed to Ignatius of Antioch[92] is one entitled *Mary the Proselyte to Ignatius*: the alleged reply of Ignatius to Mary is also extant.

*Mary to Ignatius* is of particular relevance to the discussion of women authors not simply because it purports to have been written by a woman, but because it explicitly expresses awareness of the potential offense such a letter might seem to be. Extant in Greek, the letter is relatively short, and primarily concerns the appropriateness of ordaining young men to priestly office. Mary writes to reassure Ignatius that certain men who have recently attained priestly office are well qualified, despite the fact that they are very young. In support of her claim, she offers Ignatius numerous examples from Scripture of men who, although young, were nevertheless wise and pleasing to God, and entrusted with priestly, prophetic, or royal office, including Daniel, Solomon, Josiah, and David.

At the conclusion of the letter, Mary assures Ignatius that she writes not to instruct him, but rather only to bring these testimonies to his remembrance. It would be hard not to see in this an allusion to 1 Timothy 2:12 prohibiting women from teaching men or having authority over them.

Whether the actual author was a woman seems impossible to determine on any firm grounds.[93] Either a male author wrote the letter in the name of a woman, or a woman wrote it. But there can be virtually no doubt that the letter is not an authentic epistle to the early Christian martyr, so in either case, someone fabricated a document in the name of a woman and expected it to be accepted as such. There seems little doubt that whoever wrote *Mary to Ignatius* was not, in fact, a woman named Mary living in the eastern portion of Asia Minor no later than the second century, even if the author was female. The very fact that the letter is pseudepigraphic is highly significant in light of the position the "orthodox" take in *The Debate of an Orthodox with a Montanist* about women writers. For while the "orthodox" might have meant to ban women from writing, sufficiently clever and subversive women might simply have concluded that if they wished to write, pseudepigraphy would have to be the order of the day.

It is also remarkable that the letter of *Mary to Ignatius* styles the writer Mary a proselyte of Jesus Christ, and the text of the letter is titled *Mary the Proselyte to Ignatius*. Why she is so designated is uncertain, particularly in a forgery, unless this is meant to lend some credence or some clue of identity. Possibly it is a subtle critique of 1 Timothy's dislike of proselytes as community leaders, even as the letter itself may be read as a subtle critique of 1 Timothy. Perpetua is explicitly described as a catechumen, a recent convert, and Priscilla and Maximilla may also have been new to the faith, as Eusebius's source contends.[94] Conceivably, pagan women were more likely than women born as Christians to have been educated as children and taught to write, and less likely, as converts, to accept positions such as those advocated by the Orthodox opponent of Priscilla and Maximilla, or even to convert into such churches in the first place.

Regardless of the actual identity of the author, the fact remains that someone composed a letter in the name of a woman, addressed to a bishop, which concerns issues of authority and politics in some Christian community before the sixth century. For now, it seems sufficient to suggest that the author of this work did not think a letter in the name of a woman at all implausible, provided she presented herself as cognizant of apostolic restraints on women "teaching." Further, the author of *Mary to Ignatius* depicts her as learned in Scripture and savvy in political matters; the author of *Ignatius to Mary* (possibly but not certainly the same) acknowledges her as intelligent, learned, and of the highest repute.[95] The author of another one of the pseudo-Ignatian epistles, *Ignatius to Hero*, describes her as the head of a church in her house.[96] *Mary to Ignatius* testifies to the plausibility of women's leadership in some unidentifiable Christian community associated with Asia Minor, in which women who wrote letters addressing important theological and political issues were not unknown to the author's community and not inherently unacceptable.

# — *13* —

## Women's Religious Leadership
## and Offices in Retrospect

So much of the discussion of women's leadership in early Christian communities has been oriented toward modern Christian practice that the ancient issues are sometimes obscured. We have seen that in its promulgation of social reversals and eschatologically oriented asceticism, earliest Christianity licensed alternatives for women that included participation in the public sphere often reserved, in most of Greco-Roman antiquity, for males. In its earliest renunciation of prevailing social structures, the new Christian movement advocated an egalitarian community in which, at least in theory, the prevailing social distinctions were obliterated and all might participate equally in the tasks at hand, including the leadership of the community, however that might be construed. I am not as confident as some feminist scholars that such egalitarianism was intentionally directed toward the liberation of women, but the practical effect was the same.

On the road from theory to practice, however, women's leadership rapidly encountered significant opposition. Many scholars have seen this as an inevitable consequence of the increasing desire of Christians, perhaps both male and female, to win acceptance in the Roman Empire by minimizing any appearance of social and political deviance. Those early Christian communities whose beliefs were dominated by intense eschatological expectation may have found this increasingly important as the urgency of their expectations diminished. Using the model of anthropologist Mary Douglas, we can also see that gender egalitarianism is most likely to flourish at low grid, which, together with strong group, characterized precisely those earliest Christian communities that did afford women leadership roles, and least likely to flourish at high grid, regardless of group, which is where Christian communities were increasingly located.

We have also seen that cultic leadership roles for women apparently posed no challenge to the social structures of classical Greek, classical Roman, or Greco-Roman culture. Pagan women routinely exercised the full range of reli-

191

gious offices in service to both female and male deities, and there is no record in the existing pagan literature of any debate over their legitimate ability to do so. Thus we must pursue a little further the origins and nature of much male Christian opposition to women's cultic leadership within early Christianity.

Since we cannot attribute it to pagan assumptions, can we locate its origins in Jewish perspectives? It has been a lamentable feature of some feminist Christian scholarship to blame Judaism for the misogynist patriarchal elements of Christianity, while viewing the egalitarian, potentially feminist elements of early Christianity as unique, an interpretation whose roots are firmly established in prior, largely male, Christian scholarship.[1] Certainly, we know by now that in the Greco-Roman world many Jews, Christians, and pagans (men, and probably to a large extent, women) shared a worldview that subjugated women to men as the natural order of things. If there was ever a pure egalitarian Jewish movement around Jesus of Nazareth that in modern terms could be described as feminist (and this is doubtful), its transmutation into a more restrictive and ultimately misogynist Christianity cannot be explained by the simple influence of Jewish patriarchal perspectives.

Ironically, it is even possible that the opposite is the case: some Christian women (in Asia Minor, for instance) may have learned their expectations of religious office from the examples of Jewish synagogues, as Brooten's study might suggest. But the connection between Jewish women's leadership in Greco-Roman synagogues and Christian practice is far from clear. None of Brooten's inscriptions for women synagogue officers are demonstrably pre-Christian, and the majority postdate the earliest evidence for women leaders in early churches. The Roman epitaphs of Veturia Paulla, called mother of two synagogues, and Sara Oura, possibly a synagogue elder, may date to the first century C.E. or earlier, but they may also be quite a bit later.[2] The strongest epigraphical evidence for women synagogue officers actually comes in the third and fourth centuries, if not even later, making direct borrowing from Jewish synagogues to Christian churches improbable.

Theoretically, the chronology might suggest the reverse, namely that Jewish synagogues adopted the practice of Christian churches. But the nature of Jewish-Christian relations in the first four centuries, while probably not nearly as universally hostile and antithetical as sometimes portrayed, nevertheless is likely to have precluded direct Jewish appropriation of Christian practice.

Valerie Abrahamsen has made a valuable contribution to this discussion with her suggestion that pagan women who converted to Christianity might well have brought with them the expectation of holding religious office in their new community as well.[3] Certainly, the association of converts with religious leadership is high in the literary sources, such as the story of Thecla or Lydia. None of these women are described as former priestesses or cult officers, but given the known associations between social status and cultic office in Greek cities in Asia Minor, for example, we might well expect that women like Thecla would have held such offices before they joined the Christian movement. We can also understand why subsequent Christian literature would omit such details. Of course, in the first century or so, the vast majority of Christians were converts, especially among those who adhered to strongly ascetic forms of Christianity, so at an early stage the connections between conversion and leadership may be more predictable. But Abrahamsen's suggestion should be kept in the forefront.

What, for example, would women who had formerly served in pagan cultic offices have thought of Christian restrictions on women? Would they have accepted the arguments we find in many of the church fathers? Would they rather have only, or primarily, been attracted to Christian communities which, like the one portrayed in the *Acts of Thecla*, continued to offer women positions of public participation and leadership?

As alluded to earlier, the possibility that pagan expectations affected synagogue practice should not be overlooked. Virtually all known synagogue offices have counterparts in Greco-Roman associations, and the term *synagogue* is attested as the term for a pagan association, although conceivably Jewish usage has here been transferred to the pagan milieu.[4] It is conceivable that women synagogue officers were more likely to find acceptance in Jewish communities where the incidence of conversion and pagan attachment to Judaism was high, and where, perhaps, women converts themselves successfully brought these expectations. This should not be construed as a suggestion that women's leadership in Greco-Roman synagogues can be dismissed as a pagan practice, not authentically Jewish and therefore of little relevance to our understanding of the enormous diversity of Judaism in late antiquity. Rather, it is a reminder of the extent to which Jews in the Greco-Roman Diaspora were an integral part of Greco-Roman culture.

We must also consider the likelihood that common cultural conditions in certain communities enhanced the possibilities for pagan, Jewish, and Christian women simultaneously. It is rather fascinating that in Brooten's collation of women who held the title elder, three come from the ancient city of Venusia (modern Venosa) in southern Italy, in the vicinity of those same Christian communities to whom Gelasius addressed himself in the late fifth century. Although Brooten dates them no more precisely than between the third and sixth centuries C.E., there is some reason to argue for the later as the more plausible.[5] Noting these three represent 60 percent of her *presbytera* inscriptions and that five Jewish women from Venusia bear titles of civic or religious office, Brooten suggests the Venosan (Jewish) community may have had a tradition of granting offices to women.[6] The most recently discovered inscription of a woman synagogue elder was found at Noccere, perhaps 125 kilometers from Venosa. The evidence from Gelasius suggests that such a tradition extended beyond the parameters of the Jewish community.

Yet even if Jewish women held office in larger numbers than the evidence permits us to verify, it is undeniable that women were generally excluded from the Israelite priesthood while pagan women were not.[7] Since Greek and

Roman culture can hardly be characterized as inherently less misogynist than Israelite culture (on what scale would we measure such things, and for what purpose?), I suspect that the difference lies in a radically different understanding of the nature of religion and of cultic office.

For Jewish men (and probably most Jewish women), purity appears the immediate obstacle to women's exercise of the priesthood. Women's perceived ability to pollute sacred objects, sacred space, and otherwise pure males jeopardizes men's relationship with God. But the effective dismantling of the priesthood with the destruction of the second Temple in 70 C.E. renders the issue largely moot: neither men nor women can in reality perform sacrifices and other priestly offices.[8] Subsequently, rabbinic argument against women's exercise of cultic leadership rests not on their ability to pollute (although the rabbis feared this perhaps even more than their forefathers), nor on the order of creation, nor on social conformity. Rather, as we have seen, the rabbis view cultic leadership primarily as the discharge of cultic obligations by an individual for the community. They claim that since women are exempted from such obligations in the first place, they cannot discharge them for those who are bound.[9] Unquestionably, a profound fear of female pollution continued to buttress rabbinic rejection of women's cultic leadership. But it is never offered as the explicit justification for the denial of women's public cultic performance. Conceivably, some of this is related to the fact that the focus of rabbinic cult was the Torah, which is exempt from ritual pollution.

Concern for ritual pollution was certainly not limited to Jews. Any number of pagan cult regulations require the sexual chastity and ritual cleanliness of priests and priestesses.[10] An inscription from the temple of the god Men at Sounion outside Athens reads in part:

> no-one impure is to draw near; but let him be purified from garlic and swine and woman. When members have bathed from head to foot, on the same day they are to enter. And (likewise) for ten days after (contact with) a corpse, and forty days after a miscarriage; . . . If anyone violates (these provisions) his sacrifice will be unacceptable to the god. . . . [11]

Mary Beard argues that the Vestal Virgins had to remain chaste for the entire period of their service precisely because they continually lived in Vesta's precincts, and the priestess had to be pure when she entered the temple.[12] But the systems of Greek and Roman religion do not seem to have been oriented toward protecting men from the pollution of women: rather, both sexes were vulnerable to ritual impurity and liable for purification. Of course, Jewish men could also contract ritual pollution from sources other than women, including corpses and their own sexual emissions.[13] But it is undeniable that, at least in a rabbinic worldview, the primary relationship between humans and the divine takes place between men and God, not between women and God, and the system is disproportionately concerned with protecting men from the ritual impurity of women and not vice versa. In much of pagan religion, by contrast, gender hardly disqualified women from commerce with the divine, and some even considered it particularly women's business.

If pollution is a central, if not the primary motivation for the exclusion of Israelite and later Jewish women from priestly office, we may also understand why at least some Diaspora synagogues were perfectly comfortable with women synagogue officers. First, Diaspora Jews are unlikely to have been nearly as concerned as the rabbis about ritual pollution of any sort. In Douglasian terms, the social location of Diaspora Jews, sometimes (perhaps often) well integrated into the matrix of Greco-Roman society, particularly in western Asia Minor, would not warrant adherence to strong beliefs about pollution, which function to maintain strong boundaries and separation between social groups. Second, synagogue offices whose duties are known or surmised with some degree of plausibility seem relatively unconnected to issues of purity and pollution (e.g., financial support and oversight, liaison with the local and imperial government, Jewish education, and so forth). Unfortunately, we know virtually nothing about the celebration of ritual in Diaspora synagogues, and whether the offices known to have been held by women would have carried ritual performance as some of their duties.

We do know, though, that for Christians, the arguments against women's leadership emerge early and in quite specific forms, and undoubtedly in response not just to hypotheticals, but to real women exercising real leadership. In the first and second centuries, they focus primarily on women teaching men and baptizing anyone; in later centuries, the debate extends to Christian ritual explicitly linked with priestly functions, probably including blessing and distribution of the Eucharist and forgiveness of sins. From 1 Timothy on, it is abundantly clear that the crucial issue is women's authority over men, which is seen to violate the order of creation and the self-evident rules of nature, not to mention society. The groundwork is already laid in Paul's correspondence with the Corinthians, whether or not one accords authenticity to 1 Corinthians 14:33b–36. Woman's subordination to man is demonstrated in a multiplicity of arguments, from Eve's creation after Adam, to her instrumentality in the fall, and so forth. Later Christian writers add new arguments. God intentionally excluded women from the Twelve.[14] If God had intended women to baptize, Mary would have baptized Jesus instead of John the Baptist.[15]

Yet we know now that women did baptize, teach, and exercise other priestly functions in some Christian communities. Both in antiquity and again in our own times, such women (and the men who supported them) have been denounced as heretics, and even some supposedly neutral scholarship has been content to observe that women only exercised priestly office and cultic leadership in heretical Christian churches.[16] Such arguments have contemporary implications: if only heretics did such things, they cannot be adduced as support for contemporary practice.

But the accusation of heresy is far more complex. In certain cases, such as the followers of Priscilla, Maximilla, Quintilla, and Montanus, the label of heresy or heterodoxy was assigned to them primarily and precisely because of the leadership roles they accorded to women. This is virtually explicit in the *Debate between a Montanist and an Orthodox*, when the Montanist forces the Orthodox to concede that the only thing Orthodoxy finds offensive about Priscilla and Maximilla is that they wrote books in their own names, which for the Orthodox constituted the exercise of authority over men.[17]

In fact, it may not be an exaggeration to suggest that while not all heresies accorded women authority over men, all those churches that did accord women authority over men were ultimately labeled heretical, even if there was nothing else offensive about them.

But if many Christian writers and communities display a sometimes frantic effort to impose control over women, how does that differ from Jewish and other Greco-Roman exercise of social control over women? I would suggest that the answer lies primarily in the social ramifications of Christian asceticism (including martyrdom) for women's lives that many forms of Christianity promulgated, inadvertently or otherwise. As long as Christian belief and practice advocated ascetic control of the body, a tension would exist between traditional male social control over women and the relative autonomy and freedom for women inherent in such asceticism. Christian preoccupation with orders of virgins and widows accords well with this, as do the regulations of those groups. It is, I submit, no accident that there were no orders of Christian wives. Provided they were married to Christians, wives were already sufficiently under the control of their husbands. In this regard, early Christianity differed significantly from the rest of Greco-Roman society, including Jews. Pagans and Jews could deal with unmarried women, whether virgins or widows, by marrying them off, whereas Christian adherence to asceticism and rejection of second marriage complicated their ability to do this and contributed to the growth of orders of virgins and widows.

Several ancient texts acknowledge the connections between virginity and autonomy. In his treatise *On the Veiling of Virgins*, Tertullian attempts to refute those unnamed Christians who argue that Paul did not intend to include virgins in his requirement that women cover their heads when prophesying or praying (1 Corinthians 11:2–16). His opponents claim that

> no mention of virgins is made by the apostle where he is prescribing about the veil, but that "women" only are named; whereas if he had willed virgins as well to be covered, he would have pronounced concerning "virgins" also together with the "women" named,[18]

just as Paul does in fact do when he lays down prescriptions for marriage in 1 Corinthians 7.

In the same work, Tertullian also considers whether the prohibitions against women speaking in church (1 Corinthians 14:33b–36) or teaching (1 Timothy 2:11–12) apply to virgins, concluding that they do.[19] Clearly, some Christians saw significant distinctions in the terms *virgin*, *woman*, and *widow* that exempted virgins from the regulations established for women and widows.

A catena of Origen on 1 Corinthians 14:34–35 similarly considers whether men here means only husbands. If it does, if wives are instructed to ask their husbands at home, then virgins and widows (who are not women/wives) could either speak in church or have no one to ask, no one to teach them at home. He concludes that the proper translation is "male relatives" or "the man who is her own" in some way: women are instructed here to ask their own men the answers to their questions.[20]

With the successful Christianizing of the Roman Empire, these issues seem to subside. We may hypothesize several reasons. The power the Roman church ultimately acquired allowed it to impose much more stringent controls and to destroy in one way or another those churches it considered heretical. With the Christianization of the empire, Christians may also have become less concerned to demonstrate their social orthodoxy—their orderly households, the lack of threat they posed to the empire—for they had become the empire.

Why, then, could Christians not retreat from the subordination of women to men (among other forms of social hierarchy): why could Christians not now, finally, embrace the true egalitarian qualities of the Jesus movement? The work of Mary Douglas again allows us to see some dimension of the answer. By now, strong grid characterized the experience of most Christians. For some, the constraints of strong grid were tempered with the benefits of strong group. Many others, including most women, experienced strong grid unmitigated by strong group, enduring instead the insulation and isolation of strong grid with weak group. Distant now in time and space were the social structure and attendant cosmology, rituals, and attitudes of the earliest followers of Jesus. Such egalitarianism was fundamentally antithetical to the social structure and cosmology of any high grid society and that, it seems to me, is as sufficient and persuasive as any answer.

Finally, there is conceivably some relationship between monotheism and the exclusion of women from Jewish and Christian priesthood, an exclusion that carries over to monotheistic Islam as well. When divinity is perceived to be one, and the gender of that divinity effectively presented as masculine in language, imagery, and so forth, perhaps only the sex which shares that gender is perceived as able to perform priestly functions. Conversely among the Greeks and Romans both the gods and their clergy came in two genders. But for the Greeks and Romans, there was no simple one-to-one correspondence between the gender of a deity and that of the officiants, even though it does seem that female divinities were more likely to be served by women and male divinities by men.

Interestingly, though, there seems to be some correlation between the perception of God as androgynous and the view that women could exercise Christian office. For example, many so-called gnostic Christians took Genesis 1:28 seriously and believed that since the being created in the image of God was male/female so, too, was God.[21] Unfortunately, we know so little about actual gnostics and their cultic practices and assignment of leadership that it is difficult to carry this line of reasoning much further. The work of Mary Douglas may enable us to see such correspondences as part and parcel of a particular cosmology, without requiring us to seek any direct causality from one to the other.

In any case, I know of no instances among early Christian defenses for the exclusion of women from the priesthood that reason along such lines. Instead, the arguments focus on the subordination of women to men, and on the choices that God and Jesus could have made but did not in selecting the person who baptized Jesus or choosing the apostles. In antiquity, no one argues that women cannot be priests because the priest represents Christ in the congregation, and Christ came, necessarily, in the form of a male. Only later, in the Middle Ages, does this become a key weapon in the arsenal against women's ministry, as Christians begin to argue that, contrary to Genesis, woman was not created in the image of God and therefore was unable to fulfill the office of priest.

# Epilogue:
## Toward a Theory of Women's Religions

Through the lenses of Mary Douglas's cultural theory, we have glimpsed the coherence and consonance of otherwise seemingly disparate and unrelated practices, beliefs, and symbols relevant to women's religious activities in the Greco-Roman world, as well as male attitudes toward those activities. In these final pages, I both recapitulate earlier observations and develop a little further some additional ramifications for the study of women's religions. Some readers may find my reflections here more obtuse than the preceding chapters, and too heavily laced with theoretical language. For this, I beg indulgence.

We have seen that women's religious experiences and consonant beliefs vary according to their particular social location. Gender effects subtleties of cosmology and ritual that reflect the differences between women's and men's experiences.

We have also seen that the particular constellation of strong group and weak grid is frequently the locus of increased authority and options for women, especially in the case of early Christianity. Low grid by its very nature relaxes classification and hierarchy, values achievement over ascribed status, and in general supports equality and the breakdown of social discrimination. With low grid, regardless of the accompanying degree of group, we expect to find egalitarian norms.

As I have emphasized, Douglas predicts that despite the push toward egalitarianism inherent in low grid, "the effects of the division of labour between males and females, and the extent to which valued property is inherited on lines laid by marriage alliances is crucial."[1] Thus, when coupled with strong group, weak grid ought not inherently to offer women increased options and alternatives, except where marriage, childbearing, and the transmission of valued property are negated.[2]

Beliefs in the imminent end of the world, and/or in the salvific quality of asceticism will do just that, often in tandem. If one expects the end of the current order of the world at any time, childbearing and the transmission of property become irrelevant. Childbearing can best be avoided through sexual asceticism. Even a firm belief in the value of sexual abstinence without a sense

199

of eschatological urgency will have the same effect. As we have seen, wider alternatives for women predominate in precisely those Christian communities that urgently preached asceticism and eschatology, undercutting old norms of marriage, childbearing, and property transmission. In those communities, traditional divisions of labor along gender lines come under attack, for if women do not bear children, one of the principal divisions of labor disappears, freeing women to participate in the public life of the community.

When gender is denied as a meaningful distinction, we may suggest that by virtue of the similarity of their social location, women and men will tend toward more similar perceptions of the cosmos and to shared rituals in which gender distinctions figure little or not at all. It is tempting to speculate further that the denial of gender as a meaningful category will be reflected in conceptualizations of the deity as similarly ungendered or androgynous. This is perhaps what we see in the depictions of divinity in some gnostic sources and even in Philo's Therapeutics.[3] But such suggestions are exceedingly difficult to test on the sources from antiquity, where we rarely have both constructs of divinity and details of social structure firmly under control. Douglas's model leads us to expect that the denial of gender as a significant distinction is likely to occur only at low grid and only when the social ramifications of sexuality are minimized.

The benefits of low grid and strong group are not only available to women in Christian communities with strong beliefs in asceticism and imminent eschatology. They are also available, for limited periods of time, to women whose primary social experience is that of insulation and isolation associated with strong grid and weak group, in the form of ritual flights to the diagonal opposite of weak grid and strong group. This, I propose, most accurately accounts for the characteristics and appeal of ecstatic Bacchic rituals: it also allows us to understand the gradual transformation of Bacchic rites as the social constraints of Greek women and men change significantly in the Hellenistic period. For insulated, isolated Greek women, no permanent changes in their social lives were possible. They were rarely able to change their residence, except perhaps through marriage, and even this was likely to bring them only to another village or city with similar constraints on women.

But the phenomenon of Hellenization inaugurated by Alexander the Great weakened the constraints of group and grid particularly for persons living in cities. People could travel, they could move to new cities, establish new families and new relationships, cemented often with new religions. They could, in effect, make permanent transitions to other social locations. We would expect that rites which provided only temporary respite from insulation would be less necessary as people found other ways to lessen the isolation of strong grid and weak group. Indeed, reflective of the decreased insulation and marginality of their participants, we find Bacchic rites increasingly less ecstatic and less gender segregated.

Consonant with people's ability to make real transitions from one community to another in the Hellenistic world, we find the possibility of transitions to new religions, the phenomenon we often call conversion. I suggest that such a phenomenon is possible only when it can be accompanied by transitions to new social communities, since, as Douglas allows us to see, it is exceedingly difficult, if not impossible, to sustain radical changes in cosmology without corresponding social support, which is itself much easier to obtain in a new community. This is acknowledged by Jerome in the fourth century C.E., fortunately not in such contemporary anthropological jargon, in his letter to Paula's daughter, Laeta.[4] From Jerusalem, he urges her to send him her young daughter, also named Paula, insisting that it is virtually impossible to raise a Christian virgin properly amid the temptations of the worldly city of Rome.

In the course of this study, it has become apparent that relatively few women in Greco-Roman antiquity experience high grid and high group simultaneously. Rather, this combination seems primarily to describe male social experience. Women whose male relatives experience high grid and high group are likely to share the dimension of high grid but not high group. But a few elite women do seem to experience both high grid and group, as we saw in our analysis of the rites of the Matralia.[5]

Nevertheless, the relative paucity of such women suggests that high grid/high group cosmologies exhibit little interest in women and in the feminine divine. The only exception to this will be found, perhaps, in the promulgation of male versions of rarified unreal women, who reflect the absence of contact between men and women, except in roles strongly defined by the constraints of high grid. Here we expect to find male worship of goddesses such as Athena, who despite her feminine form differs from mortal women in more than just her immortality: born from the head of Zeus, rather than the uterus of her mother, eternally aloof from sexuality, and so forth. The classic example of this may be the virgin Mary, who accomplishes something no mortal women have ever managed to do: virgin maternity.

We may now refine our observations about high grid with low group, which is, at least in Greco-Roman antiquity, still the locus of most women's social experience and that of many men as well. The more extreme the isolation of women from their high grid, high group male relatives, the more we may expect to find gender segregation in ritual as well, perhaps along the lines of gendered divisons of labor. Such a gender-segregated society will be reflected also in notions of gendered divinity. In such situations, women's worship of goddesses revered also by high grid, high group men is likely to be significantly different, as the worship of Mary by the so-called Collyridians diverged significantly from what Epiphanius considered acceptable. Conceivably, the debate over the divinity of Mary, if it is accurately represented here as women's belief opposed to that of men, may be seen as a reflection of the social distance between high grid, low group women who perceive divinity as male and female, and high grid, high group men, whose separation from women precludes this shared perception of feminine divinity. Of course, in

Epiphanius's case, things are complicated by the putative monotheism of Christianity.

But we must also take into account, as Douglas does, the fact that many men also experience the constraints of high grid with low group. Here I suspect that class is a critical factor, which Douglas obscures somewhat by lumping together women, slaves, and serfs. For free women (except perhaps those of the most elite upper classes) may still experience different separation from their male relatives than is experienced by underclass women, whose men are more likely to share their social location. We have seen, for example, that the seclusion of women in their homes, away from public intercourse, was more likely to characterize free women than freedwomen and slaves, who could rarely tolerate the economic repercussions of such seclusion. This suggests that gender may be less of a discriminating factor in the religious experiences and beliefs of freedpersons and slaves. Such persons, both women and men, may, by virtue of the similarities of their social location, tend more toward shared cosmology and ritual than upper-class persons, where gender segregation is more acute.

Yet upper-class women, often insulated and isolated from their own men and distanced from other women by virtue of the hierarchical classification of high grid, may still come to share the cosmology and ritual of others experiencing strong grid and weak group, in their fascination, for example, for magical practices and exotic foreign mysteries. The portrayal of converts to ascetic forms of early Christianity accords well with this, as does Celsus's claim that the primary targets of early Christian missionaries were the foolish, the dishonorable, and the stupid: women, children and slaves.[6]

Further, high grid, low group women segregated from their male relatives and social peers nevertheless often live within communities of other women, whose social structure and dynamics have been little studied by anthropologists and are rarely described in the sources from Western antiquity, accurately or otherwise! We might suggest that within certain limits, such women do experience a degree of high group, albeit only of other women, whose implications for ritual and cosmology remains to be explored. Based only on what we know, no better than secondhand, of their rituals and beliefs, it seems difficult to tell whether the Collyridian women, for example, may best be understood as strong grid, weak group or perhaps as weak grid and strong group.

I am also intrigued by the effects of monotheism on the cosmology and rituals of women experiencing strong grid and weak group. If in polytheistic societies, the degree of polarity of strong grid, weak group women and strong group, strong grid men is reflected in gendered divinity, what happens when monotheism is imposed on such societies? I use the word *imposed* intentionally, for with the exception of Israelite religion, we know of no ancient societies whose perception of divinity is not gendered, and there is increasing evidence that even ancient Israelite religion included a female component of divinity that took centuries to repress thoroughly.[7] If the worship of female divinities in women's rituals plays a crucial part in the reinforcement of women's social experience of insulation and isolation, what happens when such worship is artificially invalidated through the belief that God is not only one, but male? We may suggest, regrettably with insufficient evidence, that the behavior and belief of the Collyridians reflects the response of some Christian women, which will ultimately be taken up in ritual, without explicit theological articulation, by many other Christian women over the centuries, though perhaps more subversively. The male hierarchy may deny over and over again that Mary is divine, and Christian women may even explicitly articulate that when pressed, but their behavior over and over again will emulate remarkably that of centuries of women's devotion to goddesses. The same may even be ventured for Jewish women's Sabbath observance.

Although few persons in the Greco-Roman world appear to have experienced low grid with low group, we may make a few observations about this combination and its implications for women's religions. The intellectual ambiguity and tolerance of weak grid and group lead us to expect little agreement about the nature of the divine, and consequently about its gender: on this point, as on many others, persons in this location are likely to profess literal agnosticism. Low grid with low group promotes egalitarianism, as it does also with strong group, but in either combination, marriage, childbearing, and property still qualify the egalitarian impulse. Only if those things are devalued (for example, by beliefs about overpopulation) should we expect to find the possibility of real gender equality. It seems that even at low grid and low group, equality will still only be possible to the extent that marriage, children, and property are severed.

I suspect that things are more complicated than Douglas perceives here. Certainly in some quarters of late twentieth-century American society, which Douglas views as quintessentially low grid and group, it is precisely the impulse to equality that has led feminists to propose the severing of marriage, childbearing, and the transmission of property and to devalue those things because they are perceived to impinge on women's autonomy. Also problematic is the unexamined androcentrism of Douglas's theory. The universe Douglas envisions here is a male universe: it is men who do not extend their egalitarian worldview to women; it is men who view women and children as one more medium for their perpetual competition with other men in the marketplace. Whether women in low grid, low group societies share these views is beyond the scope of our study, although infinitely worth pursuing.

In any case, in antiquity, low grid and group may be useful to us as an analytic category only in the case of a few elite pagan women, whose financial autonomy, resources, and family connections allowed them to play the part of what Douglas called in her earlier works "Big Men." For these few women, whom we see not only as benefactors and patrons, but occasionally as high cultic officers, particularly in the imperial cult in the Greek cities of Asia Minor, gender may have been less relevant than for most women. But even these women seem subject to certain constraints, and the ideology used to support their roles explicitly casts these women within traditional models for women.[8]

Two final issues present themselves. The first pertains to the problem of misogynism in early Christianity. In its earliest phases, Christianity undoubtedly espoused an egalitarianism that had significant consequences for women. Yet whatever advantages early Christian communities may have offered women scarcely characterize Christianity by the end of late antiquity, and there can be no question that Christianity ultimately emerges as a major perpetrator and legitimator of misogynism in Western culture up through the twentieth century. For Christian feminists, it has become crucial to inquire whether this misogynism represents an inescapable feature of Christianity or whether it is a perversion of the core of earliest, and therefore most authentic, Christian belief.

To phrase the problem in terms of Douglas's model, even within the Greco-Roman period, we see Christian communities characterized by the combinations of high grid with high group, and low grid with high group. The former exhibit many egalitarian aspects especially with regard to gender, correlative not only with low grid, but with the presence of factors that undermine the value of marriage and property transmission, namely asceticism and eschatology. On the other hand, Christian communities that deny women major participation display predictable characteristics of high grid spread over both strong group for a few, and weak group for many, especially women. One explanation is that these different perspectives about women represent a debate between communities in two different social locations, whose resolution is accomplished when communities with strong group but weak grid fail to survive, while those with strong grid and group continue. Alternatively, it may be that Christianity evolves over time from one location to the other.

The latter model of Christian development has long been favored by many scholars and has its roots in theories about the evolution of millenarian movements, which have proposed that all such movements must ultimately acquire sufficient structure, hierarchy, and organization to guarantee their long-term survival.[9] Communities like those of earliest Christianity, with little structure for transferring leadership and authority, yet prone to fission and disagreement, could not expect to exist for very long. The price of continuity, it would seem, is the imposition of hierarchy and structure, accompanied inevitably by restrictions on women legitimized by misogynism.

Yet the preeminent Christian feminist theologian Elisabeth Schüssler Fiorenza has argued strongly against this, denying that the misogynist transformation of Christianity was inevitable if Christianity was to survive. Rather, Schüssler Fiorenza asserts that the departure from the primal egalitarianism of Christianity was artificial, and that contemporary Christianity can be restored to its original "discipleship of equals" without sacrificing its ability to continue indefinitely.[10]

Douglas suspects that an evolution from strong group and weak grid to strong group and strong grid is highly unlikely: communities do not transform themselves vertically, and movement by individuals is possible (if not probable) primarily along diagonal lines.[11] Moves from high grid and low group to low grid and high group are common enough: this is the probable path traveled by many converts to early Christianity. Douglas's model, then, opposes the traditional understanding of the inevitable transformation of millenial renewal movements such as early Christianity.

But if egalitarian low grid, high group communities do not evolve into hierarchical, high grid, high group communities, how are we to understand the transformation of Christian egalitarianism into misogynist hierarchy? Douglas's model would lead us to look for evidence that some early Christian communities were not as strongly egalitarian from their inception. Some early converts never bought wholly into an egalitarian scenario: they were never centrally located at low grid, high group, but rather perhaps on the margins of the high grid axis, where they already interpreted Christian texts and teachings in a manner consonant with their own concerns for higher grid, if not also group. This, I would argue, we have already seen in the disputes between Paul and the Corinthians spiritualists, between the Christians behind the *Acts of Thecla* and those behind 1 and 2 Timothy, and many other places as well.

The explanation for the ultimate ironic triumph of hierarchy and misogynism, then, must lie in the failure of low grid, high group communities to win the day. Douglas proposes that for the most part, such communities are unlikely to survive for long, precisely because their low level of social organization is conducive on the one hand to fission and dissolution and on the other to inadequate perpetuation.[12]

In the case of early Christianity, several key factors may be proposed. Communities characterized by an imminent sense of the end of the world and by strong sexual asceticism hampered their own continuity (with which they were little concerned) by the tendency not to reproduce biologically, but to rely on conversion to perpetuate the community. A similar problem may have existed for those Christian communities that promulgated martyrdom as the ideal Christian life, for they, too, had difficulty reproducing and continuing over long periods of time. High grid, high group Christian communities, on the other hand, were more likely to endure by their very nature, even while many of their constituents, especially women, were likely to end up insulated and isolated, if indeed they ever relinquished such conditions.

Political factors should also not be neglected. At a certain point, the tide may have turned against egalitarian Christian communities, when high grid communities ultimately obtained the real political power they needed to suppress and eliminate their egalitarian opponents. Until high grid Christians acquired the resources of the Roman Empire, their ability to coerce other Christians was limited; at the same time, the characteristics of communities at low grid and high group tended to assist their own ultimate demise. If Douglas's theory is correct, it may suggest that hopes to reform contemporary Christianity, especially those branches firmly entrenched at hierarchical high grid and high group, are inherently doomed to failure, because such branches cannot accommodate the concomitant requirement of weakened classification that egalitarianism requires.

It is clear that the greatest scope of options for women in late antiquity may be found correlative with low grid, and that, in general, the weaker the constraints of grid, the greater the options for women. In the ancient period, the advantages of low grid were available primarily in communities with strong group, whose prospects for long-term survival were never strong. We have seen, too, that weak grid by itself could increase women's alternatives and autonomy, but that ideologies which negated women's traditional social roles as wives and mothers were crucial in effecting the transition from theory to reality. We might generalize by claiming that low grid is good for women and high grid is oppressive.

Yet Douglas, who does not claim the label of feminist, argues precisely the opposite: only with strong grid can the interests of women best be served. She would concede, I think, that strong grid with weak group excludes and isolates women, but contends that strong grid with strong group provides women respect and honor, not only in their youth, but in their old age as well.[13]

With the most profound respect and admiration for Douglas's work and her ability to see connections in human culture that elude virtually all the rest of us, I dissent from this aspect of her theory. I would argue passionately that strong grid inherently confines women and serves the interests of men in opposition to women, and that only with weak grid do women stand to disengage from the confines of what many call patriarchy. In this, I come down strongly on the side of autonomy as crucial, where I understand autonomy as self-determination.

Yet unquestionably Douglas and many others would argue that the valuing of autonomy is itself a correlate of the low grid culture that I personally experience, and that the valuing of autonomy and the individual above, say, the needs of the community as a whole is not self-evident but culturally determined, if not a relatively modern conception.[14] I cannot quarrel with that assessment in theory, but in reality, it is quite evident that in those quarters where the needs of the community are put forth as surpassing those of the individual, somehow the needs and concerns and values of men appear more significant than those of women. The needs of the men are falsely represented as the needs of the community, a powerful justification for the continued suppression of women.

Yet even the demise of strong grid is insufficient to dismantle the repression of women altogether. In antiquity, it required also the promulgation of ideologies that denied the value of marriage, children, and the transmission of property, values which characterized primarily weak grid, strong group communities. In the modern world, things may not be so different, for it still seems to me that even with low grid and low group, where the opportunities for women are the strongest we ever find them, Douglas is right that the factors of marriage, childbearing, and property are still key. Only when women succeed in separating themselves from these measures of their value do they attain full autonomy.

Both for women in antiquity and for us, the price of that autonomy is high and problematic: the renunciation of sexual love and ecstasy and of their frequent correlate, children. Quite possibly, people in antiquity viewed sexuality, desire, and children so fundamentally differently than we do that we can barely comprehend their perceptions.[15] Yet we may recognize some common element in their dilemma and ours in a story in the *Acts of Thomas* whose opposite is reflected in the beliefs of Demas and Hermogenes in the *Acts of Thecla*.

According to the apocryphal *Acts of Thomas*, the apostle was compelled by the king of Andrapolis to offer a prayer in the bridal chamber of the king's daughter and her new husband. No sooner does the reluctant Thomas do so and depart than Jesus appears to the newlyweds in the form of his twin brother Thomas, urging them to abandon sexuality on the very eve of their nuptials:

> . . . know this, that if you abandon this filthy intercourse you become holy temples, pure and free from afflictions and pains both manifest and hidden, and you will not be girt about with cares for life and for children, the end of which is destruction.[16]

Jesus/Thomas then details the horrors attendant on parenthood, including the illnesses that befall children, not to mention demonic possession. Even healthy children succumb to crime of all sorts. But if the couple desists, Jesus/Thomas promises them "living children whom these hurts do not touch, and [you] shall be without care, leading an undisturbed life without grief or anxiety."[17] For their eternal salvation, the young couple assent.

Demas and Hermogenes, those enemies of Paul and Thecla in the *Acts of Thecla*, put their perspective much more succinctly. For them, the resurrection is already come "in the children whom we have."[18] In their teaching we may see awareness of the profound connection between human existence and the love of children.

In this book, and in much of my work, I have sought to recover, expose, and articulate the connections between women's lives and their religious expression. I have refrained thus far from rendering judgment on the persons behind my sources, seeing that they often had as little control over the contents of their rituals and beliefs as we have over the cosmologies which are the correlate of our own modern lives.

But if there are no theological motivations in my study of the religious beliefs and practices of women in Greco-Roman antiquity, Jewish, Christian, or otherwise, I cannot profess to have no agenda at all. I am acutely aware that the study of antiquity, particularly the study of early Christianity, and to a lesser extent the study of early Judaism, is regularly used to legitimize the stances of contemporary communities. Nor is this confined to religious communities, for ancient attitudes toward women, sexuality, and procreation are routinely used in Western culture to justify political and economic programs and to circumscribe the lives of women. While I do not personally seek the grounding of modern life in the circumstances and constraints of antiquity, I do wish to hold those who do so accountable to the realities of antiquity so far as they are accessible, even while acknowledging the limits of our ability to separate our reconstructions of the past from our constructions of the present. In particular, I wish to emphasize the extent to which the lives and options of women in antiquity were constrained by cultural constructions that I and many others do not wish to share.

So it would be disingenuous to suggest that I have no opinions about the relative merits of various ancient religions. In the end, it seems to me that not all ancient religion was good for women, except perhaps insofar as it enabled them to survive in environments that sometimes appear inherently hostile to them. I would have to agree with that same astute reviewer of *Maenads*, Ann Loades, who suggested that those preoccupied with Christianity might do well to reflect on the differences between a religion whose central myth is that of the separation and ultimate reunion of mother and daughter beloved of one another, and that of a religion whose central myth is of a father who requires the painful sacrificial death of his only son.[19]

Ironically, it seems that myths which are most gynocentric come out of ancient Greece, a society that is widely agreed to have been the most repressive and misogynist of Western antiquity. It does not resolve the dilemma this poses to see them as the reflections of an earlier, less repressive, more gynocentric society. For the most androcentric myths, those of God the father and Jesus the son, end up licensing the most equality for women, at least in theory, and occasionally even in practice. Douglas would claim, I think, that this points to a crucial distinction between gender quality and gender equality, with the latter available principally with high grid and group, and the former, only seemingly at low grid.

It might then seem that women were and are caught between a rock and a hard place. The specific belief systems that provided women in antiquity with autonomy and alternatives are enormously problematic. Ascetic and monastic women from the Therapeutics to Thecla to the desert mothers found it necessary to repudiate the body and its female associations, becoming male both in theory and in aspects of appearance in order to achieve self-determination. Splitting the body from the soul, dualist cosmologies such as those advocated by gnostics frequently provided women with alternatives denied them by those (men) who insisted on the integral connection of body and spirit. Conversely, cosmologies that value embodiment seem then to need to constrain and confine women as a necessary corollary. The notion that self-determination for women is only available at the cost of psychic self-destruction, at the cost of the repudiation of the feminine, is hardly comforting.[20]

Such dilemmas have no simple resolution. But they suggest that in the study of women's religion, there is much room for additional exploration and consideration, and in that thought I take considerable comfort.

# Abbreviations

| | |
|---|---|
| *AE* | *L'Année Epigraphique* |
| *ANF* | A. Roberts and J. Donaldson, *Ante-Nicene Christian Fathers.* |
| *ANRW* | W. Haase and H. Temporini, *Aufstieg und Niedergang der römischen Welt.* Berlin and New York: Walter DeGruyter. |
| *ANT* | M. R. James, *Apocryphal New Testament.* Oxford: Clarendon Press, 1924, rev. 1955. |
| BAGD | W. Bauer, W. Arndt, F. W. Gingrich, and F. Danker, *A Greek-English Lexicon of the New Testament and Other Early Christian Literature* (2d ed.). Chicago: University of Chicago Press, 1979. |
| *CBQ* | *Catholic Biblical Quarterly* |
| *CCCA* | M. Vermaseren, *Corpus Cultus Cybele Attidisque.* EPRO 50. Leiden: E. J. Brill, 1977. |
| *CIJ* | J.-B. Frey, *Corpus Inscriptionum Judaicarum* I, with prolegomenon by B. Lifshitz. New York, 1975. Volume II, Rome, 1936. |
| *CIL* | *Corpus Inscriptionum Latinarum.* Berlin, 1863– |
| CRINT | S. Safrai and M. Stern, *The Jewish People in the First Century. Historical Geography, Political History, Social, Cultural and Religious Life and Institutions.* Compendia Rerum Iudaicarum ad Novum Testamentum. Assen: Van Gorcum, and Philadelphia: Fortress Press, 1974. |
| EPRO | Etudes Préliminaires aux Religions Orientales dans l'Empire Romain. Leiden: E. J. Brill. |
| GCS | Die griechische christliche Schriftsteller der ersten drei Jahrhunderte. Leipzig, 1897–1941; Berlin and Leipzig, 1953; Berlin, 1954– |
| *GRBS* | *Greek, Roman and Byzantine Studies* |
| *HBD* | P. Achtemeier, ed., *Harper's Bible Dictionary.* San Francisco: Harper & Row, 1985. |
| *HTR* | *Harvard Theological Review* |
| *IEJ* | *Israel Exploration Journal* |
| *IG* | *Inscriptiones Graecae.* Berlin, 1873– |
| *IGRR* | *Inscriptiones Graecae ad res Romanas pertinentes.* Paris, 1906–1927. Chicago, 1975. |
| *ILLRP* | A. Degrassi, *Inscriptions Latinae Liberae Rei Publicae,* Florence, 1957–65. |
| *ILS* | H. Dessau, *Inscriptions Latinae Selectae.* Berlin, 1892–1916. |
| *I. Magn.* | O. Kern, *Die Inschriften von Magnesia am Meander.* Berlin, 1900. |
| *I. Syrie* | *Inscriptions grecques et latines de la Syrie.* Paris, 1929–59. |

| | |
|---|---|
| *JAAR* | *Journal of the American Academy of Religion* |
| *JBL* | *Journal of Biblical Literature* |
| *JFSR* | *Journal of Feminist Studies in Religion* |
| *JJS* | *Journal of Jewish Studies* |
| *JTS* | *Journal of Theological Studies* |
| *KJV* | *King James Version* |
| *LSAM* | F. Sokolowski, *Lois sacrées de l'Asie Mineure*. Paris: E. de Boccard, 1955. |
| *LSCG* | F. Sokolowski, *Lois sacrées des cités grecques*. Paris: E. de Boccard, 1969. |
| *LSCG Suppl.* | F. Sokolowski, *Lois sacrées des cités grecques*, supplément. Paris: E. de Boccard, 1962. |
| *Maenads* | Ross S. Kraemer, *Maenads, Martyrs, Matrons, Monastics. A Sourcebook of Women's Religions in the Greco-Roman World*. Philadelphia: Fortress Press, 1988. |
| *MAMA* | *Monumenta Asiae Minoris Antiqua*, 8 vols. Manchester, 1928– |
| *NewDocs* | Greg Horsley, *New Documents Illustrating Early Christianity*. 5 vols. North Ryde, Australia: MacQuarie University, 1981–90. |
| *NRSV* | *New Revised Standard Version* |
| *OCD* | *Oxford Classical Dictionary* (2d ed.). Oxford: Clarendon Press, 1970. |
| *OTP* | J. H. Charlesworth, *The Old Testament Pseudepigrapha*. 2 vols. Garden City, NY: Doubleday, 1983–85. |
| *PG* | J. Migne, *Patrologia Graeca*. |
| *PL* | J. Migne, *Patrologia Latina*. |
| *P. Oxy.* | *The Oxyrhynchus Papyri*, 1898– |
| *RE* | *Paulys Realencyclopädie der classichen Altertumwissenschaft* |
| *RSV* | *Revised Standard Version* |
| SBL | Society of Biblical Literature |
| *SBLSP* | *Society of Biblical Literature Seminar Papers* |
| SNTSMS | Societas Novum Testamentum Studia Monograph Series |
| SEG | Supplementum Epigraphicum Graecum. Leiden, 1923– |
| *ZPE* | *Zeitschrift für Papyrologie und Epigraphie* |

# Notes

*Chapter 1*

1. Strabo, *Geography*, 7.3.4. Similar sentiments about women's proclivities may be found everywhere from 2 Timothy 3:6–7 to Jerome, *Against Vigilantius*, 6 and his letter (133) to Ctesiphon, 4; to Juvenal's *Satire*, 6, to the famous statement of Celsus that Christians attracted "only the foolish, dishonourable and stupid, and only slaves, women and little children" (Origen, *Against Celsus*, 3:44). On the vulnerability of women to undesirable religion, see Plutarch, *The Oracles at Delphi*, 25.

2. Amy Wordelman, "Strabo on Women," defines *eudaimonia* in Strabo as "a happiness or blessedness which stems from a right or good relationship to the divine or spiritual world." *Deisidaimonia* "implies a relation to the divine or spiritual world based on fear or an overly extreme reverence. The primary characteristic distinguishing it from . . . *eudaimonia* is that *desidaimonia* governs the lives of those who do not or cannot live according to reason . . . children, the illiterate, the half-educated, women and 'any promiscuous mob'" (*Geography*, 1.2.8).

3. I do not wish here to engage in discussion about the definition of religion: there is sufficient agreement on what constituted religion in antiquity to provide more data for consideration than can ever be covered in one volume.

4. For select bibliography on late nineteenth- and early twentieth-century research on women in antiquity, see Kraemer, "Women's Religions."

5. Harnack, *Expansion*, 2:217.

6. Moore, *Judaism*, 1:328.

7. Fiorenza, "'Father,'" 305.

8. Ibid. See also Witherington, "Tendencies," which points to intriguing evidence for androcentric alterations.

9. For example, *CIJ*, 19, where Frey proposes Domitios or Domitianos to restore the fragment Domi . . .; also his restorations for *CIJ*, 62 and 71.

10. Numerous translations of the Dead Sea Scrolls are available in English, such as those by Gaster, Dupont-Sommer, and Vermes. The Nag Hammadi texts are available in Robinson, *Nag Hammadi Library* and in Layton, *Gnostic Scriptures*.

11. For an excellent detailed discussion of the transmission of ancient texts, see Reynolds and Wilson, *Scribes*.

12. For example, Hroswitha, *Callimachus*, which is based on the Acts of John.

13. Snyder, *Women Writers*. For a fuller discussion of Jewish and Christian women's authorship, see Kraemer, "Women's Authorship," as well as Cole, "Greek Women" and Pomeroy, "Technikai."

14. *The Martyrdom of Saints Perpetua and Felicitas.*

15. On the possible identity of the redactor, see below, 163.

211

16. Faltonia Betitia Proba, *Cento*; Egeria, *Diary of a Pilgrimage*.

17. See, for example, Pomeroy, "Preliminary Study" and *Women in Hellenistic Egypt*. The very recently published personal papers of a Jewish woman named Babata, who died during the Bar Kochba rebellion around 135 C.E., provide a particularly exciting illustration (Lewis, *Documents*). Regrettably, only a few of the Babata documents were published and available prior to the completion of this study (Lewis, "Papyrus Yadin" and "Provincia Arabia").

18. Cole, "Greek Women," 225.

19. See Starr, "Circulation."

20. See below, 165.

21. *PG Ser. 2*, 5: 873–80, translation mine. I doubt that it is the authorship of books as contrasted with epistles which is at stake, for the epistles of Paul and others were considered books. For a fuller discussion of the texts and the issues here, see Kraemer, "Women's Authorship," 224–26, 236–39.

22. For example, Davies, *Revolt*. On these criteria, see also Kraemer, "Women's Authorship" and Mary Lefkowitz, "Novels" in the same volume.

23. Some fruitful exceptions include the work of Gager, "Body-symbols" and *Kingdom and Community*; Meeks, *Urban Christians*; Neyrey, "Bewitched" and *Revolt*; Theissen, *Palestinian Christianity* and *Pauline Christianity*. See also Malina, *New Testament World* and *Christian Origins*; Kee, *Sociological Perspective*; and Malherbe, *Early Christianity*. An intriguing new work by Wire, *Women Prophets*, utilizes Douglas's theory for a study of Christian women prophets at Corinth.

24. For example, Fiorenza's critique of Gager and Theissen, *Memory*, 72–80.

25. Kraemer, "Attraction of Women" (also published in a less technical version as "Women of Ancient Greece"); "The Conversion of Women."

26. On deprivation theory and millenial movements, see, for example, Lanternari, *Revolt*; Gager, *Kingdom*; Burridge, *New Heaven*; Cohn, *The Pursuit of the Millenium*; Jarvie, *Revolution*.

27. For the concept of relative deprivation, see Burridge, *New Heaven*, 41–43, 48, 75–76; Gager, *Kingdom*, 59, 95.

28. Durkheim, *Elementary Forms* and *Sociological Method*.

29. Douglas has evolved this model over many years and has recently come under attack for failing to articulate a single, consistent version of the model (Spickard, "Three Versions"). I concur with Douglas's gentle response that Spickard has done no better in his attempt to summarize the theory, although I agree that it is difficult to find a concise description of grid/group theory in any one of her works. Douglas's first major articulation of grid/group theory came in *Natural Symbols* published in 1970. It was republished in 1973 in a substantially different form: all quotations here are taken from the first Vintage Books printing of the 1973 edition. She has since refined the model significantly, particularly in *Cultural Bias*. My interpretation of Douglas depends heavily on *Natural Symbols* and *Cultural Bias*, as well as on discussions with Douglas, particularly during a graduate seminar at Princeton University in 1988, which she graciously allowed me to audit.

Spickard has also written a dissertation, *Mary Douglas*, critiquing the insufficiency of grid/group as a comprehensive theoretical model. It contains a superb comprehensive bibliography of Douglas's publications and of works utilizing or critiquing Douglas.

30. *Cultural Bias*, 7.

31. Ibid., 16.
32. Ibid., 15.
33. *Essays,* 3.
34. *Natural Symbols,* 82: see also the entire chapter here on grid and group, 77–92.
35. *Cultural Bias,* 8.
36. *Essays,* 5.
37. These descriptions are condensed from Douglas's discussions in *Natural Symbols,* especially 77–92; *Cultural Bias,* 5–36; see also *Essays,* 1–8.
38. *Cultural Bias,* 21.
39. *Natural Symbols,* 117–18.
40. *Cultural Bias,* 34.
41. Ibid.
42. Ibid., 34.
43. Ibid., 34–35.
44. Feminist scholarship of the last two decades has delineated a rich and nuanced portrait of women's lives in antiquity, which offers many possibilities for interpreting the variety of women's religious activities and choices in the Greco-Roman world. See bibliography for many relevant references.
45. *Cultural Bias,* 8.
46. Ibid., 41.

## Chapter 2

1. Helen McClees (*Women,* 17) found only six dedications to Zeus by Athenian women prior to the Roman period.
2. Perlman, "Bears," 117.
3. Scholia on Aristophanes, *Lysistrata,* 645; see Perlman, "Bears," 115, n. 2 and Sale, "Arkteia."
4. Perlman, "Bears," 116.
5. Ibid., 121–23, 125.
6. In addition to Perlman, see Cole, "Maturation"; Kahlil, "Brauron"; Sourvinou, "*Lysistrata*"; Stinton, "*Iphigeneia*"; Walbank, "Artemis."
7. Perlman, "Bears," 125–26.
8. Cole, "Maturation," 242.
9. Turner, *Ritual Process,* 166–203. See also Van Gennep, *Rites of Passage.*
10. Cole, "Maturation," 243.
11. Perlman, "Bears," 127–28.
12. Perlman, 120–21; references in n. 26.
13. *LSCG Suppl.,* 115; in *Maenads,* 4.
14. *IG* II², 1514; in *Maenads,* 5. For a fascinating discussion of Greek women's ritual clothing, see Mills, "Clothing."
15. See Euripides, *Iphigeneia at Tauris,* 1464–67; Cole, "Maturation," 239.
16. Mommsen, "Ράκος," cited in Cole, "Maturation," 239, n. 33.
17. Pausanius, *Description,* 1.16.2–8; in *Maenads,* 20.
18. See Lincoln, "Persephone," 227; Arthur, "Politics," 9–10.
19. Arthur, "Politics."

20. Lincoln, "Persephone."

21. At Athens, Piraeus, Eleusis, Halimous, and Kolias. It is documented for at least thirty cities in Greece, Asia Minor, and Sicily: Farnell, *Cults*, 3: nos. 75–105.

22. See Brumfield, *Demeter*, 70–95 for the problems. The basic description comes from Rabe, *Scholia in Lucianum*, 275–76: I rely here on Brumfield's critique and analysis.

23. Diodorus, 5.4.

24. Scholia on Aristophanes, *Thesmophoriazusae*, 80; Photius, Σήνια (sic); Hesychius, Στήνια στηι ῶσαι.

25. Protrepticos, 2.17–18, cited in Lincoln, "Persephone." In the *Hymn to Demeter*, the child's name is Demophoon. Iakhos is probably a form of Dionysos/Bacchos, who was eventually associated with Demeter at Eleusis. Mylonas argued, however, that the account of Clement was largely unreliable (*Eleusis*, 304–5).

26. See H. Jones's note 1 to Strabo, 9.400, in the Loeb edition.

27. Brumfield, 84–88. See also Johansen, "Thesmophoria."

28. The only account of the ritual at the Haloa comes from a scholion to Lucian (see n. 22), 279: its source may have been Apollodorus.

29. Skov, "Priestess," 141–42.

30. Pausanius, *Description*, 2.35.6–8.; in *Maenads*, 19.

31. Pausanius, *Description*, 7.27.9–10; in *Maenads*, 22.

32. Winkler, "Laughter." For another new reading of Baubo and sexuality in antiquity, see Olender, "Baubo."

33. For example, Pomeroy, *Goddesses*; Cantarella, *Women*; Lefkowitz, *Women*; Skinner, *Creusa*; Sealey, *Women*. See also Pomeroy, "Bibliography."

34. In 1920, Helen McClees observed, "It was the women of high station that Athenian ideas of propriety would most restrict, while on account of wealth and position they had none of the freedom which the necessities of common life, as earning part of the family income, marketing, washing in the streams or at the fountains, or working in the fields, gave to the poor woman. As usual, the women of the middle class, to whom both kinds of opportunity were denied, had a life of less variety than the other two" (*Women*, 5). The division of Attic society into upper, middle, and lower classes may now be considered anachronistic, but much of McClees's perception remains pertinent!

## Chapter 3

1. Winkler, "Laughter," 189–93.

2. Ibid., 204.

3. Ibid., 205.

4. Ibid., 206.

5. Detienne, *Gardens*.

6. Winkler, "Laughter," 206.

7. Theocritus, *Idyll*, 15; excerpted in *Maenads*, 6.

8. Griffiths, "Theocritus."

9. Saller, "Marriage."

10. Griffiths, "Theocritus," 257.

11. Ibid., 247.

12. Ibid., 258.

13. Detienne, *Gardens*; Griffiths, "Theocritus," 255.

14. Griffiths, "Theocritus," 255.

15. On ritual reversal as affirmation of the social order, see Turner, *Ritual Process*, 166–203, especially 200–3.

16. Griffiths, "Theocritus," 258.

### Chapter 4

1. Much of this section utilizes and revises a portion of my doctoral dissertation (*Ecstatics*) subsequently published as "Attraction of Women." There is an extensive literature on the cult of Dionysos: for bibliography before 1976 see Kraemer, "Attraction of Women." Among subsequent studies, those especially pertinent to the study of women's worship include Cole, "New Evidence"; Henrichs, "Male Celebrant," "Greek Maenadism," and "Dionysiac Identities"; Keuls, *Sexual Politics* and "Dionysiac Ritual"; Segal, "Dionysos"; and Bremmer, "Maenadism." See also the essays in *L'association dionysiaque*.

2. On the Anthesteria and the Lenaia, see Farnell, *Cults*, 5:208–24; Deubner, *Attische Feste*, 93–134.

3. In Euripides, *Bacchae*; excerpted in *Maenads*, 1, 2. Vase paintings depict ecstatic female followers of Dionysus as early as the sixth century B.C.E. See Philippart, "Iconographie"; Lawler, "Maenads"; Keuls, "Dionysiac Ritual"; and Frontisi-Ducroux, "Ménadisme."

4. The name *maenads* comes from the Greek verb μαινομαι (to be driven mad). On the term Βακχαι as epithet, see Cole, "New Evidence," 229.

5. For example, Diodorus, 3.62–74; 3.67ff., 4.4; Philostratus, *Apollonios*, 2.9; Pausanius, *Description*, 3.24.3; Apollodorus, *Library*, 3.4.2–3. For full references, see Farnell, *Cults*, 5:85–344.

6. Dodds, "Introduction" to his edition of the *Bacchae*, xx–xxv.

7. Ibid., xxii–xxiii.

8. The passage most often cited in support of this is from Demosthenes, *On the Crown*, 259–60 (*Maenads*, 3), which explicitly refers only to Sabos. I have questioned the assumption that this is a shortened version of Sabazios: see 39–40; see also Kraemer, "'Euoi Saboi'" and "Attraction of Women," 61–63.

9. Henrichs, "Greek Maenadism."

10. For various allusions to Dionysiac madness, see Kraemer, "Attraction of Women," 59–65.

11. Kraemer, "Attraction of Women," 70. In "Male Celebrant," Henrichs points out that I and other scholars have been too quick to accept the reality of a male celebrant in the cult, with which I concur. Henrichs and others call this form of Dionysian devotion *maenadism*, or *ritual maenadism*, presumably to distinguish it from other forms of Dionysian worship, but such terminology becomes problematic when we encounter increasing evidence for male participation in such cults in the Hellenistic period.

12. Βακχων πάρα, line 1224.

13. Translation G. S. Kirk.

14. Lines 1125–36. This passage and one other (lines 734–47, the destruction of

the cattle) strongly suggest that the women enact a two-part ritual sacrifice, consisting of the tearing apart (*sparagmos*) and consumption of raw flesh (*ōmophagia*). The wording of line 1114 "First his mother started the slaughter as priestess" supports the interpretation that rending apart the body of Pentheus is a ritual sacrifice.

15. Cole, "New Evidence," also *LSCG suppl.*, 120.

16. Cole, "New Evidence," 230.

17. See below, 39–40.

18. She cites also *LSCG suppl.*, 120, a fifth-century chamber tomb inscription from Cumae: οὐ θέμις θέμις εντοῦθα κεῖσθαι ἰ μὲ τὸυ βεβαχχευμένον (it is not right for anyone to lie here who has not become a Bacchic). Cole translates this as "It is not right for anyone who has not become a *bakchos* [masc.] or *bakchē* [fem.] to lie here" ("New Evidence," 231).

19. This may pose methodological problems, since the terms Βάκχαι, and so on, may refer to these afterlife-oriented mysteries as much as anything else, but in the literary sources, this will not be such a problem, for the actual rites of the women and the exclusion of men become clearer.

20. *LSCG*, 127. Discussion in Henrichs, "Male Celebrant," 81.

21. Demosthenes, *On the Crown*, 259–60; in *Maenads*, 3.

22. On *Kittophoros* and *Liknophoros* in Dionysiac contexts, see Nilsson, *Dionysiac Mysteries*, 38–45.

23. But see Kraemer, "'Euoi Saboi.'"

24. Cole, "New Evidence," 236.

25. *LSAM*, 48; in *Maenads*, 7.

26. For example, Diodorus Siculus, 4.3.2–5; in *Maenads*, 13. Also Pausanius, *Description*, 10.4.3.

27. Henrichs, "Greek Maenadism," 149.

28. Ibid., 148; also in *Maenads*, 8.

29. *I. Magn.* 215 = *Maenads*, 9. The portion in *Maenads* is attributed to Delphi, although the inscription itself came from Magnesia because it records an oracle from Delphi, which the Magnesians sought to explain a portent.

30. On the date of the oracle, as opposed to the date of the inscription, see Henrichs, who concludes that this consultation of the oracle must have taken place between 278 and 250 B.C.E. ("Greek Maenadism," 127). Although other scholars have questioned the historicity of the oracle, Henrichs argues for its veracity (129).

31. Ibid., 133. The "scholarly" consensus is that the name refers to one who descends in lightning, an epithet sometimes applied to Zeus. Henrichs points out, of course, that the birth of Dionysos is connected with a lightning bolt of Zeus: conceivably here it applies to Dionysos himself, who "descends" in a slightly different manner in lightning!

32. See n. 26.

33. τὰς δὲ γυναῖκας κατὰ συστήματα θυσιάζειν τῷ θεῷ

34. Plutarch, *On the First Cold*, 18 (953D).

35. Plutarch, *Bravery of Women*, 13; *Maenads*, 17.

36. Plutarch, *On Isis and Osiris*, 364 E.

37. Plutarch, *Life of Alexander*, 2.1–5; *Maenads*, 18.

38. Pausanius, *Description*, 10.4.3.

39. See, for example, Cole, "New Evidence."

40. Livy, *Annals*, 39.8–19 (39.8–18 in *Maenads*, 110).

41. See, for example, Walsh, *Livy*, 13–14; 39; North, "Religious Toleration."

42. *ILS*, 18 = *ILLRP*, 511 = *CIL* I², 581: English translation in Lefkowitz and Fant, *Women's Lives*, 243 (250–52). For a detailed discussion of the differences between Livy's account and the Senate decree, see North, "Religious Toleration."

43. Livy, *Annals*, 39.15.9.

44. Lewis, *Ecstatic Religion*.

45. Ibid., 101.

46. *Bacchae*, 35–36: καὶ πᾶν τὸ θῆλυ σπέρμα Καδμείων, ὅσαι γυναῖκες ἦσαν, ἐξέμηνα δωμάτων. Conceivably, the phrase "all the female seed of the house of Camdus" excludes slaves. On the interpretation of the qualifying phrase "those who were women" see Dodds, *Bacchae*, 67. *Bacchae*, line 694, specifies that all categories of women were present: young, old, and virgins.

47. *Bacchae*, lines 695–711; in *Maenads*, 2.

48. *Natural Symbols*, especially Chapter 5, "The Two Bodies" (93–112).

49. Lewis, *Ecstatic Religion*, 85–89. Lewis suggests that in the case of sar and bori cults, male tolerance reflects "a shadowy recognition of the injustice of this contradiction between the official status of women and their actual importance to society," 88.

50. *LSAM*, 48; in *Maenads*, 7. See 41 for discussion.

51. On the participation of Cadmus and Teiresias, see Kraemer, *Ecstatics*, 48–57 and "Attraction of Women," 69–71; Evans, *God of Ecstasy*, 24–38, especially 25–27.

52. I do not know if she has said this in print: she emphasized this in her 1988 Princeton seminar.

## Chapter 5

1. See, for example, Orr, "Household Shrines" and Scullard, *Festivals*, 17–18.

2. See Chapter 8 for an extended discussion of the Vestals.

3. Plutarch, *Life of Numa*, 10; *Maenads*, 79.

4. On marriage *cum manu* and Roman marriage generally, see Gardner, *Roman Law*, 31–80.

5. Rawson cites a fourth-century Christian writer, Prudentius (*Contra orationem Symmachi*, 1.199–214), who describes a Roman mother praying before household cult objects, "Roman Family," 19. In the late first/early second century C.E. Plutarch could state unequivocally that wives should worship the gods of their husbands (*Moralia*, 140D). But Plutarch was a Greek, and his point here may have less to do with household worship than with criticizing women who participate in various religious activities opposed by their husbands. His intriguing argument here (whether or not women found it convincing) is that the gods take no pleasure in "stealthy and secret rites" performed by women (where the stealth and secrecy presumably refer to the concealment of such rites from the husband).

6. Valerius Maximus, 1.5.4. I am indebted to Amy Richlin for sharing this reference and those in notes 7, 9, 11, and 12.

7. Valerius Maximus, 2.1.6.

8. Richlin, "Rituals."

9. Aulus Gellius, *Attic Nights*, 16.16.4.

10. I leave aside the question of resident aliens on the theory that their own social customs prevailed here.

11. On the dress of matrons, Paulus ex Festus, 112L; on their special carriages, Paulus, 225L and Pompeius Festus, 282L; on the special treatment of matrons by magistrates, Pompeius Festus, 142L and Valerius Maximus, 2.1.5. On the taxation of the *ordo matronarum,* see Valerius Maximus, 8.3.3. On the Oppian law and its repeal, see Livy, *Annals,* 34.1.1–8.3; see also Pomeroy, *Goddesses,* 176–85 and below, 56–59.

12. Varro, *On the Latin Tongue,* 5.68–69; also Paulus, 131L. See also Gagé, *Matronalia* and Scullard, *Festivals,* 85–87.

13. Below, 57–58; see also Scullard, *Festivals,* 96–97.

14. Below, 55; see also Scullard, *Festivals,* 116–17.

15. Below, 61–64; 66–68; see also Scullard, *Festivals,* 149–50 on the Vestalia; 150–151 on the Matralia.

16. Below, 53–55; see also Scullard, *Festivals,* 199–201.

17. Ibid., 62–64.

18. Brouwer, *Bona Dea.*

19. Juvenal, *Satire,* 6.

20. See n. 18.

21. Collected with English translations in Brouwer, *Bona Dea,* 144–228. Brouwer observes that while Cicero is not the most exhaustive source on the goddess, she figures centrally in his writings, and Brouwer actually suggests that most subsequent literary testimony is highly influenced by Cicero's portrait: 144, 266, 398, and elsewhere. Brouwer provides a fascinating analysis of Cicero's motives, 360–70.

22. Ibid., 255–56 and notes for extensive ancient references, most of which are included in the volume.

23. Ibid., 231–53.

24. On the cult officials, see Chapter 8.

25. Plutarch, *Cicero,* 20 in Brouwer, Chapter II, no. 48 with important discussion, 361–63.

26. Clodius died during a skirmish with T. Annius Milo (a friend of Cicero's) and his armed companions. Cicero subsequently defended Milo from the resulting charges: *Pro T. Annio Milone Oratio.* See especially 27.72–73 and 31.86 for Cicero's remarks about the death of Clodius, reproduced with discussion in Brouwer, 170–72.

27. Brouwer, *Bona Dea,* 264.

28. Ibid. Elsewhere Brouwer proposes that "Cicero is the one to create an image of Bona Dea as a *locus communis* —the exponent of Roman tradition. . . ." 266.

29. Ibid., 323–99, especially 358–70 on the December festival and 370–72 on the May festival.

30. Ibid., 369.

31. Ibid., 372.

32. Pomeroy, *Goddesses,* 206–7.

33. Ibid., 206.

34. Detailed in Palmer, "Female Chastity."

35. This assessment is based on my own preliminary search for inscriptions mentioning Venus Obsequens, Venus Verticordia, Pudicitia, and Matuta. A few inscriptions to the latter are known from her temple at Satricum: *AE,* 1979, 136.

36. See, for example, Pomeroy, *Goddesses;* Hallett, *Fathers and Daughters;* Dixon, *Roman Mother;* Rawson, *Family;* Rousselle, *Porneia.* For additional references, see Pomeroy, "Bibliography."

37. On economic implications of war for women, see Pomeroy, *Goddesses,* 181; on the relative inefficacy of guardians, 151, 179–80. See also Gardner, *Roman Law,* 14–22

on guardians for women. *NewDocs* 2 (1982[1977]), 29–31 offers a useful discussion of papyrus documents in which women act without guardians.

38. On *patria potestas*, see Gardner, *Roman Law*, 5–11; Lacey, "Patria Potestas."

39. The whole affair is chronicled in Livy, *Annals*, 34.1.1–34.8.3. According to Livy, Lucius Valerius defended the position of the *matronae* who sought the repeal of the Oppian law by pointing out, among other things, that beautiful clothes and jewelry were one of the few rewards available to women, in comparison to the many offices and honors afforded (elite) men: "No offices, no priesthoods, no triumphs, no decorations, no gifts, no spoils of war can come to [women]; elegance of appearance, adornment, apparel—these are the woman's badges of honor (*insignia*); in these they rejoice and take delight; these our ancestors called the woman's world" (34.7.8–9).

40. Pomeroy, *Goddesses*, 181.

41. Valerius Maximus, 8.15.12.

42. Porte, *Les Fastes*, 391–93.

43. Pliny, *Natural History*, 7.120.

44. Ovid, *Fasti*, 4.247–348 and Livy, *Annals*, 39.14; both in *Maenads*, 128.

45. Plutarch, *Roman Questions*, 83.

46. On the Vestals, see Chapter 8.

47. Ovid, *Fasti*, 4.133–62; in *Maenads*, 12.

48. "Sub illa et forma et mores et bona fama manet" (Ovid, *Fasti*, 4.155). The mixture of poppy seed, milk, and honey was apparently a draught drunk at weddings.

49. Scullard (*Festivals*, 96) attributes the note to Verrius Flaccus.

50. Pomeroy, *Goddesses*, 208–9.

51. Stehle, "Venus." See also Pagnotta, "Il Culto."

52. Livy, *Annals*, 34.53.5–6. He also indicates (34.53.3) that a temple to Iuno Matuta was dedicated at the same time—an odd combination of the two goddesses, if correct.

53. Pomeroy, *Goddesses*, 207.

54. Livy, 10.23.3–10.

55. Livy, 10.31.9. On the Verginia dedication, see also Palmer, "Female Chastity," who argues that cults of chastity did not die out at the end of the republic.

56. Wardman, *Religion and Statecraft*, 38–39.

57. Pomeroy, *Goddesses*, 209.

58. Livy, 34.2.14–34.3.3.

59. No other sources confirm it, and some scholars suggest that Livy has confused Pudicitia here with Fortuna Virgo and related this story to explain the epithet Plebeia. See the note by B. O. Foster to the Loeb Classical Library translation of Livy, vol. 4 (1926): 442–43. See also Palmer's position (n. 55).

60. *Amores*, 1.3, 1.4, *61*. "In the case of Ovid's Corinna there can be no doubt, on the one hand, that she is synthetic, and on the other that the experiences out of which she was created were real." (Hadas, *Latin Literature*, 204–5).

62. *Amores*, 2.10.

63. *Amores*, 3.14.

64. On Isis, see Chapter 7; on Dionysos, see Chapter 6.

65. Ovid, *Fasti*, 6.473–568; in *Maenads*, 11.

66. Tertullian, *On Monogamy*, 17.

67. Plutarch, *Roman Questions*, 267D.

68. See Chapter 4, n. 5.

69. On the nurse nymphs of Dionysos: *Orphic Hymn*, 46; *Iliad*, 6.130ff.

70. Apollodorus, *Library*, 1.9.1–2; elsewhere, Apollodorus reports a myth that ties together the birth of Dionysos and the death of Ino's children, according to which Hera drove Ino's husband, Athamas, insane because he had rescued the infant Dionysos and raised him as a girl! (3.4.3). Pausanius records several versions of this myth: in one, when Ino threw herself and Melicertes into the sea, Melicertes was rescued by a dolphin. Subsequently, he was renamed and honored as Palaemon (*Description*, 1.44.7–8). Earlier, Pausanius tells the story that Ino died but was then worshipped as Leucothea (1.42.6). He also records a very different version of the tale according to which Cadmus put Semele and Dionysos into a chest, which washed up at Brasiae, where Semele was found dead, but Dionysos survived. Ino then came to Brasiae and became the nurse of Dionysos! (*Description*, 3.24.3). For other versions of Ino and her children, see also Ovid, *Metamorphoses*, 4.481–542; Hyginus, *Fables* 4 and 5.

71. Interestingly, Παλαιμον is an epithet applied by Euripides to Melicertes, the original son of Ino, *Iphigenia at Taurus*, 271: Pausanius also associates it with Melicertes (see n. 70).

72. Porte provides some useful discussion of Ovid's threefold approach to his material, presenting rites, history, and myth (*Les Fastes*, 26–30).

73. Ovid, *Fasti*, 6:559–62.

74. Paulus ex Festus, 380L. Some scholars have argued that *sororii* derives not from the Latin word for sister, *soror*, but from *sororiare*, which Festus defines as "the growth of the female breasts." Scullard proposes that early on, protection was sought for adolescent children (*Festivals*, 151).

75. τοῖς [τεκνοῖς] δε τῶν ἀδελῶν. . . . (Plutarch, *Roman Questions*, 17).

76. Livy, *Annals*, 24.47.16.

77. Livy, *Annals*, 25.7.5–6.

78. Livy, *Annals*, 6.33.5, 7.27, 28.11.1.

79. French, "Mater Matuta," 12.

80. See n. 4 on marriage. See also Rawson, "The Roman Family," especially 19–31; Pomeroy, "Married Woman"; Hallett, *Fathers and Daughters*, 90, n. 34.

81. See n. 36.

82. See Hallett, *Fathers and Daughters*, 64–69 on the concept of *filiafocality*, "daughter focus," and 76–149 for some of its central ramifications.

83. While elite Roman (male) society in many ways fits the description of strong group and strong grid (hierarchy), elements of the Roman patronage system suggest that at least a few of the Roman elite may have experienced the diagonal opposite of weak group and grid, operating as what Douglas calls "Big Men."

84. Rawson, "Roman Family," 18–19.

85. See n. 5.

86. Rawson, "Roman Family," 20. Rawson points out that women who bore their husbands' children and who were survived by their husbands were likely to have been buried in his family tomb. She observes, though, that there are some married women buried by their fathers, and suggests the need for further study.

87. Ibid., 30. In the same volume, Keith Bradley suggests that the frequent use of wet nurses among the upper classes "fulfilled for the parent a self-protective function, diminishing the degree and impact of injury in the event of loss in a society where such loss was commonly experienced" ("Wetnursing," 220). Bradley has in mind the losses caused by high infant and childhood mortality, and does not distinguish between the losses to fathers and the losses to mothers, but it would seem that in addition to the real

possibility of losing a child to illness and accident, Roman mothers had to consider the loss of a child through divorce. All of this has implications for the appeal of ascetic Christianity to upper-class women.

88. Hallett, 138–44.

89. Veyne, *Private Life*, 77–81. See also Boswell, "Abandonment."

90. According to Paulus ex Festus, women called *paelices* (singular, *paelex*) were not allowed to touch the altar of Juno, and had to sacrifice a female lamb as a penalty if they did so (248L). I owe this reference to Amy Richlin. In this text, paelices are those who "lie under those not belonging to them," that is, extramarital sexual partners. Richlin proposes that the paelex is forbidden to sacrifice to Juno, the patron goddess of marriage, because the paelex bears the brunt of the responsibility for marital discord, and also because the very nature of the paelex's sexual contact with the husband renders her impure in some sense.

91. On the specific dynamics of the mother-son conflict in ancient Greece, see the psychoanalytically oriented work of Philip Slater, *Glory of Hera*: critique in Kraemer, *Ecstatics*, 98–112 and in Arthur, "Review," 395–97.

92. Hallett, 150–92.

93. See n. 91.

94. Douglas's discussion (*Cultural Bias*, 25) on attitudes toward foreign travel allows us to hypothesize predicted attitudes toward other aspects of foreign culture, including attitudes toward foreign gods. For persons experiencing strong group and grid, foreign travel is viewed negatively and foreign space is seen as unnatural. The component of strong group, in particular, correlates with an opposition to things foreign, which presumably extends to foreign gods as well. By contrast, Douglas observes that persons experiencing weak group with strong grid have little contact with foreigners, and so entertain rather wild ideas about them. But the foreign is not inherently unnatural (as with strong group and grid) or evil (as it is with strong group and weak grid).

95. See especially Rousselle, *Porneia*, 47–62.

96. Ibid., 56–57.

97. Hallett, *Fathers and Daughters*, 83–89.

### Chapter 6

1. See Heyob, *Isis among Women*, 37–38, for references.

2. Plutarch, *On Isis and Osiris*, 12–19; in *Maenads*, 131.

3. Plutarch, *On Isis and Osiris*, 38–39.

4. See Heyob, *Isis among Women*, 54–57.

5. Apuleius, *Metamorphosis*, 11:8–17.

6. Ibid., 11.10: Viri feminaeque omnis dignitatis et omnis aetatis.

7. Ibid., 11.11.

8. Idem. Translation W. Adlington, revised by S. Gaselee.

9. The word πλοιαφεσια is an emendation: Gaselee, 568–69, n. 1.

10. *IG*, 12 Suppl., 557 = Vidman, *Sylloge*, 80.

11. Heyob, *Isis among Women*, 56–57.

12. Apuleius, *Metamorphosis* 11; excerpted in *Maenads*, 132.

13. Witt, *Isis*, 91–92.

14. *The Kyme Aretalogy*; in *Maenads*, 133.

15. *P. Oxy.*, 11.1380; in *Maenads*, 133.

16. Heyob, *Isis among Women*, 81–95, collates the data from 1,099 inscriptions in Vidman, *Sylloge*, concluding that 200, or 18.2 percent, mention women who were priestesses, members of cult associations, or ordinary devotees (81). Heyob was not counting individuals, but rather the number of inscriptions that mention women. Just before the present work went into production, I became aware of a new study that attempts a more statistically and methodologically sophisticated analysis of the presence of women in the cult of Isis. Mora (*Prosografia Isaica*) reports a total percentage of 15.6 percent for Isiac women (n = 169), with 846 men and 69 unidentified by gender. Mora's tables break down Isiac devotees by gender according to several significant categories: geographic region, chronological period, cult association membership lists (12.5 percent women), manumissions (30.9 percent), individual dedications (17.9 percent), and inventories of offerings (22.3 percent) (131–61).

17. On women in Isiac office, see Heyob, *Isis among Women*, 87–110.

18. Ibid., 80, and elsewhere.

19. Xenophon of Ephesus, *Ephesian Tale* (the story of Anthia and Habrocomes). See Griffiths, "Xenophon . . . on Isis."

20. See, for example, *ILS*, 1259–61, in *Maenads*, 124, inscriptions for Fabia Aconia Paulina, whose husband Vettius Agorius Praetextatus initiated her into all the mysteries, including those of Isis, where she held the office of priestess. See also *ILS*, 9442 = Vidman, *Sylloge* 586, discussed in Heyob, *Isis among Women*, 62–64.

21. For example, Murray, *Greek Religion*; Dodds, *Pagan and Christian*.

22 See Smith and Lounibos, *Pagan and Christian*.

23 The same concern may be raised about Mora's statistics as well.

24. See n. 15.

25. Mora also seconds the criticism by Françoise Dunand, who takes issue with Heyob's assumption that Isiac religion (and Greco-Roman religion generally) was primarily about the identification of the worshipper with the deity, suggesting that this represents a misguided retrojection of Christian identification with Christ onto non- Christian religion (*Prosopografica Isiaca*, 2, n. 9).

26. Dunand, "Hiereiai."

27. A point Heyob asserts (*Isis among Women*, 87) but never demonstrates.

28. Dunand, "Hiereiai," 352, translation mine.

29. Ibid., 374.

### Chapter 7

1. See, for example, Swidler, *Women Priests*, Paulist Press, 1977; Kendall, *Bibliography*.

2. Beard and North, *Pagan Priests*.

3. Ibid., 2.

4. McClees, *Women in Attic Inscriptions*, 5–16 (summary list, 45) identified all the women who held cultic office in the inscription from Attica collected in *IG* II and III (and in a few other sources of inscriptions as well). These volumes of *IG* were subsequently superseded by *IG²* II and III, so that McClees's citations require updating. A sampling of inscriptions may be found in Pleket, *Epigraphica*; in *Maenads*, 78, 81–83.

For women cult officers in the worship of Isis, see Heyob, *Isis among Women*, 87–110; in the worship of Bona Dea, see Brouwer, *Bona Dea*, 371–85; in service to Cybele, see Vermaseren, *Corpus Cybelae*. See also Macmullen, "Women in Public" and van Bremen, "Women and Wealth."

5. Garland, "Priests," 77.

6. Ibid., 77–78; for references, 78, n. 3.

7. Ibid., 80.

8. Beard, "Priesthood."

9. Hallett, *Fathers and Daughters*, 84; Beard, "Vestal Virgins," 14–15.

10. Hallett, *Fathers and Daughters*, 84, especially n. 29.

11. Plutarch, *Life of Numa Pompilius*, 10; in *Maenads*, 79.

12. Aulus Gellius, *Attic Nights*, I.12; in *Maenads*, 80.

13. Beard, "Sexual Status," 21.

14. Ibid., 13.

15. Ibid.

16. Ibid., 24–25.

17. Hallett, *Fathers and Daughters*, 85.

18. Ibid., 87.

19. Ibid., 89, n. 33.

20. Livy, *Annals*, 22.57.2, for the deaths of Opimia and Floronia.

21. For the deaths in 114 B.C.E., see Livy, *Periochae*, 63: see also further references in Hallett, *Fathers and Daughters*, 88, n. 32.

22. Pomeroy, *Goddesses*, 211.

23. Hallett, *Fathers and Daughters*, 87.

24. Chrysis, *IG* II², 1136; in *Maenads*, 78.

25. Pleket, *Epigraphica*, 18; in *Maenads*, 81.

26. Heyob, *Isis among Women*, 87–110; see also Dunand, "'Hiereiai'" for a discussion of women cult officers in Egyptian religion prior to its transformation in the Hellenistic period and of Hellenistic Egyptian priestesses.

27. Popescue, "Histrian Decree." For priestesses of Cybele, see Vermaseren, *Corpus Cybelae*. Interestingly, women represent about half of the persons identified as *sacerdos* of the great Idean mother of the gods in Rome and Latium: nos. 231 (*CIL*, 6.502); 235 (*CIL*, 6.508); 258 (*CIL*, 6.2257)—a *sacerdos maxima*; 291 (*CIL*, 6.2260); 360 (*CIL*, 6.2259); 423 (*CIL*, 14.371); 442 (*CIL*, 14.408); 476 (*CIL*, 10.6075); 477 (*CIL*, 6.30972); and 478 (*CIL*, 10.6074)—the same woman. The vast majority of taurobolium (bull sacrifice) performances are by men.

28. For a detailed discussion of officials in the imperial cult in Asia, see Kearsley, "Asiarchs."

29. See n 25.

30. On the expenditures entailed, see especially van Bremen, "Women and Wealth."

31. The decree for Aba of Histria, 11.22–25; discussion in Popescu, "Histrian Decree," 284.

32. Ibid., 285. Popescue consistently refers to the monetary unit as a *denarius*, and the plural as *denarii*, which is the Latin terminology. But the inscription is in Greek and uses the singular *denarion* and the plural *denaria*, which I have followed here.

33. Ibid., 286.

34. *AE*, 1971:85, cited and discussed in MacMullen, "Women in Public," 212–13.

35. Found at Assar Koy in the 1880s. Principal edition: Lanckoronski, *Städte* I–II, nos. 58, 59, 60, and 61. All but 61 in *IGRR* III, 800–2; see also G. Radot-P. Paris, "Inscriptions." Discussion in both Gordon, "Veil of Power" and van Bremen, "Women and Wealth."

36. Gordon, "Veil of Power," 228.

37. Chrysis, *IG* II², 1136; in *Maenads*, 78.

38. Tation, *CIJ*, 738; in *Maenads*, 60.

39. Popescue, "Histrian Decree," 280, 282–83.

40. Van Bremen, "Women and Wealth," 224.

41. Tation, *CIJ*, 738; in *Maenads*, 60.

42. For additional instances, see Kearsley, "Asiarchs," 186, n. 15, who emphasizes the centrality of the natal family and the father in particular determinations of social status and rank.

43. Brooten, *Women Leaders*.

44. Such views continue to be expressed on occasion, most recently by scholars such as Macmullen, "Women in Public," 214, on the high priestess of the imperial cult in Asia; against this view, see Kearsley, "Asiarchs."

45. Kearsley, "Asiarchs."

46. *I.Magn.*, 158.

47. Kraemer, "Women Elders."

48. Kearsley, "Asiarchs," 192, n. 32.

49. A lengthy treatment of the benefaction system, called *euergetism* (from the Greek) may be found in Paul Veyne, *Le pain et le cirque* (Paris: Editions du Seuil, 1976), Chapter II, 185–373 (also abridged, *Bread and Circuses*). A more recent, briefer treatment occurs in Gordon, "Veil of Power," especially 229–31. See n. 35. See also Danker, *Benefactor*.

50. Macmullen, "Women in Public," 215.

51. Ibid., 213, relying on data from a Yale dissertation (1978) subsequently published as Harl, *Politics*.

52. Van Bremen, "Women and Wealth," 233.

53. Ibid., 236–37.

54. For Artemis at Ephesus, Strabo, *Geography*, 3.4.8, 4.1.4–5, 14.1.21–26; on Cappadocian Comana, 12.2.3; and on Pontic Comana, 12.3.32–36. I am indebted for these references and an insightful preliminary discussion of Strabo to Amy Wordelman, "Strabo."

55. Strabo, *Geography*, 12.3.36.

56. See Chapter 6, 55.

57. *CIL*, 10.1549, in Brouwer, *Bona Dea* Chapter I, no. 79: dated to October 27, 62 C.E.

58. Brouwer, *Bona Dea*, 377–78.

59. Ibid., 378–83.

60. Ibid., 382.

61. Ibid., Chapter I, no. 113. Much of Brouwer's discussion of collegia in general is taken from Kornemann, *Collegium*.

62. Alexander in Lucian, *Alexander, or the False Prophet*; Apollonius of Tyana in Philostratus, *Life of Apollonius of Tyana*.

63. Pomeroy, *Goddesses*, 132; see also Snyder, *Women Writers*, 108–13. For a some-

what different view of Pythagorean women philosophers, see Waithe, *Women Philosophers*, 11–74, 127–32. See also Menage, *Women Philosophers*.

64. For example, the epistles of Crates to Hipparchia, nos. 28–33; Diogenes to Hipparchia, no. 3, in Malherbe, *Cynic Epistles*.

65. For example, Musonius Rufus, *That Women, Too, Should Study Philosopy* and *Should Daughters Receive the Same Education as Sons?*

66. An ancient brief account of the life and death of Hypatia may be found in Socrates, *Historia Ecclesiastica*, 7.15. Snyder, *Women Writers*, 113–20, provides a thorough, concise discussion of Hypatia, with useful bibliography in her notes.

67. Juvenal, *Satire*, 6; in Maenads, 25. Some other interesting examples are adduced by Lillian Portefaix in an otherwise seriously flawed study, *Sisters Rejoice*, 57.

68. For texts and discussion of Greco-Roman magic generally, a long neglected area that is now coming under serious scholarly scrutiny, see Betz, *Greek Magical Papyri*; Gager, *Curse Tablets*; Luck, *Arcana Mundi*.

69. Cole, "Male and Female."

70. *LSCG*, 65.8.

71. (Pseudo) Demosthenes, *Against Nearaea*, 75.

72. Alternatively, they may perhaps be considered within Douglas's category of "Big Men" who personally experience little constraints while exerting enormous control over others.

73. Beard, "Priesthood," 30–34.

## Chapter 8

1. Recent archaeological work in Roman Palestine supports an increasingly complex portrait of Jewish life. See, for example, Meyers, "Early Judaism" and Groh, "Jews and Christians."

2. A very useful survey of recent research may be found in Saldarini, "Rabbinic Judaism." On the utility of rabbinic sources as historical evidence, consult the many works of Jacob Neusner; Goodman, *State and Society*; Lightstone, "Judaic Context."

3. Saldarini, 440–45.

4. The Jerusalem Talmud is traditionally dated to c. 400; the Babylonian to c. 500 C.E., but the dating both of the compilation of the two Talmuds and of the traditions within them is complex and probably provisional.

5. Baskin, "Separation of Women," 7.

6. *p. Shabbat* 2, 5b, 34; *Genesis Rabbah* 17:7, cited in Baskin, "Separation of Women," 7.

7. So *m. Shabbat* 2:6, *t. Shabbat* 2:10; *b. Shabbat* 31b–32a; *p. Shabbat* 2, 5b, 34; *Genesis Rabbah* 17:8, cited in Baskin, 7–8.

8. *m. Pesah* 8:7 forbids women to assemble together to sacrifice the Pascal lamb. "And one may not constitute a fellowship group (haburah) of women, slaves and/or minors. [Such a gathering, lacking a free adult male leader, cannot qualify as a fellowship group.]" Quoted from Wegner, *Chattel or Person*, 148, emphasis original. Wegner implies that such a prohibition extended generally to religious fellowship groups of women.

9. There are, however, tantalizing references in John Chrysostom to a synagogue called "of Matrona" outside Antioch in the suburb of Daphne (*Against Judaizing*

*Christians*, 1.6.2; *Comm. on Titus*, 3.3 [*PG*, 62.679]). Whether this refers to its location (apparently near a shrine to a deity called Matrona), or to its patron, or even to its membership, is unknown: for some discussion see the notes in Harkins's translation, 22, n. 73. A more probable candidate for a female assembly may be found in the New Testament Acts of the Apostles 16:12–13. "We remained in this city [Philippi] some days; and on the sabbath day we went outside the gate to the riverside, where we supposed there was a place of prayer (προσευχή); and we sat down and spoke to the women who had come together. One who heard us was a woman named Lydia, from the city of Thyatira, a seller of purple goods, who was a worshiper of God." For discussion, see Kraemer, "Monastic Jewish Women," 367–69.

10. Josephus, *Jewish War* V.198–200; *Against Apion* II.102–4, reproduced in *Maenads*, 16.

11. *b. Kiddushin* 29a–b; in *Maenads*, 36.

12. See, for example, Meiselman, *Jewish Woman*.

13. Wegner, *Chattel or Person*, 154–55, emphasis original.

14. Jacob Neusner suggests that certain aspects of the Temple cult were not immediately abandoned, in the belief that the Temple would be restored within a reasonable amount of time, as it had been previously. Among the rites he considers candidates for some form of continuity was the ordeal of the wife suspected of adultery (*Mishnaic Law of Women*, 5:197). But it seems less likely that they would have continued after the defeat of the Bar Kochba rebellion in 135 C.E., which must have severely dampened expectations that the Temple would be restored soon: see Shukster and Richardson, "Temple and *Bet Ha-Midrash*."

15. Exodus 23:17 says only that males shall assemble before the Lord; Deuteronomy 16:16 requires all males to appear before the Lord "at the place which he will choose," which the rabbis, of course, construed to mean in Jerusalem at the Temple. See Wegner, *Chattel or Person*, 157.

16. Ibid., 156.

17. *m. Ketubot* 5:5, 5:9. See also Wegner, *Chattel or Person*, 76–77, where she points out, among other things, that a wife without slaves was expected to do the work of three!

18. Numbers 15:17–21.

19. *b. Sukkah* 28a–b exempts women, slaves, and minor children from dwelling in a *Sukkah*. But the same passage recounts the story of Shammai, who apparently thought that minor male children were still liable, for when his daughter-in-law gave birth to a baby boy during Sukkoth, he is said to have broken away some of the plaster in the roof over her head, so that the baby would be in a dwelling open to the sky, in conformity with the requirement for Sukkoth.

20. *m. Yebamot* 6.6 gives a minority opinion that women were also obligated to procreate.

21. *b. Kiddushin* 34a.

22. *b. Rosh Hashanah* 32b.

23. *m. Megillah* 2:4.

24. A. And all may be included in the quorum of seven [called to read from the Torah at Sabbath worship], even a woman or a minor.

    B. [But] one does not bring a woman to read [the Torah] in public.

    *t. Megillah* 2:11 cited in Wegner, *Chattel or Person*, 158.

25. *b. Megillah* 23a, cited and discussed in Wegner, *Chattel or Person*, 158.

26. Numbers 5:11–31.

27. Wegner suggests the suspected wife is in fact already pregnant and that miscarriage is the anticipated result if she is guilty of adultery (*Chattel or Person*, 52).

28. Ibid., 162.

29. Ibid., 161.

30. *m. Nedarim* 4:3, discussed in Wegner, *Chattel or Person*, 161.

31. See n. 14.

32. *b. Pesah* 62b.

33. For additional references on Beruriah, and discussion of the possibility that Beruriah is a fiction, see Wegner, "Image and Status," 76. See also Goodblatt, "Beruriah."

34. Rashi, Commentary on the Babylonian Talmud, *Avodah Zarah*, 18b.

35. See Boyarin, "Androcentrism."

36. *m. Gittin* 5:9.

37. *b. Gittin* 61b; *b. Hulin* 6b.

38. Wegner, *Chattel or Person*, 155.

39. *m. Ketubot* 5:5; Wegner, *Chattel or Person*, 155.

40. "[I]f the wife has carelessly forgotten the lamp [her husband has ordered her to light] the inconvenience of sitting in the dark may tempt some household member to transgress the Sabbath by kindling a light (in violation of Exod. 35:3)" (Wegner, *Chattel or Person*, 155).

41. On Jewish women in Venosa, see below, 121.

42. *CCCA* III.

43. Witt, *Isis*, 92. Similar processions were held, according to Plato, both for Athena at the port of Piraeus and for the Thracian goddess Artemis Bendis (*Republic*, 327a [and scholion]), but Jews are unlikely to have had meaningful contact with Greek practices in Plato's day. For a lengthy treatment of the symbol of the menorah and of light among Jews in the Greco-Roman period, see especially Goodenough, *Jewish Symbols*, 4:71–98.

44. Cakes were routinely part of the sacrificial offerings to Greek and Roman goddesses and gods. Cakes offered to Demeter were central in the Thesmophoria, and the festival of μεγαλάρτιος to Demeter on the island of Delos was the festival of the great bread loaves. See also 32 on the cakes offered to Adonis, and 62 on the toasted cakes offered at the Matralia to the Mater Matuta. Stephen Benko of California State University at Fresno has amassed numerous references to the practice of sacrificing cakes to goddesses, which he was kind enough to share in an unpublished manuscript.

45. Jeremiah 7:18, 44:15–25.

46. See Pope, *Song of Songs*, 222, 378–79.

47. *b. Ketubot* 62b and *j. Ketubot* 30 b (to *m. Ketubot* 5:11). That a wife must eat with her husband every Shabbat evening is taken as a euphemism that she must have sex with him. See Goldenberg, "Jewish Sabbath," 424, n. 51. By contrast, *Jubilees* 50 (second century B.C.E.) explicitly forbids sex on the Sabbath.

48. Interestingly, the connections between the lighting of candles, marital intercourse, and veneration of the feminine divine were not lost on E. R. Goodenough, who termed traditional rabbinic explanations regarding the sin of Eve "irrelevant" (*Jewish Symbols*, 4:96, n. 152). On this constellation, he wrote, "In this rite [the lighting of Sabbath candles] all the hope which men for ages have found in God, in light and in the Great Mother, still carries on as a value in Judaism, little as Jews trained in the rab-

binic tradition have been taught consciously to recognize it" (Ibid., 96–97); and also remarked, "[i]t still seems to me no coincidence that the ancient ritualistic use of sex still survives in the old requirement that on the evening of a Sabbath or festival (that is, after the wife has lighted the lights), the husband must have intercourse with her. I should guess that it is with his wife as the Light of God that he has relations" (Ibid., 98, n. 155). Regrettably, Goodenough still focused on the significance of the feminine divine from a male perspective and gave little thought to the implications of his perceptions for Jewish women and women's Judaism(s).

49. The most well-known instance of this was the Vestal Virgins, who remained chaste for their thirty years of service. See Chapter 7, 81–84. Under certain circumstances, Israelite males were also expected to abstain from sexual activity. In Exodus 19:15, Moses instructs the Israelites to avoid women in order to be pure when they appear before God, although Susan Niditch has suggested to me that this may be intended quite literally, rather than euphemistically. Sexual abstinence was also expected during certain military actions (1 Sam. 21:4–6).

50. In *Maenads*, 26.

51. Leviticus 15 describes the circumstances under which bodily emissions render men and women ritually impure and prescribes the necessary cleansing procedures and offerings: 15:1–18 concern the impurity that results from male discharges; 15:19–30 detail the impurity conveyed by female discharge. Leviticus 18:19 forbids men from sex with menstruating women, as does 20:18. Leviticus 12 deals with the ritual impurity conferred by childbirth (seven days plus thirty-three after the birth of a son; fourteen days plus sixty-six after the birth of a daughter).

52. See Cohen, "Menstruants," for an excellent interpretation of menstrual purity within the framework of multiple forms of polluting discharges, sexual and otherwise, for both women and men. Leviticus 15 specifies the circumstances under which both men and women are rendered ritually impure from bodily emissions, including semen and menstrual blood.

53. *Maenads*, 113.

54. *CIJ*, 391. Whether this accurately reflects the differences in ages between the children is uncertain, for epitaphs have a tendency to round off ages to certain preferred numbers, and many scholars are dubious about their reliability. However, it is generally agreed that parents were more likely to know the accurate ages of their children than they were to know their own, so that children's ages are more likely to be reliable than those of adults. Here, of course, the parents and children are all dead, complicating the matter, but the adults who commissioned the inscription might well have known the accurate ages of the children.

55. In correspondence, relying especially on *b. Hullin* 109b: "Yaltha once said to R. Nahman, "Observe, for everything that the Divine Law has forbidden us it has permitted us an equivalent: . . . it has forbidden us intercourse during menstruation but it has permitted us the blood of purification [*dam tohar*: pure blood]." Susan Niditch of Amherst College, also in correspondence, concurs that the rabbis must have in mind some similar interpretation of Leviticus, particularly since *m. Niddah* 4:6 treats the case of a woman who is miscarrying during the days of purification for a daughter.

56. Chrysostom, *Against Judaizing Christians*, 7.1.3, 7.1.5.

57. *b. Yebamot 63*, discussed in Wegner, "Image," 80–81. Also *b. Nedarim* 20a.

58. Wegner, "Image," 80–81.

59. Rousselle, *Porneia*, 37. Her study contains an excellent discussion of theories of

contraception, especially Chapter 2, "The Bodies of Women." See also Blayney, "Theories of Conception."

60. *m. Rosh Hashanah* 3:8.

61. Wegner, *Chattel or Person*, 147: see also 146–56.

62. Ibid., 147, 153.

63. Above, 97–98.

64. Neusner, *Mishnaic Law of Women*, 5:13–14; Wegner, *Chattel or Person*, 5–6, 17; "Image," 77; Baskin, "Rabbinic Separation," 131.


*Chapter 9*

1. *b. Sotah 22b*; in Maenads, **28.**

2. See, for example, Kraemer, "Women in the Diaspora," "Non-Literary Evidence," "Hellenistic Jewish Women"; Brooten, "Jewish Women's History."

3. Brooten, *Women Leaders.*

4. *I. Syrie*, 1322–27, 1329, 1332, 1335, 1336 = *Maenads*, **65.**

5. *CIL*, 8.12457a = Brooten, *Women Leaders*, 161, no. 22 (also in Goodenough, *Jewish Symbols*, 3: Fig. 894; 2:91–100).

6. *CIJ*, 738 = *Maenads*, **60.** See also above, 85.

7. See Brooten, "Did the Ancient Synagogue Have a Women's Gallery or Separate Women's Section," in *Women Leaders*, 103–38.

8. Suetonius, *Augustus*, 44–5; cited in Rawson, "Roman Family," 31.

9. *Thecla*, 32.

10. Tertullian, *On the Veiling of Virgins*, 9, 2–3.

11. *Constitutions of the Holy Apostles*, 2.57; in *Maenads*, **109.**

12. *CIJ*, 748.

13. See n. 7.

14. See above, 101–4 on menstrual purity.

15. *CIJ*, 476 = *Maenads*, **120.**

16. On Judith, see Craven, *Artistry and Faith*; Nickelsburg, *Jewish Literature*, 105–9. Zeitlin speculates on the seeming oddity that the deeds of the pious heroine Judith are excluded from the canon by the rabbis, while those of the considerably more problematic Esther are left in (in Enslin, *Judith*, 21–26). Additions to the book of Esther (see Nickelsburg, 172–75) attempted to address the apparent lack of piety on the part of the Persian Jewish queen.

17. See Zeitlin, in Enslin, *Judith*, 14.

18. See Nickelsburg, *Jewish Literature*, 106–7.

19. *Against Judaizing Christians*, 2.3.3–2.3.6; 4.7.3; in *Maenads*, **31.** Harkins, in his translation, considers the festival of trumpets Rosh Hashanah. Ford ("Montanism") argued that the festival of trumpets was Yom Kippur and thought that Chrysostom's targets were actually Montanists, in a creative, but largely erroneous study.

20. *Against Judaizing Christians*, 1.2.7. *m. Ta'anith* 4.8 asserts that women in Jerusalem wore borrowed white clothing and danced in the vineyards on Yom Kippur. Ford ("Montanism") connects both the white clothing and the dancing with mourning, and attempts to connect them, via the cult of Cybele, with the Montanists.

21. On the shofar at Rosh Hashanah, see Hoenig, "City-Square and Synagogue."

22. *Maenads*, **54.**

23. *CIJ* 849.

24. Juvenal, *Satire*, 6; in **Maenads, 25.**

25. D'Angelo, "Women in Luke-Acts," 460.

26. *m. Sotah* 22b.

27. For the references to Lilith, see Louis Ginsburg, *Legends*, 5:87–88; also 1:65–66. For more recent discussion, see Umansky, "Lilith"; Patai, *Hebrew Goddess*, 207–45; Phillips, *Eve*, 39–40, with additional references 180, n. 5. For a feminist retelling of Lilith, see also Plaskow, "The Coming of Lilith."

28. *Testament of Job*, 46.1–52.2; in *Maenads*, 119. On *Testament of Job*, see van der Horst, "The Role of Women," also Kraemer, "Monastic Jewish Women," 462–63.

29. Translated (except for Chapters 22–29) in *Maenads*, 113.

30. *Aseneth*, 4:8.

31. Some of the most important textual work on *Aseneth* has been done in recent years by Christoph Burchard. For his translation of a preliminary reconstruction of the text, introduction, and extensive bibliography, see *OTP*, 2:177–247. See also Burchard, "Die jüdische Asenethroman." An earlier study by Marc Philonenko, based on a different reconstruction of the text, still has much merit: *Joseph et Aséneth*. A recent dissertation on Aseneth contains extensive bibliography and a helpful discussion of the textual problems: Chesnutt, *Joseph and Aseneth*. For a summary of recent scholarly discussion, see Robert Doran, "Narrative Literature," in Kraft and Nickelsburg, 290–93; see also Nickelsburg, *Jewish Literature*, 258–63.

32. *Aseneth*, 8:4, translation mine, using the text of Philonenko.

33. Ibid., 8:5 in Burchard, *OTP*.

34. Ibid., 8:11 in Philonenko's text; 8:9 in Burchard.

35. Ibid., 21:18–21, abridged, from Burchard.

36. Lefkowitz, "Novels."

37. Ibid., 213.

38. *Aseneth*, 2:10–11. This and subsequent references follow Philonenko's text.

39. Ibid., 10:2–8.

40. Ibid., 17:4–5.

41. Ibid., 15:7–8.

42. Crucial sections are extracted in *Maenads*, 14. This section is adapted from my article "Monastic Jewish Women."

43. "The vocation of these philosophers is at once made clear from their title of Therapeutae (masculine plural) and Therapeutrides (feminine plural)." (*On the Contemplative Life*, 2.)

44. Ibid., 34–35. There is a wonderful provocative discussion of the diet of Christian male monastics in Roman Egypt and its relation to the suppression of desire, which may well apply to the Therapeutics as well: see Rousselle, *Porneia*, 160–78.

45. *On the Contemplative Life*, 30–33.

46. Ibid., 19.

47. Ibid., 68–69.

48. Ibid., 80. On whether women sang individually, see Kraemer, "Monastic Jewish Women," 346–47.

49. *On the Contemplative Life*, 86–88.

50. Ibid., 68–69.

51. On the general question of women's education, see Cole, "Read and Write?" and Pomeroy, "Technikai kai Mousikai."

52. See Kraemer, "Monastic Jewish Women," 350–51; also Pomeroy, "Women

in Roman Egypt" and *Women in Hellenistic Egypt*; also Kraemer, "Non-Literary Evidence," 94.

53. When he lists those whom philosophers abandon to take up the contemplative life, husbands are conspicuous by their absence, whereas wives are specifically noted (*On the Contemplative Life*, 13). "Wife" here is the only gender-specific one in the list: all the others, at least in the grammatical constructions of Hellenistic Greek, could connote both males and females.

54. *On the Contemplative Life*, 68.

55. On Philo's allegorical methods, see *On Allegorical Interpretation*; Goodenough, *Introduction to Philo*, 139ff. For Philo's own statement about the veracity of both literal and allegorical interpretation, see *On the Migration of Abraham*, 89. On male and female, see Baer, *The Categories Male and Female*.

56. *Questions and Answers on Exodus* 1:8.

57. *Questions and Answers on Genesis* 4:15. *On the Cherubim*, 50.

58. Kraemer, "Monastic Jewish Women," 355–56.

59. *The Special Laws*, 3.169–75 = *Maenads*, 15.

60. This argument, which is developed at length in my dissertation (*Ecstatics and Ascetics*) and several subsequent articles, depends heavily on the work of Burridge, *New Heaven* and Lewis, *Ecstatic Religion*.

61. On the decline of the birthrate, see Hopkins, "Contraception" and Noonan, *Contraception*, 23–46. Additional discussion of the issues of contraception, abortion, and so forth, may be found in Cantarella, *Pandora's Daughters*, 135–70.

62. See Gardner, *Women in Roman Law*, 20–22 and 31–116; Raditsa, "Augustus' Legislation" and *NewDocs* 2 (1977 [1982]), 29–32.

63. Rousselle, *Porneia*, 45.

64. Pomeroy, *Goddesses*, 166.

65. *On the Special Laws*, 3.62. Numbers 5:28, on which this is based, says only that if the woman is innocent, she will conceive children.

66. *On the Special Laws*, 3.35.

67. *Questions and Answers on Genesis* 4:145 (= Genesis 24:67). The paraphrase is taken from Goodenough, *Introduction to Philo*, 143–44.

68. Goodenough's arguments about the significance of desire for the mother in ancient religious beliefs, Jewish and otherwise, are elaborated in *Jewish Symbols*, 4:25–62: in Neusner's abridged edition, 72–74.

69. Philonenko, 104–5; Delcor, "Joseph et Asénath."

70. While the identification of *Aseneth* as a Therapeutic text seems feasible, I sometimes think that the *Testament of Job*, or at least the portion about the hymning daughters, would make as much sense in a Montanist context. Such an hypothesis would seem to render *Testament of Job* inappropriate as evidence for Jewish women. But if, as some scholars think, Montanism may have significant links to some forms of Judaism in Asia Minor, locating *Testament of Job* in a Montanist context would not immediately have this effect. See below, 164.

71. Brooten, *Women Leaders*.

72. Rufina (*CIJ*, 741 = *Maenads*, 84) and Theopempte *CIJ*, 756 = *Maenads*, 86.

73. Sophia of Gortyn (*CIJ*, 731c = *Maenads*, 86).

74. Peristeria (*CIJ*, 696b = *Maenads*, 87).

75. Beronike (*CIJ*, 581), Mannine (*CIJ*, 590), and Faustina (*CIJ*, 597) *Maenads*, 88.

76. As yet unpublished: found at Noccere near Pompeii by Marisa Conticello de'Spagnolis.

77. Rebeka (*CIJ*, 692); Makaria (*SEG*, 27 [1977], 1201) = *Maenads*, 88; Eulogia (Kraemer, "A New Inscription from Malta"); also in *Maenads*, 89.

78. See Miller in Schürer, *History of the Jewish People*, III.1: 87–107 for a detailed discussion of the internal organization of Jewish communities, including religious offices.

79. Ibid., 107–37. Among others, see also Tcherikover, *Hellenistic Civilization*; Cohen, *Maccabees to the Mishnah* 104–23; Smallwood, *Jews under Roman Rule*; and Applebaum, "Legal Status" and "Organization" on the nature of Jewish communities as political and social entities.

80. On *collegia*, see above, 88. On the common terminology, see *NewDocs*, 4 (1979 [1987]), no. 113.

81. Brooten, *Women Leaders*. On head of the synagogue, 27–30; on leader, 37–39; on elder, 46–55.

82. Ibid., 28.

83. Ibid.

84. Ibid., 54.

85. *CIJ*, 738 = *Maenads*, 60; see above, 85.

86. See Leon, *Jews of Ancient Rome*, 135–66.

87. The connotations of the term *Jew* are not obvious: see Kraemer, "On the Meaning of the Term 'Jew.'"

88. Lifshitz, *Donateurs et Fondateurs*, 33 = *MAMA*, 6.264; also in Brooten, *Women Leaders*, 158, no. 6, with English translation, and Ramsay, *Cities and Bishoprics*, 673.

89. The actual term in both inscriptions, Tation's and Julia Severa's, is οἶκος, which can mean not simply a house, but a dwelling generally or a public building.

90. Conceivably, the Julia Severa inscription only means that the synagogue renovated a building originally built by Julia Severa, but not necessarily as a synagogue (as we know was the case for the synagogue building at Sardis).

91. Ramsay, *Cities and Bishoprics*, 647, no. 550; see the discussion of the synagogue inscription in Lifshitz, *Donateurs*, 35, n. 4; see also the new Schürer, *History of the Jewish People*, III.1:31. The most reasonable assumption is that Julia Severa was high priestess in the imperial cult of Asia. See Kearsley, "Asiarchs." See also the discussion in Lifshitz.

92. Lifshitz, *Donateurs et Fondateurs*, 35.

93. Ramsay, *Cities and Bishoprics of Phrygia*, 650.

94. For example, Poppaea, wife of Nero (in Josephus, *Antiquities*, 20.186–96); or Plotina, wife of Trajan (*Acts of Isidorus* Recension C, cols 2.21–37). For discussion, see among others, Smallwood, "Poppaea Sabina." For references on the general question of Jewish sympathizers, see Kraemer, "Monastic Jewish Women," 367–68, n. 34.

95. This issue has come to the fore now, though, with the discovery of an important donative inscription from Aphrodisias, a major city in western Asia Minor in the Roman period, published in Reynolds and Tannenbaum, *Jews and Godfearers*.

96. Veturia Paulla (*CIJ*, 523); Marcella (*CIJ*, 496); Simplicia (*CIJ*, 166).

97. Coelia Paterna (*CIJ*, 639).

98. Brooten, *Women Leaders*, 64–65.

99. *CIJ*, 523.

100. *CIJ*, 606, 619d.

101. Brooten, *Women Leaders,* 63.

102. Ibid.

103. Ibid., 69–70.

104. Pleket (*Epigraphica,* 31) and Robert (*Documents de l'Asie Mineure,* 86, n. 6) both considered the title mother of the city to be honorary.

105. *CIJ,* 523.

106. *CIJ,* 741. For a more detailed discussion of Rufina, see Kraemer, "Women in the Diaspora."

107. See Kraemer, "On the Meaning of the Term 'Jew.'"

108. Brooten, *Women Leaders,* 10, considers it most likely that this is the imperial treasury, with which Millar concurs (Schürer, *History of the Jewish People,* III.1:106. Conceivably it is the treasury of the local *hieron.*

109. *Tituli Asiae Minoris* 3 (1941): 448 = *Maenads,* 58.

110. Fuks, "Freedpersons."

111. See Cohen, "Conversion to Judaism" and "Crossing the Boundary."

112. Cohen, "The Matrilineal Principle."

113. See, for example, Gager, *The Origins of Anti-Semitism;* Cohen, "Crossing the Boundary."

114. Although numerous sources from the Greco-Roman period denounce the marriage of Jews to non-Jews (*Testament of Levi,* 14:5–8; *Testament of Job,* 45:4; *Jubilees,* 30:11–17; Philo, *The Special Laws,* 3.29; Josephus, *Antiquities,* 8.191–93) the marriage of Jews to non-Jews is attested in Josephus, *Antiquities,* 20.141–44 and in Acts 24:24. Several inscriptions strongly suggest the possibility of intermarriage between Jews and non-Jews, but none is conclusive.

115. Douglas, *Cultural Bias,* 34.

116. It might be interesting to examine contemporary rationales for the observance of menstrual purity, to see whether we find divergent explanations that differ predictably according to the matrix locations of those who proffer them. Here I have in mind arguments that menstrual purity laws are healthier, promote good marriages, enhance fertility, and so forth. Similar analysis might be offered for contemporary explanations about the origins of kashrut or of circumcision as healthier for both men and women.

117. See, for example, *m. Sanhedrin* 4:3–4; *t. Sanhedrin* 8:1–2.

118. Douglas, "Grid/Group Analysis," 6.

*Chapter 10*

1. Origen, *Against Celsus,* 3.44.

2. See above, 3; 211, n.1.

3. Origen, *Against Celsus,* 3.49.

4. For references, see Kraemer, "Women's Religions."

5. For example, Cameron, "Male nor Female."

6. D'Angelo, "Women in Luke-Acts."

7. Ibid., 443–51.

8. See, for example, Koester, *Introduction,* 2:310. It is within the realm of possibility that the patron of Luke-Acts was in fact not a man named Theophilus, as Luke 1:3 and Acts 1:1 currently read, but a woman named Theophile. Although there is no tex-

tual support for such a reading, there are other instances in which ambiguous names become clarified as male, as in the case of Junia(s) in Romans 16:7. See 136. In this particular instance, the difference between the masculine and feminine vocation is simply the change from ε to η, a frequent change in ancient manuscripts that could have occurred under numerous circumstances. I am indebted to Robert Kraft for this suggestion. A woman patron would accord well with a portrait of women which, in antiquity, would have seemed quite favorable, however we construe it today. It would be consistent with Luke's particular emphasis on women patrons, especially the figures of Joanna and Susanna in Luke 8:3. We might even speculate whether the detailed portrait of Lydia in Acts 16:13–15 points to Luke's actual experience of women patrons, all the more so because Lydia's apparent autonomy (evidenced in her lack of a husband, father, or identified patron of her own) does not accord well with Luke's overall program to portray Christian women as socially respectable.

9. Mark 16:1–8; Matthew 28:1–8; Luke 24:1–11; John 20:1–18; paralleled also in *The Gospel of Peter* 9:35–13:57; *Acts of Pilate* 13:1–3.

10. John 20:1–19.

11. Mark 16:9–11. The scholarly consensus is that the verses in Mark represent a later addition.

12. Luedemann, *Opposition*, 46–52. On the possibility that Cephas should not be identified with Peter, see Ehrman, "Cephas."

13. Beare, *Earliest Records*, 241. More recently, Crossan argues not only that the empty tomb tradition is late, but that it is the creation of the author of Mark, *Gospels*, 160.

14. See n. 13 for Crossan's claim that Mark concocted the empty tomb story. Crossan's language suggests, though, that he believes all our versions of the empty tomb story go back to Mark, but not necessarily that the whole construct is Markan.

15. In the Gospel of Peter, there are no resurrection appearances to women. By the time Mary of Magdala and her female companions go to the tomb and find it empty, the resurrection has been observed by a multitude of men (9:35–13:57), greatly muting the potential impact of the women's discovery of the tomb. For a fascinating reconstruction of the history of these traditions, including the Gospel of Peter, see Crossan, *Gospels*, 159–64, who regrettably does not address the specific feminist implications.

16. Fiorenza, *In Memory of Her*. See also Kraemer, "Women's Religions," for additional references.

17. Fiorenza, *In Memory of Her*, 138.

18. Beare, though, argues that Matthew knows two versions of the story, which accounts for these particular differences, as well as the inclusion of the sheep saying: *Earliest Records*, 132.

19. In Matthew 10:5–6, Jesus instructs his disciples to "Go nowhere among the Gentiles and enter no town of the Samaritans, but go rather to the lost sheep of the house of Israel."

20. Mark 12:41–44; Luke 7:11. The omission of this story from Matthew is consonant with Matthew's version of the beatitudes in 5:3: "Blessed are the poor in spirit, for theirs is the kingdom of heaven."

21. Luke 2:36–38.

22. Luke 7:11–17.

23. In the Greek, the term *widow* occurs eight times in Luke, three times in Acts, twice in Mark, and not at all in Matthew. It is also absent from the Gospel of John. In

Acts, the use of the term is ambiguous. In 9:36–41, where Peter raises from the dead a disciple named Tabitha, widow may signify Christian women as a whole, or it may designate a more specific group within the broader community. In 6:1, where Luke relates conflict over daily distributions to widows, it is again unclear whether all women are intended, or only a smaller group known as widows, and if the latter, whether the group comprised any and all women whose husbands were dead, or whether it had some other connotation.

24. Mark 15:40–41; Joses is called Joseph in the Matthean parallel: Matthew 27:55–56. Not inconceivably, this is in fact a reference to the mother of Jesus, in accord with Matthew 13:55: "Is not this the carpenter's son? Is not his mother called Mary? And are not his brothers James and Joseph and Simon and Judas?" Curiously, Winsome Munroe skips right over this issue: *HBD*, 610–11.

25. Mark 1:29–31; Matthew 8:14–15; Luke 4:38.

26. Matthew 27:55–56.

27. The long version of Mark repeats this, but almost certainly depends here on Luke.

28. Meeks, *Urban Christians*, 55–63.

29. Ibid., 72.

30. Ibid., 52–55, 72–73.

31. Ibid., 70–71.

32. See Brooten, "Junia."

33. Meeks, *Urban Christians*, 57.

34. Sopater in Acts 20:4 is said to be the son of Pyrrhus, and the parentage of Timothy is given in Acts 16:1, without specific names.

35. In the *NRSV*, "a mother to me also."

36. Meeks, *Urban Christians*, 60: see also above, 86–88 and 120–21, on the use of "Mother" as a patronage term.

37. Meeks, *Urban Christians*, 60–61.

38. On the possible implications of the order of spouses' names, see Flory, "Women Precede Men."

39. Meeks, *Urban Christians*, 57.

40. Meeks, *Urban Christians*, 63.

41. See D'Angelo, "Women in Luke-Acts."

42. See D'Angelo, "Women Partners," 73–74, 77–81.

43. As several recent studies point out, the Roman concept of marriage was itself sometimes elusive. There was no required marriage ceremony or any mandatory marriage documents, although there were three necessary conditions for a legally valid marriage: legal capacity, age, and consent of the necessary parties. See Gardner, *Roman Law*, 31–50; also Rawson, "Roman Family," 19–24. See also Veyne, "The Roman Empire," in Veyne, *Private Life*, 33–70, especially 61, where he observes that slaves had neither wives nor children, in the process assuming that slaves were by definition male!

44. Among the multitude of studies on the formation of the Jesus movement, see especially the excellent brief treatments by Fredriksen, *Jesus to Christ*, and Segal, *Rebecca's Children*.

45. For example, Mark 1:16–20 = Matthew 4:18–22 (and see Luke 5:1–11): Mark 10:17–21 = Matthew 19:16–21 = Luke 18:18–22; also Mark 10:28 = Matthew 10:28 = Luke 18:28. See also *The Gospel of Thomas*, 55, 101.

46. Q 9:57–60 (60–61) [Luke 9:57–62 = Matthew 8:19–22]; Q 12:(49) 51–5

[Matthew 10:34–36 = Luke 12:49–53]; Q 14:26–27; 17:30 [Luke 14:26–27; 17:30 = Matthew 10:37–39; see also *The Gospel of Thomas*, 55, 101]; also Matthew 10:21. See also Mark 13:12–13 = Matthew 10:21–22 = Luke 21:16–17. For reconstructions and analysis of Q, see Kloppenborg et al., *Q Thomas*; Kloppenborg, *Q Parallels*; Kloppenborg, *Formation of Q*. See also Amy-Jill Levine, "Feminist Observations."

47. Luke 14:26. Ironically, this new translation from the *NRSV* attempts to eliminate gender-specific language, but the reference to wives unparalleled by any mention of husbands makes clear the perspective of the author. A less harsh wording of this saying occurs in Matthew 10:34–38; paralleled also in *The Gospel of Thomas*, 55, 101.

48. 1 Corinthians 7; see also Wire, *Corinthian Women Prophets*.

49. And from Matthew 19:10–12 which attributes to Jesus the saying that some persons (presumably males are meant here) make themselves eunuchs (and therefore presumably celibate) for the sake of the kingdom of heaven. This may be interpreted as an endorsement of such abstinence, although not a requirement, since Jesus says also that not all can receive these sayings on marriage but only those to whom it is given.

50. Rousselle, *Porneia*.

51. See, for example, Pagels, *Adam*, 17.

52. 1 Timothy 2:13–15.

53. Fiorenza, *In Memory of Her*, 130–40.

54. Levine, "Feminist Observations."

55. For example, Gager, *Kingdom and Community*; Meeks, *Urban Christians*; Theissen, *Palestinian Christianity* and *Pauline Christianity*. A few scholars of early Christianity have utilized Douglas's theory, with varying degrees of sophistication. Among the more productive are Gager, "Body-symbols"; Owen and Isenberg, "Bodies"; Neyrey, "Bewitched," "Witchcraft," and *Revolt*; see also Wire, *Women Prophets*. Less effective is Malina, *Christian Origins*. See also Atkins, *Grid-Group Analysis*.

56. See n. 45 and n. 46.

57. Acts 1:13–14.

58. On the absence of the use of father, see Fiorenza, "Father," and *In Memory of Her*, 147–51; see also D'Angelo, "Abba, Father."

59. See above, 87–88.

60. Fiorenza, *In Memory of Her*, 245–342. See also Wire, who suggests that the conflict with Paul over head covering is in part a conflict about public/private dichotomies. Corinthian women prophets do in the "public" space of the ecclesia what they should only do in the private space of their homes: go uncovered (*Corinthian Women Prophets*, 183).

61. Rosaldo, "Overview."

62. This description may, however, emanate from Luke's desire to show that Christians attempted to remain within the Jewish community, but were unable to do so through no fault of their own, and therefore have little historical value.

63. Meeks, *Urban Christians*, 158.

64. Schüssler Fiorenza, *In Memory of Her*, 165–66.

65. On the subject of Christian women's participation in ritual meals, see Corley, "Greco-Roman Meals."

66. On food laws in particular, see Douglas, "Abominations" and Carroll, "Abominations."

67. Additionally, it may be significant that the laws governing women's ritual purity are found in Leviticus 15:19–30 (menstruation and other flows of blood) and Leviticus 12 (childbirth), that is, in that portion of the law given to Moses after the golden calf

incident in Exodus 32, at least according to the chronology of the biblical text. Ultimately, Christians came to view this second presentation of the law as God's punishment on Israel for their idolatrous worship of the golden calf. This portion of the law was therefore seen to be temporary and not binding on Christians. See Thielman, *Plight*, 75, who finds such an interpretation in Pseudo-Philo's *Liber Antiquitatum Biblicarum*, dated perhaps before 100 C.E.

68. Fiorenza and D'Angelo point out, though, that this story is not quite as positive as it may appear, for the serving role of Martha, whose neglect by Mary Jesus approves, is in fact the table ministry. Luke's real point, as D'Angelo observes, is that "women who have been discouraged from ministry have chosen the better part," "Luke-Acts," 455; see also Fiorenza, *In Memory of Her*, 165.

69. For example, 1 Corinthians 14, where Paul argues for the superiority of prophecy over glossolalia, but this is still within the framework of the centrality of personal religious experience: that is, after all, the center of Paul's own claim to authority.

70. 1 Corinthians 5:9 makes it clear that there is at least one prior letter.

71. Wire, *Corinthian Women Prophets*, especially 181.

72. Walker, "1 Corinthians 11:2–16" argues that the entire passage has been interpolated. Murphy-O'Connor, "Non-Pauline Character" and "Sex and Logic," defends the integrity of these verses; see also "Once Again." Cope, "1 Cor 11:2–16" agrees that the verses have been inserted, but disagrees with Walker's identification of three separate interpolations. Trompf defends and expands the arguments of Walker, "Attitudes toward Women." Wire accepts the arguments of Murphy-O'Connor that the passage is genuine. For an excellent summary of the discussion and additional bibliography, see Wire, *Corinthian Women Prophets*, 220–23.

73. See Wire; see also Neyrey, "Body Language."

74. See Thompson, "Portraits." Ancient testimony to head coverings for respectable pagan women includes Plutarch, *Quaest. Rom*, 267a and *Apophth. Lac.*, 232c; Dio Chrysostom, *33rd Discourse*. Thompson sets the question squarely within the context of social conventions in Roman Corinth. In addition, sanctuary regulations from a temple of Despoina in the third century B.C.E. demonstrate the prior existence of such cultic regulations for women and men: "Nor (let it be permissible to enter) for women with their hair braided, nor for men with their heads covered." *IG* V, 2 [1913] 514, reprinted in Guarducci, *Epigrafia*, 4.20. Text, translation, and discussion in *New-Docs* 4 (1979[1987]) 25, pp. 108–9. See also Apuleius, *Metamorphoses*, 11.10 on the hairstyles/head coverings of Isiac initiates.

75. Douglas, *Natural Symbols*, 101. Douglas suggests that this distance is easier to perceive in table manners than in dress and grooming.

76. Ibid., 102.

77. At the same time as we see some critique of that—see Thompson, "Portraits."

78. See Oster, "When Men Wore Veils." D'Angelo has recently argued, though, that Paul is not equally concerned with male comportment ("Unveiled Women").

79. See Douglas, *Natural Symbols*, 99–100: "A study of anti-ritualism must focus on the expression of formality and informality. It seems not too bold to suggest that where role structure is strongly defined, formal behavior will be valued. . . . Formality signifies social distance, well-defined, public, insulated roles. Informality is appropriate to role confusion, familiarity, intimacy. Bodily control will be appropriate where formality is valued, and most appropriate where the valuing of culture above nature is most emphasized."

80. Ibid., 102.

81. Ibid., 109.

82. Wire, *Corinthian Women Prophets*, 183.

83. The *RSV* reads: "As in all the churches of the saints, *the women* should keep silence in the churches." The remainder is the same, except for the phrase "as the law *also* says," which was previously translated "as *even* the law says." The *KJV* reads: "Let your women keep silence in the churches: for it is not permitted unto them to speak; but they are commanded to be under obedience, as also saith the law. And if they will learn any thing, let them ask their husbands at home: for it is a shame for women to speak in the church."

84. Extant manuscripts embed the story of Thecla and Paul within other apocryphal stories of Paul, known collectively as the *Acts of Paul*. Many of these stories, including that of Thecla, probably initially circulated individually. Titling the tale *The Acts of Paul and Thecla* gives primacy to the figure of the apostle (and parallels the titles of numerous ancient romances). Yet in the actual story, Thecla is clearly the central character, hence my preference for the title *Acts of Thecla*.

85. 1 Timothy 2:8–15.

86. 1 Timothy 5:5 defines the real widow as one who prays night and day. That such activity is to be private is suggested by 1 Timothy 5:13, which upbraids widows for "gadding about from house to house."

87. MacDonald, *Legend*.

88. Ibid., 54–77.

89. *Thecla*, 18.

90. Ibid., 20.

91. Ibid., 35.

92. Ibid., 34.

93. Ibid., 43. There is an alternate ending to the tale of Thecla in which a band of would-be rapists pursue Thecla into a cave. Praying to God, she disappears into the very rock of the cave, eluding the men, but dying (unviolated!) in the process. In *ANT*, 281.

94. Tertullian, *On Baptism*, 17.

95. MacDonald, *Legend*, 59–66, especially 59.

96. 2 Timothy 4:10.

97. 2 Timothy 2:17–18.

98. The term *Asia* normally refers to the Roman province of Asia, in what is now modern Turkey.

99. *Thecla*, 14.

100. 1 Timothy 4:13.

101. *Thecla*, 25.

102. τοὺς δὲ βεβήλους καὶ γραώδεις μύθους παραιτοῦ.

103. MacDonald, *Legend*, 34–53.

104. Kraemer, "Conversion"; Davies, *Revolt*; Burrus, *Chastity*. Additional references in *Maenads*, 407.

105. MacDonald, *Legend*, 14–15, 102.

106. Tertullian believes the presbyter to have been male. A woman presbyter named Ammion lived in Phrygia in the mid third century—*Maenads*, 94. On Christian women elders, see below, 183–85.

107. MacDonald asserts, though, that this is originally an oral tale: *Legend*, 17–18.

108. Kraemer, "Conversion," 304.

109. At one point in the *Acts of Andrew*, Maximilla's husband entreats her to return to their marital life: "If you would be the woman you once were, living together with me as we are accustomed to—sleeping with me, having sex with me, conceiving children with me—I would treat you well in every way" (*Acts of Andrew*, 36 (4), from a new edition and translation: MacDonald, *Andrew*, 371). Aside from this conditional statement, nothing in the text identifies Maximilla as a mother.

110. The model of defective or inaccessible standards of worth is derived from the work of Kenelm Burridge, *New Heaven*, who did not consider its implications for women in any detail. See Kraemer, *Studies*, 85–97: "Conversion," especially 301–2, n. 11.

111. Schüssler Fiorenza, *In Memory of Her*, 182–83.

### Chapter 11

1. 2 Timothy 3:6–7.
2. Jerome to Ctesiphon, Letter 133, 4.
3. Ancient sources on the New Prophecy and Montanism are conveniently collected in de Labriolle, *Sources* (Greek and Latin with French translation facing) and most recently in Heine, *Montanist Oracles*. Selections relevant to the study of Montanist women are excerpted in *Maenads*, 101–6.

  The scholarly literature on Montanism is extensive, much of it done in the early part of this century, for example, Labriolle, *La Crise*; Schepelern, *Der Montanismus*. Interesting but improbable is Ford, "Jewish-Christian Heresy." More important recent studies include Strobel, *Das heilige Land*; Barnes, "Chronology of Montanism"; Powell, "Tertullianists and Cataphrygians." Several recent dissertations have focused on the movement, including McGinn-Moorer, *The New Prophecy*; Goree, *Montanism*; and Klawiter, *The New Prophecy*.

4. Eusebius, *History of the Church* 5.16.7–10; in *Maenads*, 104.
5. Ibid., 5.18.3.
6. Ibid., 5.16.4.
7. Ibid., 5.16.17.
8. Ibid., 5.16.10.
9. Hippolytus, *Refutation of All Heresies*, 8.12; in *Maenads*, 102.
10. Eusebius, *History of the Church*, 5.18.5.
11. "Anonymous," cited in Eusebius, *History of the Church*, 5.16.18.
12. Powell, "Tertullianists," 44–46.
13. Eusebius, *History of the Church*, 5.16.13.
14. Powell, "Tertullianists," 48.
15. Ibid., 42–43.
16. *Perpetua*, 1.
17. Ibid.
18. Ibid., 6.
19. Ibid., 4.
20. Ibid., 7–8.
21. Ibid., 10.
22. Ibid., 11–13.
23. Lefkowitz, "St. Perpetua's Martyrdom."

24. See Kraemer, "Conversion."

25. Tertullian, *On Fasting.*

26. *On Monogamy, On Chastity, On Modesty.* On the criteria for distinguishing Tertullian's Montanist works, see among others Barnes, *Tertullian,* 42–48.

27. Powell, "Tertullianists," 42, n. 45, where he suggests that "[f]or devout women to abandon cohabitation was in no sense a Montanist innovation."

28. Epiphanius, *Medicine Box,* 49; in *Maenads,* 103. Ambrosiaster claims that the Cataphrygians argued from 1 Timothy 3:8–11 that women could be deacons, which he attempts to refute by pointing out that in Acts 6:3ff no women were among the seven commissioned by the apostles to take on the responsibilities of table fellowship. *Commentary on 1 Timothy 3.11:* in Labriolle, *Les Sources,* 109; also in Heine, *Montanist Oracles,* 126–29.

29. Tertullian, *On the Soul,* 9; in *Maenads,* 101.

30. Tertullian, *On Baptism,* 17. *On the Veiling of Virgins,* 9, and *The Prescription against Heretics,* 41, also oppose women teaching and baptizing, but without specific reference to the example of Thecla.

31. See n. 26.

32. *On Flight in the Face of Persecution.*

33. "Anonymous" in Eusebius, *History of the Church,* 5.16.13.

34. See Barnes, *Tertullian,* 265, who doubts the veracity of this attribution: it is regrettably asserted without discussion of Wilson-Kastner in her introduction to *A Lost Tradition,* xi, but not by Rader in the introduction to her translation of *Perpetua* in the same volume, who refers simply and more judiciously to the "redactor."

35. Epiphanius, *Medicine Box,* 48.8.8. On the source question, see Heine, *Montanist Oracles,* x; "The Role of the Gospel of John," 3–11.

36. Epiphanius, *Medicine Box,* 48.9.1.

37. Ibid., 48.9.

38. Ibid., 48.14, 49.1.

39. Ibid., 49.2.

40. See below, 185–86.

41. Epiphanius, *Medicine Box,* 49.2.

42. Jerome to Marcella, Letter 41, 4. As far as I can tell, Jerome is the sole source of the claim that Montanus was a Cybelean priest, which renders it historically tenuous.

43. Schepelern saw a connection here, although he generally argued against the derivation of Montanism from the worship of Cybele. See Powell, "Tertullianists," 47.

44. Apollonius in Eusebius, *History of the Church,* 5.18.11. This interpretation was shared by Stephen Benko in the draft of an unpublished study. Various scholars, however, read this passage differently: some see it as a description of Maximilla and Priscilla; others see it as a reference to false male prophets.

45. See Powell, "Tertullianists," 46–47.

46. Ibid., 46.

47. Ford, "Jewish Christian Heresy."

48. *Perpetua,* 4.9: *et clamavit me et de caseo quod mulgebat dedit mihi quasi buccellam; et ego accepi iunctis manibus et manducavi et universi circumstantes dixerunt: Amen.* Musurillo's translation, used in *Maenads,* unfortunately obscures the reference to cheese (*caseus*): "He called me over to him and gave me, as it were, a mouthful of the milk he was drawing," influenced, I think, by the reference in 4.8 that the man was milking sheep. Rosemary Rader's translation (in P. Wilson-Kastner; see n. 34) preserves

the reference to cheese, but obscures some of the difficulties as well: "Then he beck-oned me to approach and gave me a small mouthful of the cheese he was making. I accepted it with cupped hands and ate it." A key term here is *mulgebat* for what the man was doing to the substance from the sheep—he was milking it, not making cheese, but in fact, as he milked it, a cheeselike substance was produced. This is demonstrated by the fact that Perpetua chews it—*manducavi*—rather than drinks it. But the sub-stance is sufficiently amorphous that Perpetua must hold it in her cupped hands, rather than eat it with her fingers. Elements such as this instantaneous transformation of the milk into a soft cheese buttress the claim that *Perpetua* recounts real dreams or visions, not literary creations, for real dreams have precisely such oddities.

49. Powell, "Tertullianists," 48. The metaphor of milk, of course, has a long history in ancient religion, Christian and otherwise. Milk figures prominently in the food of the blessed in Greek mysteries and is employed by Paul as a metaphor for the teachings one can give to neophyte Christians. In a sheep-based economy, cheese and yogurt are important dietary staples. Cheese offerings are attested for the worship of Artemis at Sparta (see *OCD*, 127).

50. *The Debate between an Orthodox and a Montanist.*

51. Eusebius, *History of the Church*, 15.16.5: the text does not state explicitly that the opponents were present, but that seems quite feasible; 25.5–7 mentions a written discussion (ἐγγράφως διαλεχθείς) between Gaius, bishop of Rome, and Proclus, a leader of the Cataphrygians.

52. ὥστε καὶ βίβλους ἐξ ὀνόματος αὐτῶν (Heine, *Montanist Oracles*, 124).

53. A similar argument connecting the writing of books with the exercise of author-ity over men is made by Didymus of Alexandria, cited in Labriolle, *Les Sources*, 160; Heine, *Montanist Oracles*, 144–47.

54. Epiphanius, *Medicine Box*, 78.23, 79; also in *Maenads*, 29 and 30.

55. See, for example, Carroll, *Virgin Mary*, 46.

56. Epiphanius, *Medicine Box*, 79.1.

57. Ibid., 78.23.

58. Ibid., 79.3; Jeremiah, 7:18, 44:15–25.

59. Benko, *Goddesses*, 35–36.

60. Abrahamsen, "Women at Philippi." Her work focused solely on Philippi, and she suggests that we may attribute some conflict between Paul and the Philippians to precisely such expectations. A study by Portefaix, *Philippian Women*, on first-century C.E. Philippian women and their reception of Christianity is regrettably seriously flawed.

61. Gibson, *Christians*, 132, 136; Labriolle, *Les Sources*, no. 152; see also Strobel, *Das heilige Land*.

62. *Maenads*, 94.

63. Gibson, *Christians*, 143–44.

64. Ibid., 136. Also in Gibson, "Montanist Epitaphs."

65. Gibson, *Christians*, 143.

66. Who read inscriptions, and what inscriptions say about the literacy of those who commission them is a matter of some recent debate: see, for example, MacMullen, "Epigraphic Habit"; Harris, "Literacy and Epigraphy."

67. In Gibson, *Christians*, no. 29, one family has ten children; one has five; no. 15 has six.

68. In Gibson, *Christians*, no. 28, a man named Sosthenes died at age thirty having been married for three years. Regrettably, the inscription does not mention the age of the wife.

69. See below, 185–86.

70. See 84–88 and 118–23 for discussion of numerous priestesses, patrons, and synagogue leaders from Asia Minor.

71. I suppose Paul's instructions in 1 Corinthians, 11:3–16 could be read to mean that women prophets at Corinth are married.

72. Heine, "The Role of the Gospel of John."

73. *Shepherd of Hermas*, Vision 1:2.2, Vision 2:1.3. The woman who appears to Hermas is dressed "ἐν ἱματισμῷ λαμπροτάτῳ": the figure who appears to the Montanist prophetess came (both use the Greek ἦλθε[ν]) "ἐν στολῇ λαμπρᾷ" (Epiphanius, *Medicine Box*, 49.1).

74. Pagels, *Gnostic Gospels*, 3–32.

75. This is somewhat less crucial if Powell is correct that the movement was never excessively eschatological in the first place.

76. See n. 29.

77. Excerpted in *Maenads*, 107, 108, 109.

78. Again, the male pronoun is intentional. I cannot conceive of a female author for the *Constitutions* any more than for 1 Timothy.

79. *The Constitutions* does concede that women prophesied of old, naming Miriam, Deborah, Huldah, and Judith: also Mary, the mother of Jesus, Elizabeth, Anna, and the daughters of Philip. Since the text purports to be written by the Apostles themselves, it cannot know of women's prophecy much later, and even cleverly says, "in *our own time*, [emphasis added] the daughters of Philip . . ." But the Corinthians were in fact contemporaries of the surviving disciples of Jesus.

80. It is discussed briefly in *Constitutions*, 8.1 and 8.2.

81. See n. 75.

82. See Shaw, "Age of Roman Girls"; Saller, "Men's Age."

83. See Carolyn Osiek, "Widow as Altar."


## Chapter 12

1. See above, 104.

2. The major work on this subject is Fiorenza, *In Memory of Her*. Newer studies include Wire, *Women Prophets*; D'Angelo, "Women Partners" and "Women in Luke-Acts."

3. D'Angelo, "Women in Luke-Acts," 452. The letters of Paul also attest the financial support of women, such as Phoebe in Romans 16:1.

4. 1 Corinthians 12:28–29.

5. 1 Corinthians 11:3–16: see above, 146–49.

6. Romans 16:7. On Junia, see especially Brooten, "Junia . . ."

7. Above, 146–49.

8. On early cult, see Meeks, *Urban Christians*, 140–63.

9. MacDonald, *Legend*, 86–89.

10. Acts 18:2ff, where she is called Priscilla. Paul calls her Prisca (1 Cor. 16:19; Rom. 16:3).

11. Acts 18:18, 25. In Romans 16:3, Paul also greets Prisca before Aquila. For some perspective on the significance of the word order of spouses' names, see Flory,

"Where Women Precede Men."

12. Acts 18:25.

13. Fiorenza, *In Memory of Her*, 172.

14. Revelation, 2:20–23.

15. Acts 16:14–15. See Fiorenza, *In Memory of Her*, 160–204; Kraemer, "Monastic Jewish Women," 367–68; Horsley, "The Cities of Revelation"; *NewDocs* 3 (1978[1983]) 53–55.

16. Revelation 2:20.

17. Collins, "Women's History," 83.

18. That the author does not know 1 Timothy may not be surprising, since the scholarly consensus would date Revelation earlier than the first Pastoral. That the author does not cite 1 Corinthians 14:33b–36 may be more significant.

19. Transmitted in Latin in the *Adumbrationes*, GCS 17, 215: *Secunda Iohannis epistola, quae ad virgines scripta est, simplicissima est. Scripta vero est ad quandam Babyloniam, Electam nomine, significat autem electionem ecclesiae sanctae.* Cited in Brown, *Epistles of John*, 646. Clement is known, though, for some strange ideas about ancient authorship.

20. See, for just a few examples, Dodd, *Johannine Epistles*, 143–45; Bultmann, *Johannine Epistles*, 107–8; Kümmel, *New Testament*, 313; Johnson, *New Testament*, 505.

21. For example, in the *Shepherd of Hermas*, vision 4 (also other visions).

22. See Kümmel, *New Testament*, 313 for references.

23. Fiorenza, *In Memory of Her*, 248–49. Another example of a woman head of a church may be that of Nympha in Colossians 4:15.

24. Especially Hippolytus, who says that the Montanists "magnify these wretched women above the Apostles and every gift of Grace." *Refutation*, 8.12; in *Maenads*, 102.

25. Ibid.

26. See above, 166.

27. Powell, "Tertullianists and Cataphrygians," 44.

28. Gibson, "Montanist Epitaphs." Also in Gibson, *Christians for Christians*, 136 (in *Maenads*, 94). The inscription is now at Uşak but its exact provenance is unknown.

29. Gibson, *Christians for Christians*, 136.

30. Tertullian, *On the Soul*, 9; in *Maenads*, 101.

31. *Perpetua*, 4; in *Maenads*, 53.

32. *Perpetua*, 16, 18.

33. Ibid., 13.

34. Klawiter, "Montanism," 251, 254.

35. Ibid., 254.

36. Hippolytus, *Apostolic Tradition*, 10.1, cited in Klawiter, 254.

37. See above, 160.

38. Pregnant women sometimes received a reprieve until they gave birth. Presumably this reflects less Roman sentimentality than Roman law regarding the ownership of children. Since the unborn child belonged to its father, or if the mother was a slave, to her owner, to execute a pregnant (Christian) woman was to deprive an innocent party of his property. Correspondence between the Emperor Trajan and Pliny the Younger, governor of Bithynia in Asia Minor, gives us some insight into Roman policies and procedures with regard to Christians in the early second century C.E. (Pliny, *Epistles*, 10.96 and 97).

39. Eusebius, *History of the Church*, 5.1.3–63; in *Maenads*, 123.

40. Klawiter, "Montanism," 256.

41. Tertullian, *On Baptism*, 17. See above, 152..

42. Gryson (*Ministry of Women*, 18) thinks this tentativeness suggests it was not much of a problem, but Tertullian's vigorous attack on Thecla suggests otherwise.

43. Tertullian, *On the Veiling of Virgins*, 9. His concern here is to determine whether the prohibitions on women extend to virgins, concluding that they do. Offerings here presumably means a Eucharist.

44. *Didascalia*, 3.9.1–3.

45. Excerpted in *Maenads*, 107–9.

46. *Constitutions of the Holy Apostles*, 3.9; in *Maenads*, 108.

47. Ibid.

48. For references, see διάκονος in *Lidell-Scott*; BAGD; *NewDocs* 4 (1979[1987]), 122.

49. Romans 16.1–2.

50. For an excellent discussion, with many illustrations of ancient usage, see *NewDocs* 4 (1979[1987]), 122.

51. See, for example, Fiorenza, *In Memory of Her*, 181–83.

52. The feminine form does occur in Greek and is not used here.

53. 1 Timothy 3:1–7 on bishops, 3:8–13 on deacons, 5:3–16 on widows, and 5:17–22 on elders.

54. Pliny, *Epistles*, 10.96.8.

55. See above, 88.

56. The *Constitutions* also give the following prayer to be said at the ordination of a deaconess: "O Eternal God, the Father of our Lord Jesus Christ, the Creator of man and of woman, who didst replenish with the Spirit Miriam, and Deborah, and Anna and Huldah, who didst not disdain that Thy only begotten Son should be born of a woman; who also in the tabernacle of the testimony, and in the temple, didst ordain women to be keepers of Thy holy gates,—do Thou now also look down upon this Thy servant, who is to be ordained to the office of a deaconess, and grant her Thy Holy Spirit, and 'cleanse her from all filthiness of flesh and spirit' [2 Cor. 7.1] that she may worthily discharge the work which is committed to her to Thy glory, and the praise of Thy Christ, with whom glory and adoration be to Thee and the Holy Spirit for ever. Amen." (Translation, *ANF*, 8.20).

57. *Maenads*, 95.

58. They may have begun "as a shouted acclamation of the people, designed to flatter a wealthy citizen." Jones, "Three Foreigners," 223–24.

59. Guarducci, *Epigrafica*, 4.368–70.

60. *Maenads*, 96.

61. IG III² x. 3527.

62. Gryson, *Ministry of Women*, 153 (notes 151–58) lists the following: from Axylos, Nunes, Strateges, Pribu, and Matrona [*MAMA* 1: 323b, 324, 326, 383]; Masa, Aurelia Faustina, Paula from Laodicea Combusta [*MAMA* 1:178, 194, 226]; an Encratite deaconess, Elaphia in Nevinne (*MAMA* 1: p.xxv); Timothea at Corycos (*MAMA* 3: 744) and Arete at Aphrodisias, in Grégoire, *Recueil*, 1, n. 258.

63. The inscription of Domna comes from the general region of the Montanists with names attested frequently for that area, but there is nothing specific to suggest any connection with Montanist or any kind of heterodoxy.

64. *Maenads*, 100.

65. Regrettably, even some of those in *Maenads.*

66. Epiphanius, *Medicine Box,* 49.2; in *Maenads,* 103.

67. Rossi, "Priesthood," a translation and discussion of Otranto, "Sacerdozio femminile." I am indebted to Professor Rossi for bringing this material to my attention and for sharing the draft of her translation and commentary in advance of its publication.

68. Decree 26: 376–77 in Thiel. Translation is mine from the Latin: *Nihilominus impatienter audivimus, tantam divinarum rerum subisse despectum, ut feminae sacris altaribus ministrare firmentur, cunctaque non nisi virorum famulatui deputata sexum, cui non competunt, exhibere.*

69. In Labriolle, *Sources,* 226–30.

70. See Gryson, *Ministry of Women,* 44–74, 100–8.

71. *Quod non oportet eas quae dicuntur presbyterae vel presidentes in Ecclesiis ordinari.*

72. *Hae quae presbyterae dicebantur, praedicandi, iubendi, vel docendi . . . officium sumpserat.* Ep. 8 (*PL* 134, 114).

73. Ferrua, "Note su Tropea."

74. Flavia Vitalia, a presbyter from Salona in Dalmatia, in Bulic, "Iscrizione inedita." She sells a cemetery plot to a man named Theodosious, apparently in her official capacity as presbyter. Also a fragment [*sace*]*rdotae* (*CIL* 3.14, 900); a *presbiterissa* from Ippona, *AE* 1953:107; and a graffito from Poitiers in Gaul, *Martia presbyteria. . .* (*CIL* 13, 1183). Not inconceivably, one or more of these might be Jewish.

75. See above, 178.

76. *NewDocs* 1 (1976[1981]), 79.

77. Ibid.

78. *AE* 1975:454; in *Maenads,* 93.

79. Epiphanius, *Medicine Box,* 49; in *Maenads,* 103.

80. Ibid., 49.3.

81. Ibid., 78.23 in *Maenads,* 29; Ibid., 79; in *Maenads,* 30.

82. Ibid., 79.1

83. Ibid., 79.2.

84. The actual regulation of widows comes from 1 Timothy. This is one of the only cases where Epiphanius uses a noun to designate priesthood, and he uses feminized forms: ἱερίσσα and πρεσβυτέρας: *Medicine Box,* 79.4.

85. Chrysostom, *On the Priesthood,* 3.9.

86. Ibid., 3.9.1–4.

87. Clement, *Stromateis,* 3.6.53, 3–4.

88. See above, 144.

89. *First Homily on "Salute Priscilla and Aquila"* 3 (*PG* 51, 191D–192C), excerpted and translated in Gryson, *Ministry of Women,* 81–82.

90. One of Chrysostom's many letters to Olympias is translated in *Maenads,* 74; an ancient *Life of Olympias* is given in 75. Elizabeth Clark has written extensively on Chrysostom's friendship with Olympias: for references, see the entry on Chrysostom in "About the Authors and Sources," *Maenads,* 400–1.

91. Nagel, "Lettre Chrétienne."

92. This section is excerpted from my article "Women's Authorship." Ignatius is usually dated to the late first/early second century, on the basis of the references in Polycarp. Joly, *Dossier d'Ignace,* has suggested, however, that the references to Ignatius in Polycarp's letter to the Philippians are interpolated, requiring a rethinking of the

actual date of Ignatius, whom he would place considerably later. For a summary of these problems and additional bibliography, see Schoedel, *Ignatius*, especially 4–7; and Schoedel, "Letters of Ignatius." It appears that having been universally rejected as authentically Ignatian, the alleged correspondence between this Mary and Ignatius has been deemed of little further interest, even though Roberts and Donaldson observed over a century ago that such texts were worthy sources of later Christian conflicts and communities: *ANF*, 1:106, n.1.

93. For a fuller treatment of these problems, see Kraemer, "Women's Authorship," 236–39.

94. Eusebius, *History of the Church*, 5.16.6.

95. *Ignatius to Mary*, 1,4.

96. *Ignatius to Hero*, 5.

## Chapter 13

1. See Plaskow, "Blaming the Jews."

2. *CIJ*, 523 (Veturia Paulla) and *CIJ*, 400 (Sara Oura).

3. Abrahamsen, "Women at Philippi." Portefaix, *Sisters Rejoice*, also considers the perspectives pagan women would have brought to Christianity, but her work has many flaws.

4. See, for example, *NewDocs* 4 (1979[1987])), **113**, especially 219–20; *NewDocs* 3 (1978[1986]), 43; *NewDocs* 1 (1976[1981]), 5.

5. Principally the use of *shalom* in *CIJ*, 597, the epitaph of Faustina, but this can be somewhat circular.

6. Brooten, *Women Leaders*, 44: one more in Kraemer, "New Inscription."

7. There are some enigmatic passages in the Hebrew Bible and in later Jewish literature which may suggest that Israelite women did once serve as priests, or did at least exercise cultic office in ancient Israel: a brief but thorough discussion of these sources may be found in Brooten, *Women Leaders*, 83–95. Some scholars have considered the probability that Miriam was a priest and that a priestly tradition associated with Miriam continued for some time, but was eventually expunged. Some of this tradition and conflict may be reflected in Numbers 12. See Bird, "Women in Israelite Cultus" and Burns, *Miriam*.

8. See above, 98.

9. See above, 104.

10. See Beard, "Vestal Virgins," 12–13.

11. *IG²* II 1366 , reprinted with commentary in *NewDocs* 3 (1978[1983]), 6.

12. Beard, "Vestal Virgins," 13.

13. This is abundantly clear in Leviticus 15, which spells out the circumstances under which both women and men are rendered unclean from sexual and other bodily fluids.

14. Above, 187.

15. Above, 187.

16. For instance, Geoffrey de Ste. Croix, in the as yet unpublished 1988 Townsend lectures.

17. See above, 165.

18. Tertullian, *On the Veiling of Virgins*, 4.

19. Ibid., 9.

20. Jenkins, "Origen," 41, no. 74.

21. See Pagels, *Gnostic Gospels*, especially Chapter 4: King, *Feminine in Gnosticism*.

## Epilogue

1. Douglas, *Cultural Bias*, 34.

2. Ibid., 35.

3. On gnosticism and the feminine divine, see King, *Feminine in Gnosticism*; on the Therapeutics, see above, 113–17.

4. Jerome to Laeta, Letter 107; in *Maenads*, 70.

5. Above, 61–69.

6. Origen, *Against Celsus*, 3.44.

7. See 248, n. 7.

8. See above, 87–88.

9. On this apparent routinization of charisma, see Weber, *Sociology*, 104–105: for its specific application to early Christianity, see Gager, *Kingdom*, 67–88.

10. Fiorenza, *In Memory of Her*, 80–84.

11. Douglas offered this interpretation during a seminar at Princeton University in the spring of 1988: I do not know whether she has articulated this detail in print.

12. Douglas suggests that they are not inherently doomed to disappear and under certain circumstances, they may continue more or less indefinitely. *Natural Symbols*, 102–3.

13. These observations were also made at the 1988 seminar.

14. See Spickard, *Relativism*, which critiques Douglas on the grounds that the centrality of the individual in Douglas's model is not a cultural universal. Conceivably, this is a reflection of the location on the matrix from which the model is necessarily generated. Ironically, this might suggest that the model reflects the experience and perspective of low grid and group. When I asked her about this, Douglas thought that the model could only be generated at certain locations on the grid, but we did not discuss which. Douglas tends to align herself with strong grid and group, although I think that all her years in academia belie that somewhat. She might disagree.

15. For example, Winkler, *Constraints of Desire*; Halperin et al., *Before Sexuality*.

16. *Acts of Thomas*, I.12.

17. Ibid.

18. *Acts of Thecla*, 14; in *Maenads*, 114.

19. Loades, "Review."

20. I am indebted to Daniel Boyarin of the University of California at Berkeley for a discussion shortly before this book went into production. He suggested that one of the central concerns of ancient rabbis was precisely a refutation of dualist cosmologies and the insistence on the integrity of body and soul, expressed particularly in their affirmation of sexuality. But regrettably, such an affirmation does not lead to a correlate affirmation of women.

# Ancient Sources and Translations

Unless otherwise indicated, text and translation of Greek and Latin authors, including Philo and Josephus, may be consulted in the *Loeb Classical Library*. Christian authors and texts not listed here may be found in the *Ante-Nicene Fathers* or the *Nicene and Post-Nicene Fathers*: some are also available in the *Loeb*. An asterisk indicates that part or all of the text may be found in English translation in *Maenads*.

For translations of major rabbinic sources, see H. Danby, *The Mishnah* (Oxford: Clarendon Press, 1933); Jacob Neusner, *The Mishnah. A New Translation* (London and New Haven: Yale University Press, 1988); I. Epstein, *The Babylonian Talmud* (London: Soncino, 1935–52); Jacob Neusner, *The Talmud of Babylonia: An American Translation* (Brown Judaic Studies, 1980–84); and J. Neusner, *The Tosefta* (New York: Ktav, 1977).

More detailed discussion and bibliography for many sources may be found in *Maenads*, 393–417 (About the Authors and Sources).

*Acts of Isidorus*
H. Musurillo, *Acts of the Pagan Martyrs*. Oxford: Oxford University Press, 1954. Reprinted New York: Arno Press, 1979.

*Acts of Pilate*
F. Scheidweiler in E. Hennecke and W. Schneemelcher, eds., *New Testament Apocrypha*. Trans. R. McL. Wilson. Philadelphia: Westminster Press, 1963, I:179–87. Robert W. Funk, *New Gospel Parallels*. Philadelphia: Fortress Press, 1985, I:444–70.

*Acts of Thecla*
M. Bonnet and R. Lipsius, *Acta Apostolorum Apocrypha*. Reprinted Darmstadt: Wissenschaftliche Buchgesellschaft, 1959, I:235–72.

*Apuleius, Metamorphoses*
J. Gwyn Griffiths, *The Isis-Book (Metamorphoses XI): Apuleius of Madauros*. EPRO 39. Leiden: E. J. Brill, 1975.

Clement of Alexandria, *Stromateis*
H. Chadwick and J. E. L. Oulton, *Alexandrian Christianity*. Library of Christian Classics. Philadelphia: Westminster Press, 1954.

*Conversion and Marriage of Aseneth*
M. Philonenko, *Joseph et Aséneth. Introduction, texte critique, traduction, et notes.* Leiden: E. J. Brill, 1968.
C. Burchard, in A.-M. Denis, avec la collaboration d'Yvonne C. Burchard, in Janssens et le concours du CETEDOC, Concordance Grecque des Pseudepigraphies d'ancien Testament. Louvain-la-Neuve: Université catholique de Louvain, Institute Orientaliste, 1987.


*Cynic Epistles*
Abraham J. Malherbe, *The Cynic Epistles: A Study Edition.* SBL Sources for Biblical Study 12. Missoula, MT: Scholars Press, 1977.


*Dead Sea Scrolls*
A. Dupont-Sommer, *The Essene Writings from Qumran.* Trans. Geza Vermes. Cleveland: World, 1961.
Theodore H. Gaster, *The Dead Sea Scriptures in English Translation with Introduction and Notes.* Revised and enlarged. Garden City, NY: Anchor Books, 1964.
Geza Vermes, *The Dead Sea Scrolls in English* (3d ed.). London: Penguin, 1987; New York: Viking Penguin, 1987.


*Debate between a Montanist and an Orthodox*
G. Ficker, "Widerlegung eines Montanisten." *Zeitschrift für Kirchengeschichte* 26 (1905): 446–63.
Greek text with French translation in P. de Labriolle, *Les Sources de l'histoire du Montanisme.* Fribourg: Université de Fribourg, 1913, 93–108.
Greek text with English translation in Ronald J. Heine, *Montanist Oracles and Testimonia.* North American Patristic Society 14. Macon, GA: Mercer University Press, 112–27.


*The Decree for Aba of Histria*
Pleket, *Epigraphica,* 33–35.


*Didascalia Apostolorum*
F. X. Funk, *Didascalia et Constitutiones Apostolorum I.* Munich: Paderborn. 1905.
*Didascalia Apostolorum in Syriac.* Edited and translated by Arthur Vööbus. Louvain: Secrétariat du Corpus Scriptorum Christianorum Orientation, 1979.


Egeria, *Diary of a Pilgrimage*
P. Maraval. *Journal de voyage (itinéraire): Egérie.* Sources chrétiennes 296. Paris: Editions du Cerf, 1982.
English translation in George Gingras, *Egeria, Diary of a Pilgrimage.* Ancient Christian Writers 38. New York and Ramsey, NJ: Newman Press, 1970.


*Epiphanius, *Medicine Box*
*Griechische Christliche Schriftsteller* 25 (1915); 31 (1980); 37 (1985).
Partial translation in Frank Williams, *The Panarion of Epiphanius of Salamis: Book I (Sects 1–46).* Nag Hammadi Studies 35. Leiden: E. J. Brill, 1987.

*Epistles of Gelasio*
A. Thiel, *Epistulae Romanorum pontificiem genuinae*. Hildesheim: Georg Olms, 1974, 360–79.

*Euripides, Bacchae
E. R. Dodds, *Euripides Bacchae* (2d ed.). Oxford: Oxford University Press, 1960.
English translation in G. S. Kirk, *The Bacchae. A Translation with Commentary*. Englewood Cliffs, NJ: Prentice Hall, 1970.

Festus, Sextus Pompeius, *Epitome of Verrius Flaccus' De Significatu Verborum*.
W. M. Lindsay, *Sexti Pompei Festi De verborum significatu quae supersunt cum Pauli epitome*. Leipzig, 1913. Reprinted Hildesheim, Georg Olms, 1965.

Flaccus, Verrius, *De Significatu Verborum*
See Festus.

*Genesis Rabbah*
H. Freedman and M. Simon, *Mishnah Rabbah*. London and New York: Soncino, 1983, vols. 1–2.
J. Neusner, *Genesis Rabbah: The Judaic Commentary to the Book of Genesis*. A New American Translation. Brown Judaic Studies, 104–6. Atlanta: Scholars Press, 1985.

*Gospel of Peter*
C. Maurer, in E. Hennecke and W. Schneemelcher, eds., *New Testament Apocrypha*. Trans. R. McL. Wilson. Philadelphia: Westminster Press, 1963, I:179–87.

*Homeric Hymn to Demeter*
N. J. Richardson, *The Homeric Hymn to Demeter*. Oxford: Clarendon Press, 1974.

Harpocration, Valerius, *Lexicon in decem oratores Atticos*
W. Dindorf, Groningen: Bouma's Boekhuis N.V., 1969.

Hesychius, *Lexicon*
K. Latte, *Hesychii Alexandrini Lexicon*. 1953–66.

Hroswitha, *Callimachus*
H. Homeyer, *Hrotsvitae Opera*. Munich: Paderborn; Vienna: Schoningh, 1970.
English translation in Larissa Bonfante, *The Plays of Hrotswitha of Gandersheim*. New York: New York University Press, 1979.

Hyginus, *Geneaologiae*
H. J. Rose, Leiden, 1934.

John Chrysostom, *Against Judaizing Christians*
Paul Harkins, *Saint John Chrysostom, Discourse against Judaizing Christians*. Fathers of the Church 68. Washington, DC: Catholic University of America Press, 1979.
*Comm. on Titus PG* 62.679.

*Jubilees*
*OTP* II: 35–142. *AOT* 1–39.

\* *The Kyme Aretalogy*
W. Peek, *Der Isishymnus von Andros und verwandte Texte.* Berlin: Weidmann, 1930.

\* *Letter of the Churches of Lyons and Vienne*
Eusebius, *History of the Church*, 5.1.
H. Musurillo, *Acts of the Christian Martyrs.* Oxford: Clarendon Press, 1972, 62–85.

\* *The Martyrdom of Saints Perpetua and Felicitas*
H. Musurillo, *Acts of the Christian Martyrs.* Oxford: Clarendon Press, 1972, 106–31.

*Mary the Proselyte to Ignatius*
*PG* Ser. 2, 5:873–80.
*ANF* 1:120–23.
Greek with Latin also in William Cureton, *Corpus Ignatianum.* London: Francis and
John Rivington, 1849, 119–27.

Musonius Rufus
*That Women, Too, Should Study Philosophy*
*Should Daughters Receive the Same Education as Sons?*
Cora E. Lutz, "Musonius Rufus: 'The Roman Socrates.'" *Yale Classical Studies* 10
(1947): 3–150.

Origen, *Against Celsus*
Henry Chadwick, *Origen Contra Celsum.* Cambridge: Cambridge University Press,
1953. Reprinted 1986.

*Orphic Hymns*
Apostolos N. Athanassakis, *The Orphic Hymns.* SBL Text and Translations 12, Greco-
Roman Religions 4. Missoula MT: Scholars Press, 1977.

Paulus Diaconus ex Festus
See Festus.

Photius, *Lexicon*
S. A. Naber. Leiden: Brill, 1864–65; Christos Theodoridis. Berlin and New York:
de Gruyter, 1982–.

Proba, *Cento*
Elizabeth Clark and Diane Hatch, *The Golden Bough, The Oaken Cross: The Virgilian
Cento of Faltonia Betitia Proba.* Chico: Scholars Press, 1981.

*Scholia on Aristophanes*
L. M. Positano, D. Holwerda, and W. J. W. Koster, *Scholia in Aristophanem*, Groningen: Wolters-Noordhoff, Amsterdam: Swets & Zeitlinger, 1960–62.

*Scholia in Lucianum*
H. Rabe, Leipzig, 1906.

*Testament of Levi*
*OTP* I:788–95; *AOT*, 526–37.

* *Testament of Job*
*OTP* I:829–68; *AOT*, 617–48.

*Theocritus, *Idyll*
A. F. S. Gow, *Theocritus.* Cambridge: Cambridge University Press, 1950.

Valerius Maximus, *Factorum ac dictorum memorabilium libri IX.*
C. Kempf, Teubner, 1888.

Xenophon of Ephesus, *Ephesian Tale*
G. Dalmeda, *Xénophon d'Ephèse. Les Ephésiaques.* Paris: Budé, 1924.
English translation in B. P. Reardon, ed., *Collected Ancient Greek Novels.* Berkeley: University of California Press, 1989.

# Bibliography

Abrahamsen, Valerie. "Women at Philippi: The Pagan and Christian Evidence." *JFSR* 3 (1987) 2:17–30.

Achtemeier, Paul, ed. *Harper's Bible Dictionary.* San Francisco: Harper & Row, 1985.

Amaru, Betsy Halpern. "Portraits of Biblical Women in Josephus' *Antiquities.*" *JJS* 39 (1988) 2:143–70.

Applebaum, Shimon. "The Legal Status of the Jewish Communities in the Diaspora." CRINT 1:420–63.

———. "The Organization of the Jewish Communities in the Diaspora." CRINT 1:464–503.

Arthur, Marilyn B. "Politics and Pomegranates: An Interpretation of the Homeric Hymn to Demeter." *Arethusa* 10 (1977):7–47.

———. "Review Essay: Classics." *Signs: Journal of Women in Culture and Society* 2 (1976):382–403.

*L'association dionysiaque dans les sociétés anciennes.* Actes de la table ronde organisée par l'Ecole française de Rome. Rome: Ecole française de Rome, 1986.

Atkins, Robert Alan. *The Integrating Function of Adoption Terminology Used by the Apostle Paul: A Grid-Group Analysis.* Ph.D dissertation, Northwestern University, 1987.

Baer, Richard. *Philo's Use of the Categories Male and Female.* Leiden: E. J. Brill, 1971.

Balch, David. L. *Let Wives Be Submissive: The Domestic Code in 1 Peter.* SBL Monograph Series 26. Chico, CA: Scholars Press, 1981.

———. "1 Cor 7:32–35 and Stoic Debates about Marriage, Anxiety and Distraction." *JBL* 102 (1983) 3:429–39.

Barnes, Timothy. "The Chronology of Montanism." *JTS* N.S. 21 (1970) 2:403–8.

———. *Tertullian. A Historical and Literary Study.* Oxford: Clarendon Press, 1971.

Baskin, Judith. "The Separation of Women in Rabbinic Judaism," in Ellison Findly and Yvonne Haddad, eds., *Women, Religion and Social Change.* Albany: State University of New York Press, 1984, 3–18.

Beard, Mary. "Priesthood in the Roman Republic," in Beard and North, *Pagan Priests*, 19–48.

———. "The Sexual Status of the Vestal Virgins." *Journal of Roman Studies* 70 (1980):12–27.

———. and North, John, eds. *Pagan Priests. Religion and Power in the Ancient World.* Ithaca, NY: Cornell University Press, 1990.

Beare, Francis W. *The Earliest Records of Jesus.* Oxford: Basil Blackwell, 1962.

Benko, Stephen. *The Virgin Goddess in Pagan and Christian Tradition.* Unpublished manuscript.

Betz, Hans Dieter, ed. *The Greek Magical Papyri in Translation.* Vol. I: *Texts.* Chicago: University of Chicago Press, 1986.

Bird, Phyllis. "The Place of Women in the Israelite Cultus," in Paul D. Hanson, Patrick D. Miller, and S. Dean McBride, eds., *Ancient Israelite Religion: Essays in Honor of Frank M. Cross.* Philadelphia: Fortress Press, 1987, 397–419.

Blayney, Jan. "Theories of Conception in the Ancient Roman World," in Rawson, *The Family in Ancient Rome,* 230–36.

Boswell, John. "Expositio and Oblatio: The Abandonment of Children and the Ancient and Medieval Family." *American Historical Review* 89 (1984) 1:10–33.

———. *The Kindness of Strangers: The Abandonment of Children in Western Europe from Late Antiquity to the Renaissance.* New York: Pantheon, 1988.

Boyarin, Daniel. "Reading Androcentrism against the Grain: Women, Sex and the Study of Torah." *Poetics Today* 12 (1991) 1:29–53.

Bradley, Keith. "Wet-nursing at Rome: A Study in Social Relations," in Rawson, *The Family in Ancient Rome,* 201–29.

van Bremen, Riet. "Women and Wealth," in A. Cameron and A. Kuhrt, eds., *Images of Women in Antiquity.* Detroit: Wayne State University Press, 1985, 223–42.

Bremmer, J. N. "Greek Maenadism Reconsidered." *ZPE* 55 (1984):267–86.

Brooten, Bernadette J. "Early Christian Women and Their Cultural Context," in Adela Yarbro Collins, ed., *Feminist Perspectives on Biblical Scholarship.* Chico: Scholars Press, 1985, 66–91.

———. "Jewish Women's History in the Roman Period: A Task for Christian Theology," in G. W. E. Nickelsburg and George W. MacRae, eds., *Christians among Jews and Gentiles. Essays in Honor of Krister Stendahl on His Sixty-Fifth Birthday.* Philadelphia: Fortress Press, 1986, 22–30 (= *HTR* 79 (1986):3–4).

———. "Junia . . . Outstanding among the Apostles," in L. and A. Swidler, eds., *Women Priests: A Catholic Commentary on the Vatican Declaration.* New York: Paulist Press, 1977, 141–44.

———. *Women Leaders in the Ancient Synagogue.* Brown Judaic Studies 36. Chico: Scholars Press, 1982.

Brouwer, H. H. J. *Bona Dea. The Sources and a Description of the Cult.* EPRO 110. Leiden: E. J. Brill, 1989.

Brown, Peter. *The World of Late Antiquity, A.D. 150–750.* New York: Harcourt Brace Jovanovich, 1971.

———. *The Body and Society: Men, Women and Sexual Renunciation in Early Christianity.* New York: Columbia University Press, 1988.

Brown, Raymond. *The Epistles of John.* Anchor Bible. Garden City, NY: Doubleday, 1982.

Bulic, F. "Iscrizione inedita." *Bulletino di Archeologia e Storia Dalmatia* 37 (1914): 107–11.

Bultmann, Rudolph. *The Johannine Epistles; A Commentary on the Johannine Epistles.* Trans. R. Philip O'Hara with Lane McGaughy and Robert Funk. Hermeneia—A Critical and Historical Commentary on the Bible. Philadelphia: Fortress Press, 1973.

Burchard, Christoph. "Die jüdische Asenethroman und seine Machwirkung. Von Egeria zu Anna Katharina Emmerick odor von Moses aus Aggel zu Karl Kerényi." *ANRW* II.20.1:543–667.

Burns, Rita J. *Has the Lord Indeed Spoken Only through Moses? A Study of the Biblical Portrait of Miriam.* SBL Dissertation Series 84. Atlanta: Scholars Press, 1987.

Burridge, Kenelm. *New Heaven, New Earth: A Study of Millenarian Activities.* New York: Schocken, 1969.

Burrus, Virginia. *Chastity as Autonomy: Women in the Stories of the Apocryphal Acts.* Studies in Women and Religion 23. Lewiston and Queenston: Edwin Mellen Press, 1987.

Cameron, Averil. "Neither Male nor Female." *Greece and Rome,* 27 (1980):60–68.

———— and Kuhrt, Amelie, eds. *Images of Women in Antiquity.* Detroit: Wayne State University Press, 1985.

Cantarella, Eva. *Pandora's Daughters: The Role and Status of Women in Greek and Roman Antiquity.* Baltimore and London: Johns Hopkins University Press, 1987.

Carroll, Michael P. *The Cult of the Virgin Mary. Psychological Origins.* Princeton: Princeton University Press, 1986.

————. "One More Time: Leviticus Revisited." *Archives européennes de sociologie* 19 (1978):339–46. Reprinted in Bernhard Lang, ed., *Anthropological Approaches to the Old Testament.* Philadelphia: Fortress Press and London: SPCK, 1985, 117–26.

Chadwick, Henry. *Origen Contra Celsum.* Cambridge: Cambridge University Press, 1953. Reprinted 1986.

Chesnutt, Randall. *Conversion in Joseph and Aseneth: Its Nature, Function and Relation to Contemporary Paradigms of Conversion and Initiation.* Ph.D. dissertation, Duke University, 1986.

Clark, Elizabeth A., and Hatch, Diane, eds. *The Golden Bough, the Oaken Cross: The Virgilian Cento of Faltonia Betitia Proba.* Chico: Scholars Press, 1981.

Cohen, Shaye J. D. "Conversion to Judaism in Historical Perspective: From Biblical Israel to Postbiblical Judaism." *Conservative Judaism* 36 (1983) 4:31–45.

————. "Crossing the Boundary and Becoming a Jew." *HTR* 82 (1989) 1:13–33.

————. *From the Maccabees to the Mishnah.* Philadelphia: Westminster Press, 1987.

————. "Menstruants and the Sacred in Judaism and Christianity," in Sarah B. Pomeroy, ed., *Women's History, Ancient History.* Chapel Hill: University of North Carolina Press, 1991, 273–99.

————. "The Origins of the Rabbinic Law." *Association for Jewish Studies Review* 10 (1985) 1:19–53.

Cohn, Norman. *The Pursuit of the Millenium: Revolutionary Millenarians and Mystical Anarchists of the Middle Ages* (2d ed., revised and expanded). New York and Oxford: Oxford University Press, 1970.

Cole, Susan Guettel. "Could Greek Women Read and Write?" *Women's Studies* 8 (1981):129–55. Reprinted in Foley, *Reflections of Women in Antiquity,* 219–45.

————. "Male and Female in the Greek Leges Sacrae." Unpublished paper read at the annual meeting of the American Philological Association, Baltimore, 1988.

————. "New Evidence for the Mysteries of Dionysos." *GRBS* 21 (1980) 3:223–38.

————. "The Social Function of Rituals of Maturation: The Koureion and the Arkteia." *ZPE* 55 (1984):233–44.

Collins, Adela Yarbro. "Women's History and the Book of Revelation." *SBLSP* 26 (1987):80–91.

Conzelmann, Hans. *A Commentary on the First Epistle to the Corinthians.* Hermeneia—A Critical and Historical Commentary on the Bible. Philadelphia: Fortress Press, 1975.

Cope, Lamar. "1 Cor 11:2–16: One Step Further." *JBL* 97 (1978) 3:435–36.

Corley, Kathleen, "Were the Women Around Jesus Really Prostitutes? Women in the Context of Greco-Roman Meals." *SBLSP* 28 (1989), 48/–521.

Craven, Toni. *Artistry and Faith in the Book of Judith.* SBL Dissertation Series. Chico: Scholars Press, 1983.

Crossan, John Dominic. *Four Other Gospels. Shadows on the Contours of the Canon.* San Francisco: Harper & Row, 1985.

Cureton, William, ed. *Corpus Ignatianum.* London: Francis and John Rivington, 1849.

Dahl, Nils. "Paul and the Church at Corinth According to 1 Corinthians 1:10–4:21," in W. F. Farmer, C. F. D. Moule, and R. R. Niebuhr, eds., *Christian History and Interpretation: Studies Presented to John Knox.* Cambridge: Cambridge University Press, 1967, 313–35.

D'Angelo, Mary Rose. "*Abba* and 'Father': Imperial Theology and the Traditions about Jesus." *JBL* in press.

———. "A Critical Note: John 20:17 and the Apocalypse of Moses 31." *JTS* 41 (1990): 529–36.

———. "Women in Luke-Acts: A Redactional View." *JBL* 109 (1990) 3:441–61.

———. "Women Partners in the New Testament." *JFSR* 6 (1990) 1:65–86.

———. "Women's Heads as Sexual Members: Paul and the Unveiled Women of Corinth." Paper presented at the Annual Meeting of the Society of Biblical Literature, New Orleans, November, 1990.

Danker, Frederick J. *Benefactor: Epigraphic Study of a Graeco-Roman and New Testament Semantic Field.* St. Louis, MO: Clayton, 1982.

Davies, Stevan. *The Revolt of the Widows. The Social World of the Apocryphal Acts.* Carbondale: Southern Illinois University Press, 1980.

Delcor, M. "Un roman d'amour d'origine thérapeute: Le Livre de Joseph et Asénath." *Bulletin de Littérature Ecclesiastique* 63 (1962):3–27.

Detienne, Marcel. *The Gardens of Adonis: Spices in Greek Mythology.* Trans. Janet Lloyd. Atlantic Highlands, NJ: Humanities Press, 1977.

Deubner, Ludwig. *Attische Feste.* Berlin: Verlag Heinrich Keller, 1932.

Dixon, Suzanne. *The Roman Mother.* Norman: Oklahoma University Press, 1988.

Dodd, C. H. *The Johannine Epistles.* The Moffatt New Testament Commentary. London: Hodder and Stoughton, 1946.

Dodds, E. R. Euripides *Bacchae* (2d ed.). Oxford: Oxford University Press, 1960.

———. "Maenadism in the Bacchae," in *The Greeks and the Irrational.* Sather Classical Lectures 25. Berkeley: University of California Press, 1951.

Doran, Robert. "Narrative Literature," in Kraft and Nickelsburg, *Early Judaism and its Modern Interpreters,* 290–93.

Douglas, Mary. "The Abominations of Leviticus," in *Purity and Danger: An Analysis of Concepts of Pollution and Taboo.* London: Routledge and Kegan Paul, 1966, 41–57. Reprinted in Bernhard Lang, ed., *Anthropological Approaches to the Old Testament.* Philadelphia: Fortress Press and London: SPCK, 1985, 100–16.

———. "The Background of the Grid Dimension: A Comment." *Sociological Analysis. A Journal in the Sociology of Religion* 50 (1989) 2:171–76.

———. *Cultural Bias.* Occasional paper no. 35 of the Royal Anthropological Institute of Great Britain and Ireland. London: 1978. Reprinted in *In the Active Voice.* London: Routledge and Kegan Paul, 1982.

———. "Introduction to Grid/Group Analysis," in M. Douglas, ed., *Essays in the Sociology of Perception.* London: Routledge and Kegan Paul, 1982.

———. *Natural Symbols. Explorations in Cosmology.* London: Barrie and Rockcliffe; New York: Pantheon Books, 1970. Reprinted 1973.

Dunand, Françoise. "Le Statut des 'Hiereiai' en Egypte romaine," in M. B. de Boer and T. A. Eldridge, eds., *Hommages à Maarten J. Vermaseren*. Leiden: E. J. Brill. I: 352–74.

Durkheim, Emile. *The Elementary Forms of Religious Life*. London: Allen and Unwin, 1915. First published as *Les Formes élémentaires de la vie religieuse*, 1912.

———. *The Rules of Sociological Method*. New York: Free Press, 1965. First published as *Les Règles de la méthode sociologique*, 1912.

Ehrman, Bart D. "Cephas and Peter." *JBL* 109 (1990) 3:463–74.

Emmett, Alanna. "Female Ascetics in the Greek Papyri." *Jahrbuch der österreichischen Byzantinistik* 32 (1982) 2:507–15.

Enslin, Morton. *The Book of Judith*. Edited with a general introduction and appendices by Solomon Zeitlin. Leiden: E. J. Brill for Dropsie University, 1972.

Evans, Arthur. *The God of Ecstasy*. New York: St. Martin's Press, 1988.

Ewing, Anne Hickey. *Women of the Roman Aristocracy as Christian Monastics*. Studies in Religion 1. Ann Arbor, MI: UMI Research Press, 1987.

Farnell, Lewis R. *Cults of the Greek City States*. Reprinted Chicago: Aegean Press, 1971.

Ferrua, A. "Note su Tropea." Cited incompletely in Otranto "Note."

Fiorenza, Elisabeth Schüssler. *In Memory of Her: A Feminist Theological Reconstruction of Christian Origins*. New York: Crossroad, 1983.

———. "'You are not to be called Father': Early Christian History in a Feminist Perspective." *Cross Currents* 30 (1979):301–23.

Flory, Marleen Boudreau. "Where Women Precede Men: Factors Influencing the Order of Names in Roman Epitaphs." *The Classical Journal* 79 (1984) 3:216–24.

Foley, Helene P., ed. *Reflections of Women in Antiquity*. New York: Gordon and Breach, 1981.

Ford, J. Massingberd. "Was Montanism a Jewish-Christian Heresy?" *Journal of Ecclesiastical History* 17 (1966):145–58.

French, Valerie. "The Cult of the Mater Matuta." Paper presented at the 1987 Berkshire Conference on Women's History.

Fredriksen, Paula. *From Jesus to Christ. The Origins of the New Testament Images of Jesus*. New Haven: Yale University Press, 1988.

Frontisi-Ducroux, Françoise. "Images du ménadisme féminin: Les vases des "Lénéennes," in *L'association dionysiaque dans les sociétés anciennes*, 165–76.

Fuks, G. "Where Have All the Freedpersons Gone?" *JJS* 36 (1985):25–32.

Funk, Robert W., ed. *New Gospel Parallels*. Philadelphia: Fortress Press, 1985.

Gagé, Jean. *Matronalia: Essai sur les dévotions et les organisations culturelles des femmes dans l'ancienne Rome*. Collection Latomus LX. Brussels: Berchem, 1963.

Gager, John G. "Body-symbols and Social Reality: Resurrection, Incarnation and Asceticism in Early Christianity." *Religion* 12 (1982):345–63.

———. *Kingdom and Community. The Social World of Early Christianity*. Englewood Cliffs, NJ: Prentice Hall, 1975.

———. ed. *Curse Tablets and Binding Spells from the Ancient World*. New York and Oxford: Oxford University Press, in press.

Gardner, Jane. *Women in Roman Law and Society*. Bloomington and Indianapolis: Indiana University Press, 1986.

Garland, Robert. "Priests and Power in Classical Athens," in Beard and North, *Pagan Priests*, 75–91.

Gibson, Elsa. *The "Christians for Christians" Inscriptions of Phrygia. Greek Texts, Translation and Commentary*. Harvard Theological Studies 32. Missoula: Scholars Press, 1978.

*Bibliography*

———. "Montanist Epitaphs at Uşak." *GRBS* 16 (1975):433–42.

Ginsburg, Louis. *Legends of the Jews.* Philadelphia: Jewish Publication Society, 1925.

Goldenberg, Robert. "The Jewish Sabbath in the Roman World." *ANRW* II.19.1:414–47.

Goodblatt, David. "The Beruriah Traditions." *JJS* 26 (1975):68–85.

Goodenough, Erwin R. *An Introduction to Philo Judaeus* (2d ed. rev.). Oxford: Basil Blackwell, 1962.

———. *Jewish Symbols in the Greco-Roman Period.* Bollingen Series XXXVII. Princeton: Princeton University Press, 13 vols., 1953–68. Reprinted, abridged in one volume, ed. Jacob Neusner. Princeton: Princeton University Press, 1988.

Goodman, Martin. *State and Society in Roman Galilee, A.D. 132–212.* Totowa, NJ: Rowman and Allanheld, 1983.

Gordon, Richard. "The Veil of Power: Emperors, Sacrificers and Benefactors," in Beard and North, *Pagan Priests,* 201–31.

Goree, William B., Jr. *The Cultural Bases of Montanism.* Ph.D. dissertation, Baylor University, 1980.

Grégoire, H. *Recueil des inscriptions grecques chrétiennes de l'Asie Mineure.* Paris: Leroux, 1922. Reprinted Amsterdam: A. M. Hakkert, 1968.

Griffiths, Frederick. "Home before Lunch: The Emancipated Woman in Theocritus," in Foley, *Reflections of Women in Antiquity,* 247–73.

Griffiths, J. Gwyn. "Xenophon of Ephesus on Isis and Alexandria," in M. B. de Boer and T. A. Eldridge, eds., *Hommages à Maarten J. Vermaseren.* Leiden: E. J. Brill, 1978. I: 409–37.

Groh, Dennis E. "Jews and Christians in Late Roman Palestine: Towards a New Chronology." *Biblical Archaeologist* 51 (1988):80–98.

Gryson, Roger. *The Ministry of Women in the Early Church.* Trans. Jean Laporte and Mary Louise Hall. Collegeville, MN: Liturgical Press, 1976.

Guarducci, Maria. *Epigrafica Greca.* Rome: Istituto poligrafico dello Stato, Libreria dello Stato, 4 vols, 1967–78.

Hadas, Moses. *A History of Latin Literature.* New York and London: Columbia University Press, 1952.

Hallett, Judith. *Fathers and Daughters in Roman Society: Women and the Elite Family.* Princeton: Princeton University Press, 1984.

Halperin, David M., Winkler, John J., and Zeitlin, Froma I., eds. *Before Sexuality: The Construction of Erotic Experience in the Ancient Greek World.* Princeton: Princeton University Press, 1990.

Harl, K. W. *Civic Coins and Civic Politics in the Roman East A.D. 180–275.* Berkeley: University of California Press, 1987.

von Harnack, Adolph. *The [Mission and] Expansion of Christianity in the First Three Centuries.* Trans. James Moffatt. New York: G. P. Putnam's Sons, 1904–5. Reprinted, New York: Books for Libraries Press, 1972.

Harris, W. V. "Literacy and Epigraphy, I." *ZPE* 52 (1983):87–111.

———. *Ancient Literacy.* Cambridge, MA: Harvard University Press, 1989.

Heine, Ronald J. *Montanist Oracles and Testimonia.* North American Patristic Society 14. Macon, GA: Mercer University Press, 1989.

———. "The Role of the Gospel of John in the Montanist Controversy." *The Second Century* 6 (1987/88):1–19.

Henrichs, Albert. "Male Intruders among the Maenads: The So-Called Male Celebrant," in Harold D. Evjen, ed., *Mnemai. Classical Studies in Memory of Karl K. Hulley.* Chico: Scholars Press, 1984, 69–91.

———. "Greek Maenadism from Olympias to Messalina." *Harvard Studies in Classical Philology* 82 (1978):121–60.

———. "Changing Dionysiac Identities," in B. F. Meyers and E. P. Sanders, eds., *Jewish and Christian Self-Definition.* Vol. III: *Self-Definition in the Graeco-Roman World.* Philadelphia: Fortress Press, 1982, 137–60; notes 213–36.

Heyob, Sharon Kelly. *The Cult of Isis among Women in the Greco-Roman World.* EPRO 51. Leiden: E. J. Brill, 1975.

Hoenig, Sidney B. "City-Square and Synagogue." *ANRW* II.19.1: 448–76.

Holmberg, Bengt. *Paul and Power: The Structure of Authority in the Primitive Church as Reflected in the Pauline Epistles.* Philadelphia: Fortress Press, 1980.

Hopkins, Keith. "Contraception in the Roman Empire." *Comparative Studies in History and Society* 8 (1965):124–51.

James, M. R. *The Apocryphal New Testament.* Oxford: At the Clarendon Press, 1924. Reprinted (corrected) 1969.

Jenkins, Claude. "Origen on 1 Corinthians." *JTS* 10 (1908): 29–51.

Johansen, J. Prtyz. "The Thesmophoria as a Women's Festival." *Temenos* 11 (1975): 78–87.

Johnson, Luke. *An Interpretation of the New Testament.* Philadelphia: Fortress Press, 1986.

Joly, Robert. *Le Dossier d'Ignace d'Antioche.* Université Libre de Bruxelles 69. Brussels: Editions de l'Université, 1979.

Jones, C. P. "Three Foreigners in Attica," *Phoenix* 32 (1978):222–28.

Kahil, Lilly. "L'Artémis de Brauron: Rites et Mystère." *Antike Kunst* 20 (1977):86–98.

Kearsley, R. A. "Asiarchs, *Archiereis* and the *Archiereiai* of Asia." *GRBS* 27 (1986): 183–92.

Kendall, Patricia A. *Women and the Priesthood: A Selected and Annotated Bibliography.* Philadelphia: The Episcopal Diocese of Philadelphia, 1976.

Keuls, Eva C. "Male-Female Interaction in Fifth-Century Dionysiac Ritual as Shown in Attic Vase Painting." *ZPE* 55 (1984):287–97.

———. *The Reign of the Phallus: Sexual Politics in Ancient Athens.* New York: Harper & Row, 1985.

King, Karen, ed. *Images of the Feminine in Gnosticism.* Philadelphia: Fortress Press, 1988.

Kirk, G. S. *The Bacchae. A Translation with Commentary.* Englewood Cliffs, NJ: Prentice Hall, 1970.

Klawiter, Frederick C. *The New Prophecy in Early Christianity: The Origin, Nature and Development of Montanism, A.D. 165–220,* Ph.D. dissertation, University of Chicago, 1975.

———. "The Role of Martyrdom and Persecution in Developing the Priestly Authority of Women in Early Christianity: A Case Study of Montanism." *Church History* 49 (1980) 3:251–61.

Kloppenborg, John S. *The Formation of Q: Trajectories in Ancient Wisdom Collections.* Philadelphia: Fortress Press, 1987.

———. *Q Parallels. Synopsis, Critical Notes & Concordance.* Sonoma, CA: Polebridge Press, 1988.

——— et al. *Q Thomas Reader.* Sonoma, CA: Polebridge Press, 1990.

Koester, Helmut. *Introduction to the New Testament.* Foundations and Facets: New Testament. 2 vols. Philadelphia: Fortress Press, 1982.

Kornemann, E. "*Collegium,*" *RE* IV I (1900) cols. 380–480.

Kraft, Robert A., and Nickelsburg, George W. E., eds., *Early Judaism and Its Modern Interpreters, The Bible and Its Modern Interpreters,* Vol 2. Douglas A. Knight, ed. Atlanta: Scholars Press, 1986.

Kraemer, Ross S. "The Conversion of Women to Ascetic Forms of Christianity." *Signs: Journal of Women in Culture and Society* 6 (1980) 2:298–307. Reprinted in Judith M. Bennet, Elizabeth A. Clark, Jean O'Barr, B. Anne Vilen, and Sarah Westphal-Wihl, eds., *Sisters and Workers in the Middle Ages.* Chicago: University of Chicago Press, 1989, 198–207.

———. "Ecstasy and Possession: The Attraction of Women to the Cult of Dionysos." *HTR* 72 (1979):55–80.

———. "Ecstasy and Possession: Women of Ancient Greece and the Cult of Dionysos," in Rita Gross and Nancy Falk, eds., *Unspoken Worlds: Women's Religious Lives in Non-western Cultures.* New York: Harper & Row, 1980, 53–69. Reprinted as *Unspoken Worlds: Women's Religious Lives.* Belmont, CA: Wadsworth Press, 1989, 45–55.

———. *Ecstatics and Ascetics. Studies in the Functions of Religious Activities for Women in the Greco-Roman World.* Ph.D. dissertation, Princeton University, 1976.

———. "'Euoi Saboi' in Demosthenes' *De Corona*: In Whose Honor Were the Women's Rites?" SBLSP 20 (1981):229–36.

———. "Hellenistic Jewish Women: The Epigraphical Evidence." SBLSP 25 (1986):183–200.

———. "Jewish Women in the Diaspora World of Late Antiquity," in Judith Baskin, ed., *Jewish Women in Historical Perspective.* Detroit: Wayne State University Press, 1991, 43–67.

———. "Monastic Jewish Women in Greco-Roman Egypt: Philo on the Therapeutrides." *Signs: Journal of Women in Culture and Society* 14 (1989) 1:342–70.

———. "A New Inscription from Malta and the Question of Women Elders in Diaspora Jewish Communities." *Harvard Theological Review* 78 (1985)3–4:431–38.

———. "Non-Literary Evidence for Jewish Women in Rome and Egypt," in Marilyn B. Skinner, guest editor, *Rescuing Creusa: New Methodological Approaches to Women in Antiquity = Helios* 13 (1986) 2:85–101.

———. "Women's Authorship of Jewish and Christian Literature in the Greco-Roman Period," in Amy-Jill Levine, ed., *Jewish Women in Hellenistic Literature,* Septuagint and Cognate Studies. Atlanta: Scholars Press, 1991, 221–42.

Kümmel, Werner George, ed. *Introduction to the New Testament* (14th ed., rev.). Trans. A. J. Mattill, Jr. Nashville: Abingdon Press, 1966.

de Labriolle, P. *Les Sources de l'histoire du Montanisme.* Fribourg: Université de Fribourg, 1913.

———. *La Crise Montanist.* Paris: E. Leroux, 1913.

Lacey, W. K. "Patria Potestas," in Rawson, *The Family in Ancient Rome,* 121–44.

Lanckoronski, Graf. *Städte Pamphyliens und Pisidiens* I–II. Vienna, 1890–92.

Lanternari, Victor. *The Revolt of the Oppressed: A Study of Modern Messianic Cults.* New York: New American Library, 1965.

Lawler, Lillian. "The Maenads: A Contribution to the Study of the Dance in Ancient Greece." *Memoirs of the American Academy in Rome* 6 (1927):69–112.

Layton, Bentley R. *The Gnostic Scriptures.* Garden City, NY: Doubleday, 1987.

261

Lefkowitz, Mary R. "The Motivations for St. Perpetua's Martyrdom." *JAAR* 44 (1976):417–21.

———. *Women in Greek Myth*. Baltimore and London: Johns Hopkins University Press, 1986.

———, and Fant, Maureen B., eds. *Women's Lives in Greece and Rome: A Source Book in Translation*. Baltimore: Johns Hopkins University Press, 1982.

Leon, Harry J. *The Jews of Ancient Rome*. Philadelphia: Jewish Publication Society, 1960.

Levine, Amy-Jill. "Who's Catering the Q Affair? Feminist Observations on Q Paraenesis." *Semeia* 50 (1990):145–61.

Lewis, I. M. *Ecstatic Religion*. Harmondsworth: Penguin, 1971. Reprinted with a new introduction. New York and London: Routledge, 1989.

———. "Spirit Possession and Deprivation Cults." *Man*, N.S. 1 (1966) 3:307–29.

Lewis, Naphtali. *The Documents from the Bar Kochba Period in the Cave of Letters: Greek Papyri*. Judean Desert Studies 2. Jerusalem: Israel Exploration Society, Hebrew University, Shrine of the Book, 1989.

———. Katzoff, Ranon, and Greenfield, Jonas. "*Papyrus Yadin* 18." *Israel Exploration Journal* 37 (1987) 4: 229–50.

———. "Two Greek Documents from Provincia Arabia." *Illinois Classical Studies* 3 (1978):100–114.

Lifshitz, Baruch. *Donateurs et Fondateurs dans les synagogues juives. Répertoires des dédicaces grecques relatives à la construction et à la réfection des synagogues*. Paris: J. Gabalda et Cie, 1967.

Lightstone, Jack. "Christian Anti-Judaism in Its Judaic Mirror: The Judaic Context of Early Christianity revised," in Stephen G. Wilson, ed., *Separation and Polemic. Anti-Judaism in Early Christianity* 2. Studies in Christianity and Judaism 2. Waterloo, Ontario: Wilfred Laurier University Press, 1986, 103–132.

Lincoln, Bruce. "The Rape of Persephone: A Greek Scenario of Women's Initiation." *HTR* 72 (1979) 3–4: 223–35. Reworked in Lincoln, *Emerging from the Crysalis: Studies in Rituals of Women's Initiation*. Cambridge, MA: Harvard University Press, 1981, 71–90.

Loades, Ann. "Review of R. S. Kraemer, 'Maenads, Martyrs, Matrons, Monastics,'" *Religious Studies Review* 15 (1989) 4:353.

Luck, Georg. *Arcana Mundi: Magic and the Occult in the Greek and Roman Worlds*. Baltimore and London: Johns Hopkins University Press, 1985.

Luedemann, Gerd. *Opposition to Paul in Jewish Christianity*. Trans. M. Eugene Boring. Minneapolis: Fortress Press, 1989.

Lutz, Cora E. "Musonius Rufus: 'The Roman Socrates.'" *Yale Classical Studies* 10 (1947):3–150.

MacDonald, Dennis Ronald. *The Legend and the Apostle: The Battle for Paul in Story and Canon*. Philadelphia: Westminster Press, 1983.

———. *The Acts of Andrew and The Acts of Andrew and Matthias in the City of the Cannibals*. Texts and Translations 33. Christian Apocrypha Series 1. Society of Biblical Literature. Atlanta: Scholars Press, 1990.

MacDonald, Margaret Y. *The Pauline Churches: A Socio-Historical Study of Institutionalization in the Pauline and Deutero-Pauline Writings*, SNTSMS 60. Cambridge: Cambridge University Press, 1988.

MacMullen, Ramsay. "The Epigraphic Habit in the Roman Empire." *American Journal of Philology* 103 (1982):233–46.

————. "Woman in Public in the Roman Empire." *Historia* 29 (1980):208–18.

McClees, Helen. *A Study of Women in Attic Inscriptions.* New York: Columbia University Press, 1920.

McGinn-Moorer, Sheila Elizabeth. *The New Prophecy of Asia Minor and the Rise of Ecclesiastical Patriarchy in Second Century Pauline Traditions.* Ph.D. dissertation, Northwestern University, 1989.

Malherbe, Abraham. *Social Aspects of Early Christianity.* Baton Rouge: Louisiana State University, 1977.

Malina, Bruce. *Christian Origins and Cultural Anthropology.* Atlanta: John Knox Press, 1986.

————. *The New Testament World: Insights from Cultural Anthropology.* Atlanta: John Knox Press, 1981.

Meeks, Wayne. *The First Urban Christians: The Social World of the Apostle Paul.* New Haven: Yale University Press, 1983.

Meiselman, Moshe. *Jewish Woman in Jewish Law.* New York: Ktav, 1978.

Menage, Gilles. *The History of Women Philosophers.* Trans. Beatrice H. Zedler. Lanham, MD: University Press of America, 1984. Originally written c. 1690–92.

Meyers, Eric. "Early Judaism and Christianity in the Light of Archaeology." *Biblical Archaeologist* 51 (1988):69–79.

Mills, Harrianne. "Greek Clothing Regulations: Sacred and Profane?" *ZPE* 55 (1984):255–65.

Mommsen, A. "Ράκος" aus attischen Inschriften." *Philologus* 58 (1899):343–47.

Moore, George Foote. *Judaism in the First Centuries of the Christian Era: The Age of the Tannaim.* 3 vols. Cambridge, MA: Harvard University Press, 1927.

Mora, Fabio. *Prosopografia Isiaca. II. Prosopografia Storica E Statistica Del Culto Isiaco.* EPRO 113. Leiden: E. J. Brill, 1990.

Murphy-O'Connor, Jerome. "1 Corinthians 11:2–16 Once Again." *CBQ* 50 (1988) 2:265–74.

————. "The Non-Pauline Character of 1 Corinthians 11:2–16?" *JBL* 95 (1976): 615–21.

————. "Sex and Logic in 1 Corinthians 11:2–1." *CBQ* 42 (1980):482–500.

Murray, Gilbert. *Five Stages of Greek Religion.* New York: Columbia University Press, 1925. Reprinted New York: AMS Press, 1978.

Mylonas, George E. *Eleusis and the Eleusinian Mysteries.* Princeton: Princeton University Press, 1961.

Nagel, M. "Lettre chrétienne sur papyrus (provenant de milieux sectaires du IVe siècle?)" *ZPE* 18 (1975):317–23.

Neusner, Jacob. *A History of the Mishnaic Law of Women.* Studies in Judaism in Late Antiquity 33. 5 vols. Leiden: E. J. Brill, 1980.

Neyrey, Jerome H. "Bewitched in Galatia: Paul and Cultural Anthropology." *CBQ* 50 (1988):72–100.

————. "Body Language in 1 Corinthians," *Semeia* 35 (1986):129–70.

————. *An Ideology of Revolt: John's Christology in Social-Science Perspective.* Philadelphia: Fortress Press, 1988.

————. "Witchcraft Accusations in 2 Cor 10–13: Paul in Social Science Perspective." *Listening* 21 (1986) 2:160–70.

Nickelsburg, G. W. E. *Jewish Literature between the Bible and the Mishnah: A Historical and Literary Introduction.* Philadelphia: Fortress Press, 1981.

Nilsson, M. *The Dionysiac Mysteries in the Hellenistic and Roman Ages.* Lund: C. K. Gleerup, 1957. Reprinted New York: Arno Press, 1975.

Noonan, John T., Jr. *Contraception: A History of Its Treatment by the Catholic Theologians and Canonists.* New York: Mentor. Reprinted 1967.

North, J. A. "Religious Toleration in Republican Rome." *Proceedings of the Cambridge Philological Society* 25 (1979):85–103.

Olender, Maurice. "Aspects of Baubo: Ancient Texts and Contexts," in Halperin, Winkler, and Zeitlin, *Before Sexuality,* 83–109.

Orr, D. "Roman Domestic Religion: The Evidence of the Household Shrines." *ANRW* II.16.2:1556–91.

Ortner, Sherry B. "Is Nature to Culture as Female Is to Male?" in Michelle Rosaldo and Louise Lamphere, eds., *Woman, Culture and Society.* Stanford: Stanford University Press, 1974, 67–88.

Osiek, Carolyn. "The Widow as Altar: The Rise and Fall of a Symbol." *The Second Century* 3 (1983):159–69.

Oster, Richard E. "When Men Wore Veils to Worship: The Historical Context of 1 Corinthians 11:14." *New Testament Studies* 34 (1988) 4:481–505.

Otranto, Giorgio. "Note sul sacerdozio femminile nell'antichita in margine a una testimonianza di Gelasio I." *Vetera Christianorum* 19 (1982):341–60.

Owen, Dennis, and Isenberg, Sheldon R. "Bodies Natural and Contrived." *Religious Studies Review* 3 (1977) 1:1–17.

Pagels, Elaine. *Adam, Eve and the Serpent.* New York: Random House, 1988.

———. "Christian Apologists and the 'Fall of the Angels': An Attack on Roman Imperial Power?" *HTR* 78 (1985):3–4, 301–25.

———. *The Gnostic Gospels.* New York: Random House, 1974.

Pagnotta, Maria Antonietta. "Il culto di Fortuna Virile e Venere Verticordia nei Riti delle Calende di Aprile a Roma." *Annali Della Facoltà di Lettere e Filosofia,* Università degli Studi di Perugia 16 (1978/79) n.s. II, 144–56.

Palmer, Robert. "Roman Shrines of Female Chastity from the Caste Struggle to the Papacy of Innocent I." *Rivista storica dell 'antichità* 4 (1974):294–309.

Patai, Raphael. *The Hebrew Goddess.* New York: Ktav, 1967.

Peradotto, John, and Sullivan, J. P., eds. *Women in the Ancient World: The Arethusa Papers.* Albany, NY: State University of New York Press, 1984.

Perlman, Paula. "Plato *Laws* 833C-834D and the Bears of Brauron." *GRBS* 24 (1983):115–30.

Petersen, W. L. "Can *arsenokoitai* be translated by 'Homosexuals'? (1 Cor. 6:9, 1 Tim. 1:10)." *Vigiliae Christianae* 40 (1986) 2:187–91.

Philippart, H. "Iconographie des Bacchantes d'Euripide." *Revue Belge de Philologie et d'Histoire* 9 (1930):5–72.

Phillips, J. A. *Eve: The History of an Idea.* San Francisco: Harper & Row, 1984.

Philonenko, Marc. *Joseph et Aséneth. Introduction, texte critique, traduction et notes.* Leiden: E. J. Brill, 1968.

Plaskow, Judith. "Blaming Jews . . . for the Birth of Patriarchy." *Lilith* 7 (1980): 11–13.

———. "The Coming of Lilith: Toward a Feminist Theology," in R. Ruether, ed., *Religion and Sexism: Images of Women in the Jewish and Christian Traditions.* New York: Simon & Schuster, 1974, 341–43. Reprinted in C. Christ and J. Plaskow, eds., *Womanspirit Rising: A Feminist Reader in Religion.* San Francisco: Harper &

Row, 1979, 198–209.

Pleket, H. W. *Texts on the Social History of the Greek World. Epigraphica* vol II. Leiden: E. J. Brill, 1969.

Pomeroy, Sarah B. *Goddesses, Whores, Wives and Slaves: Women in Classical Antiquity.* New York: Schocken, 1975.

———. "Technikai kai Musikai: The Education of Women in the Fourth Century and in the Hellenistic Period."*American Journal of Ancient History* 2 (1977):51–68.

———. *Women in Hellenistic Egypt: From Alexander to Cleopatra.* New York: Schocken Books, 1984.

———. "Women in Roman Egypt: A Preliminary Study Based on Papyri," in Foley, *Reflections of Women in Antiquity,* 303–22.

———, with Kraemer, Ross S., and Kampen, Natalie. "Selected Bibliography on Women in Classical Antiquity," in Peradotto and Sullivan, *Women in the Ancient World,* 317–72.

Pope, Marvin. *Song of Songs.* Anchor Bible. Garden City, NY: Doubleday, 1977.

Popescue, E. M. "The Histrian Decree for Aba (second century of our era)." *Dacia* n.s. IV (1960): 273–96.

Porte, Danielle. *Etiologie Religieuse dans Les Fastes d'Ovide.* Paris: Société d'Edition "Les Belles Lettres," 1985.

Portefaix, Lilian. *Sisters Rejoice. Paul's Letter to the Philippians and Luke-Acts as Received by First Century Philippian Women.* Coniectanea Biblica New Testament Series 20. Stockholm: Almqvist and Wiksell International, 1988.

Powell, Douglas. "Tertullianists and Cataphrygians." *Vigiliae Christianae* 29 (1975):33–54.

Raditsa, Leo Ferrero. "Augustus' Legislation Concerning Marriage, Procreation, Love Affairs and Adultery." *ANRW* II.13:278–339.

Radot, G., and Paris, P. "Inscriptions de Syllion en Pamphylie." *Bulletin de Correspondance Hellenique* 13 (1889):486–97.

Ramsay, William M. *Cities and Bishoprics of Phrygia.* Oxford: Clarendon Press, 1895–97. Reprinted New York: Arno Press, 1975.

Rawson, Beryl, ed. *The Family in Ancient Rome: New Perspectives.* Ithaca, NY: Cornell University Press, 1986.

———. "The Roman Family," in B. Rawson, *The Family in Ancient Rome,* 1–57.

Reynolds, Joyce, and Tannenbaum, Robert. *Jews and Godfearers at Aphrodisias. Greek Inscriptions with Commentary.* Cambridge Philological Society Supplementary Vol. 12. Cambridge: Cambridge Philological Society, 1987.

Reynolds, L. D., and Wilson, N. G. *Scribes and Scholars: A Guide to the Transmission of Greek and Latin Literature,* (2d ed., revised and enlarged). New York and Oxford: Oxford University Press, 1974.

Richlin, Amy. "Rituals of the Body: Roman Women's Religion." Unpublished paper delivered at Princeton University and elsewhere in the fall of 1989.

Robert, Louis. *Documents de l'Asie Mineure Méridionale: Inscriptions, monnaies et géographie.* Geneva: Droz, 1966.

Robertson, N. "Greek Ritual Begging in Aid of Women's Fertility and Childbirth." *Transactions of the American Philological Association* 113 (1983):143–69.

Rohrbach, Richard L. "Social Location of 'Thought' as a Heuristic Construct in New Testament." *Journal for the Study of New Testament* 30 (1987):103–19.

Rosaldo, Michelle. "Introduction: A Theoretical Overview," in Michelle Rosaldo Rosaldo and Louise Lamphere, eds., *Woman, Culture and Society*. Stanford: Stanford University Press, 1974, 17–42.

Rossi, Mary Ann. "Priesthood, Precedent and Prejudice: On Recovering the Women Priests of Early Christianity. G. Otranto's 'Notes on the Female Priesthood in Antiquity.'" *JFSR* 7 (1991) 1: 73–94.

Rousselle, Aline. *Porneia. On Desire and the Body in Antiquity*. London: Basil Blackwell, 1988.

Saldarini, Antony. "Reconstructions of Rabbinic Judaism," in Kraft and Nickelsburg, *Early Judaism and Its Modern Interpreters*, 437–77.

Sale, W. "The Temple Legends of the Arkteia." *Rheinisches Museum für Philologie* 118 (1975):265–84.

Saller, Richard P. "Men's Age at Marriage and Its Consequences in the Roman Family." *Classical Philology* 82 (1987)21–34.

Schepelern, Wilhelm. *Der Montanismus und die Phrygischen Kulte*. Tubingen: J. C. B. Mohr, 1929.

Schilling, Robert. *Rites, Cultes, Dieux de Rome*. Paris: Editions Klincksieck, 1979.

———. *La religion romaine de Vénus depuis les origines jusqu'au temps d'Auguste*. Paris: E. de Boccard, 1954.

Schoedel, William R. *Ignatius of Antioch*. Hermeneia—A Critical and Historical Commentary on the Bible. Philadelphia: Fortress Press, 1985.

———. "Are the Letters of Ignatius of Antioch Authentic?" *Religious Studies Review* 6 (1980) 3:196–201.

Schürer, Emil. *The History of the Jewish People in the Age of Jesus Christ*. Edited and revised by Matthew Black, Geza Vermes, Fergus Millar, and Martin Goodman. 3 vols. Edinburgh: T & T Clark, 1986.

Scroggs, Robin. "Paul and the Eschatological Woman." *JAAR* 40 (1972):283–303.

———. "Paul and the Eschatological Woman: Revisited." *JAAR* 42 (1974):532–37.

Scullard, H. H. *Festivals and Ceremonies of the Roman Republic*. Ithaca, NY: Cornell University Press, 1981.

Sealey, Raphael. *Women and Law in Classical Greece*. Chapel Hill and London: University of North Carolina Press, 1988.

Segal, Alan F. *Rebecca's Children: Judaism and Christianity in the Roman World*. Cambridge, MA: Harvard University Press, 1986.

Segal, Charles. "The Menace of Dionysos: Sex Roles and Reversal in Euripides' *Bacchae*," in Peradotto and Sullivan, *Women in the Ancient World*, 195–212.

Shaw, Brent. "The Age of Roman Girls at Marriage: Some Reconsiderations." *Journal of Roman Studies* 77 (1987):30–46.

Shukster, Martin B., and Richardson, Peter. "Temple and *Bet Ha-Midrash* in the Epistle of Barnabas," in Stephen G. Wilson, ed., *Separation and Polemic. Anti-Judaism in Early Christianity*, vol. 2. Waterloo, Ontario: Wilfred Laurier University Press, 17–31.

Skinner, Marilyn B., guest ed. *Rescuing Creusa: New Methodological Approaches to Women in Antiquity (= Helios 13* (1986)2).

Skov, G. E. "The Priestess of Demeter and Kore and her Role in the Initiation of Women at the Festival of the Haloa at Eleusis." *Temenos* 11 (1975):136–47.

Slater, Philip. *The Glory of Hera. Greek Mythology and the Family*. Boston: Beacon Press, 1968.

Smallwood, E. Mary. *The Jews under Roman Rule from Pompey to Diocletian: A Study in Political Relations.* Leiden: E. J. Brill, 1981.

———. "The Alleged Jewish Tendencies of Poppaea Sabina." *Journal of Theological Studies* N.S. 10 (1959):329–35.

Smith, Robert C., and Lounibos, John. *Pagan and Christian Anxiety: A Response to E. R. Dodds.* Lanham, MD: University Press of America, 1984.

Snyder, Jane McIntosh. *The Woman and the Lyre: Women Writers in Classical Greece and Rome.* Carbondale: Southern Illinois University Press, 1989.

Sourvinou, Christiane. "Aristophanes, *Lysistrata,* 641–47." *Classical Quarterly* 65 (1971):339–42.

Spickard, James V. "A Guide to Mary Douglas' Three Versions of Grid/Group Theory." *Sociological Analysis. A Journal in the Sociology of Religion* 50 (1989) 2:151–70.

———. *Relativism and Cultural Comparison in the Anthropology of Mary Douglas.* Ph.D. dissertation, Graduate Theological Union (Berkeley, CA), 1984.

Starr, Raymond J. "The Circulation of Literary Texts in the Roman World." *Classical Quarterly* 37 (1987): 213–23.

Stehle, Eva. "Venus, Cybele and the Sabine Women: The Roman Construction of Female Sexuality." *Helios* 16 (1989) 2:143–64.

Stinton, T. C. W. "Iphigeneia and the Bears of Brauron." *Classical Quarterly* 26 (1972):12–14.

Stroble, August. *Das heilige Land der Montanisten.* Berlin and New York: Walter de Gruyter, 1980.

Szemler, G. J. "Priesthoods and Priestly Careers in Ancient Rome." *ANRW* II.16.3:2314–31.

Tcherikover, Victor. *Hellenistic Civilization and the Jews.* Trans. S. Applebaum. Philadelphia: Jewish Publication Society, 1966.

Theissen, Gerd. *Sociology of Early Palestinian Christianity.* Philadelphia: Fortress Press, 1978.

———. *The Social Setting of Pauline Christianity.* Philadelphia: Fortress Press, 1982.

Thompson, Cynthia. "Portraits from Roman Corinth: Hairstyles, Headcoverings and St. Paul." *Biblical Archaeology* 51 (1988) 2:99–115.

Trompf, G. W. "On Attitudes toward Women in Paul and Paulinist Literature: 1 Cor 11:3–16 and Its Context." *CBQ* 42 (1980):196–215.

Turner, Victor. *The Ritual Process: Structure and Anti-Structure.* Ithaca, NY: Cornell University Press, 1969. Reprinted 1977.

Tyson, Joseph B. *The New Testament and Early Christianity.* New York: Macmillan, 1984.

Umansky, Ellen. "Lilith." *Encyclopedia of Religion,* Mircea Eliade, ed. New York: Macmillan, 1987, 8:554–55.

van der Horst, Piet. "The Role of Women in the Testament of Job." *Nederlands Theologisch Tijdschrift* 40 (1986):273–89.

Vermaseren, M. J. *Corpus Cultus Cybelae Attidisque.* EPRO 50 Leiden: E. J. Brill, 1977–78.

Veyne, Paul. *Bread and Circuses: Historical Sociology and Political Pluralism.* Abridged with introduction by Oswyn Murray. Trans. Brian Pearce. London: Penguin Press, 1990. Originally published as *Le pain et le cirque.* Paris: Editions du Seuil, 1976.

———, ed. *A History of Private Life.* Vol 1: *From Rome to Byzantium.* Trans. Arthur Goldhammer. Cambridge, MA and London: Belknap Press of Harvard University Press, 1987.

Vidman, L. *Sylloge Inscriptionum Religionis Isaicae et Sarapiacae*. Berlin: de Gruyter, 1969.

Waithe, Ellen. "Hypatia of Alexandria," in E. Waithe, ed., *History of Women Philosophers*. Vol. 1: *Ancient Women Philosophers, 600 B.C.–500 A.D.* Dordrecht, Boston, and Lancaster: Martinus Nijhoff, 1987.

Walbank, M. B. "Artemis Bear-Leader." *Classical Quarterly* 31 (1981):276–81.

Walker, William O. "1 Corinthians 11:2–16 and Paul's Views Regarding Women." *JBL* 94 (1975):94–110.

Walsh, P. G. *Livy, His Historical Aims and Methods*. Cambridge: Cambridge University Press, 1963.

Wardman, Alan. *Religion and Statecraft among the Romans*. London: Granada, 1982.

Weber, Max. *Sociology of Religion*. Boston: Beacon Press, 1963. Originally published 1922.

Wegner, Judith. *Chattel or Person: The Status of Women in the Mishnah*. New York and Oxford: Oxford University Press, 1988.

———. "The Image and Status of Women in Classical Rabbinic Judaism," in Judith Baskin, ed., *Jewish Women in Historical Perspective*. Detroit: Wayne State University Press, 1991, 68–93.

Wilson-Kastner, Patricia, et al. *A Lost Tradition: Women Writers of the Early Church*. Lanham, MD: University Press of America, 1981.

Winkler, John J. "The Laughter of the Oppressed: Demeter and the Gardens of Adonis," in *The Constraints of Desire: The Anthropology of Sex and Gender in Ancient Greece*. New York and London: Routledge, 1990, 188–209.

Witherington, Ben, III. "Anti-feminist Tendencies of the 'Western' Text in Acts." *JBL* 103 (1984) 1:82–84.

Witt, R. E. *Isis in the Greco-Roman World*. Ithaca, NY: Cornell University Press, 1971.

Wordelman, Amy. "Strabo on Women, the Ephesian Artemis, and the Goddesses of Comana." Paper presented to the Women in the Biblical World section of the Society of Biblical Literature, Anaheim, 1989.

Wright, D. F. "Homosexuals or Prostitutes? The Meaning of *arsenokoitai* (1 Cor. 6:9, 1 Tim. 1:10)." *Vigiliae Christianae* 32 (1984) 2:125–53.

———. "Translating ΑΡΣΕΝΟΚΟΙΤΑΙ (1 Cor. 6:9, 1 Tim. 1:10)." *Vigiliae Christianae* 41 (1987) 4:396–98.

# *Index*